METHODS IN MOLECULAR BIOLOGY™

Series Editor
John M. Walker
School of Life Sciences
University of Hertfordshire
Hatfield, Hertfordshire, AL10 9AB, UK

For other titles published in this series, go to
www.springer.com/series/7651

Zebrafish

Methods and Protocols

Edited by

Graham J. Lieschke*, Andrew C. Oates†, and Koichi Kawakami‡

** Cancer and Haematology Division, Walter and Eliza Hall Institute of Medical Research, Parkville, VIC, Australia*

† Max Planck Institute of Molecular Cell Biology and Genetics, Dresden, Germany

‡ Division of Molecular and Developmental Biology, National Institute of Genetics, Mishima, Japan and

‡ Department of Genetics, The Graduate University for Advanced Studies (SOKENDAI), Mishima, Japan

Humana Press

Editors
Graham J. Lieschke
Cancer and Haematology Divison
Walter and Eliza Hall Institute of Medical Research
Parkville, VIC
Australia

Andrew C. Oates
Max Planck Institute of Molecular Cell Biology
and Genetics,
Dresden
Germany

Koichi Kawakami
Division of Molecular and Developmental Biology
National Institute of Genetics
Mishima
Japan
and
Department of Genetics
The Graduate University for Advanced Studies (SOKENDAI)
Mishima
Japan

ISSN: 1064-3745 ISSN: 1940-6029 (eBook)
ISBN: 978-1-62703-899-7 ISBN: 978-1-60327-977-2 (eBook)
DOI: 10.1007/978-1-60327-977-2
Springer Dordrecht Heidelberg London New York

Softcover re-print of the Hardcover 1st edition 2009

Printed on acid-free paper

Springer is part of Springer Science+Business Media (www.springer.com)

Preface

In the last 20 years, research activity using the zebrafish *Danio rerio* has increased dramatically. Their contribution to modern genetic and molecular research originates with their use as a vehicle for testing ideas concerning the genetic basis of vertebrate brain formation and function at the University of Oregon's Institute for Neuroscience. Their research use has expanded into their becoming a leading model system for understanding the basic genetics, cell biology, and physiology of vertebrate development and human disease states in hundreds of labs around the world. It has been a heady time for the little fish! There are good reasons for this rapid rise of popularity, both practical and technical. Practically, zebrafish are easy and inexpensive to keep, breed, and raise, and—similar to yeast, mice, and fruit flies—zebrafish like being around humans. Technically, the genetic tractability, embryonic accessibility, and imaging potential of the zebrafish are, in our opinion, the features that have tempted so many people to push the boundaries of zebrafish research so far in such a short time. Although each model organism has its strengths and weaknesses, we now regard zebrafish as sitting alongside mouse, worm and fruit fly as key animal model systems in modern biology.

There are already a number of excellent books and papers dealing with zebrafish experimental techniques, which begs the question—why another one? In choosing the contributions to this book, we were guided by three principles as we sought to make sure that this volume made a useful contribution to the field. First, because of the rapid development of techniques and reagents, we looked for material that was not yet well known or widely distributed. Second, we sought experience from newer labs with approaches that had not received exposure. Third, we tried to avoid duplicating familiar, well tested, and trusted material. The material in this volume is organized loosely along three strengths of the zebrafish: genetic modification, accessibility for manipulation, and ease of in vivo live imaging.

With a nearly complete sequenced genome, with significant genetic homology to that of humans, and with ease of mutagenesis and housing of sufficient numbers to enable forward genetic screens, the zebrafish is a natural candidate for genetic analysis of biologic processes. Chapters 1 and 2 describe dense chemical and retroviral mutagenesis, Chapter 3 covers resource-efficient haploid screening, and Chapter 4 discusses effective cryopreservation of zebrafish sperm for the precious mutants harvested from these techniques.

External fertilization and the production of large numbers of embryos from each mother have made practical the microinjection of lineage dyes, mRNA for protein overexpression, and DNA for transgenesis, as well as the transplantation of cells for genetic cell-autonomy studies. It has also made possible large-scale screens for gene expression using *in situ* hybridization, and enhancer traps. Part II of this volume develops these themes, describing the use of transposons in Chapters 5 and 6, or homologous recombination in bacterial artificial chromosomes in Chapter 7 to modify zebrafish chromosomal DNA for transgenic analysis of gene expression, as well as efficient single-copy transgenesis in

Chapter 8. Having thus created reporter strains of zebrafish with fluorescently-labeled cells, a novel method of ablating these cells specifically with nitroreductase allows their role in the organism to be tested, and is discussed in Chapter 9. Such cellular-level precision is also found in Chapter 10, which focuses on the focal electroporation of dyes or DNA into cells deep within the fish. However, sometimes a slightly larger specific region of the embryo must be manipulated, and zebrafish surgical techniques along the lines of those utilized in chick experimental embryology are presented in Chapter 11. Having plentiful embryonic material also facilitates the use of microarrays to analyze mRNA expression. Chapters 12 and 13 describe their synthesis and use for the zebrafish. The recent emergence and importance of microRNA biology has been underscored by pioneering work in the zebrafish; Chapter 14 outlines methods for validating microRNA targets in vivo.

It is perhaps the optical transparency of the zebrafish embryo that has most tipped the balance in its favor. In this volume, we included chapters showcasing methods that most labs with access to the equipment of a modern biology department can use. Chapter 15 describes a protocol for following tissue-scale morphogenesis simultaneously in multiple embryos that allows for the estimation of precision and variability. The striking beauty and power of single-cell resolution in the living zebrafish is seen in Chapters 16 and 17, which focus on imaging the early immune system using laser confocal scanning microscopy and the deeper cells of the gastrula using two-photon. The significant technical challenges of imaging the late-developing gut are tackled in Chapter 18 with a range of methods that include principles with application to other larval organ systems. Finally, Chapter 19 presents methods for achieving the lofty goal of following every cell in an organ, or indeed an entire organism, during development.

We hope that these chapters not only meet experimental needs that already exist, but also that they might inspire approaches that were not previously considered, and finally that they might give close insight and perspective into the emerging literature. The editors would like to thank John Walker and the staff at Humana Press and Springer for their continuous assistance, and the authors for their hard work and flexibility.

Parkville, Australia — *Graham J. Lieschke*
Dresden, Germany — *Andrew C. Oates*
Mishima, Japan — *Koichi Kawakami*

Contents

Part IV Analyzing Gene Expression

Part V Imaging

Contributors

LOLA BAJARD • *Max Planck Institute of Molecular Cell Biology and Genetics Dresden, Germany*

ISAAC H. BIANCO • *Department of Anatomy and Developmental Biology, University College London, London, UK*

MICHAEL BRAND • *Biotechnology Center and Center of Regenerative Therapies Dresden, University of Technology, Dresden, Germany*

SHAWN M. BURGESS • *Genome Technology Branch, National Human Genome Research Institute, National Institutes of Health, Bethesda, MD*

CARRIE CARMICHAEL • *Zebrafish International Resource Center, University of Oregon, Eugene, OR*

LARA CARVALHO • *Max Planck Institute of Molecular Cell Biology and Genetics, Dresden, Germany*

JONATHAN D.W. CLARKE • *MRC Centre for Developmental Neurobiology, King's College London, London, UK*

KATHY CROSIER • *Department of Molecular Medicine and Pathology, School of Medical Sciences, The University of Auckland, Auckland, New Zealand*

PHIL CROSIER • *Department of Molecular Medicine and Pathology, School of Medical Sciences, The University of Auckland, Auckland, New Zealand*

EDWIN CUPPEN • *Hubrecht Institute for Developmental Biology and Stem Cell Research, Utrecht, The Netherlands*

EWART DE BRUIJN • *Hubrecht Institute for Developmental Biology and Stem Cell Research, Utrecht, The Netherlands*

HARMA FEITSMA • *Hubrecht Institute for Developmental Biology and Stem Cell Research, Utrecht, The Netherlands*

MARIA VEGA FLORES • *Department of Molecular Medicine and Pathology, School of Medical Sciences, The University of Auckland, Auckland, New Zealand*

ZHIYUAN GONG • *Department of Biological Sciences, National University of Singapore, Singapore*

CHRIS HALL • *Department of Molecular Medicine and Pathology, School of Medical Sciences, The University of Auckland, Auckland, New Zealand*

JOAN K. HEATH • *Colon Molecular and Cell Biology Laboratory, Ludwig Institute for Cancer Research, Royal Melbourne Hospital, Parkville, VIC, Australia*

CARL-PHILIPP HEISENBERG • *Max Planck Institute of Molecular Cell Biology and Genetics, Dresden, Germany*

LEAH HERRGEN • *Max Planck Institute of Molecular Cell Biology and Genetics, Dresden, Germany*

BENJAMIN M. HOGAN • *Hubrecht Institute for Developmental Biology and Stem Cell Research, Utrecht, The Netherlands*

Li-En Jao • *Cell and Developmental Biology, Vanderbilt University, Nashville, TN*

Hong Jiang • *Department of Molecular, Cell and Developmental Biology, University of California Los Angeles, Los Angeles, CA*

R. Krishna Murthy Karuturi • *Genome Institute of Singapore, Agency for Science Technology and Research, Singapore*

Koichi Kawakami • *Division of Molecular and Developmental Biology, National Institute of Genetics, Mishima, Japan, Department of Genetics, The Graduate University for Advanced Studies (SOKENDAI), Mishima, Japan*

Hiroshi Kikuta • *Division of Molecular and Developmental Biology, National Institute of Genetics, Mishima, Japan*

Siew Hong Lam • *Department of Biological Sciences, National University of Singapore, Singapore*

Judith E. Layton • *The Walter and Eliza Hall Institute of Medical Research, Parkville, VIC, Australia*

Graham J. Lieschke • *Cancer and Haematology Division, Walter and Eliza Hall Institute of Medical Research, Parkville, VIC, Australia*

Shuo Lin • *Department of Molecular, Cell and Developmental Biology, University of California Los Angeles, Los Angeles, CA*

Sinnakarupan Mathavan • *Genome Institute of Singapore, Singapore*

Sean G. Megason • *Department of Systems Biology, Harvard Medical School, Boston, MA*

Andrew C. Oates • *Max Planck Institute of Molecular Cell Biology and Genetics, Dresden, Germany*

Adam C. Parslow • *Colon Molecular and Cell Biology Laboratory, Ludwig Institute for Cancer Research, Royal Melbourne Hospital, Parkville, VIC, Australia*

Michael J. Parsons • *Department of Surgery, Johns Hopkins School of Medicine, Baltimore, MD*

Luke Pase • *Cancer and Haematology Division, Walter and Eliza Hall Institute of Medical Research, Department of Medical Biology, University of Melbourne, Parkville, VIC, Australia*

Alexander Picker • *Biotechnology Center and Center of Regenerative Therapies Dresden (CRTD) University of Technology, Dresden, Germany*

Harshan Pisharath • *Department of Comparative Medicine, Johns Hopkins School of Medicine, Baltimore, MD*

Olivier Pourquié • *Stowers Institute for Medical Research, Howard Hughes Medical Institute, Kansas City, MO*

Daniela Roellig • *Max Planck Institute of Molecular Cell Biology and Genetics (MPI-CBG), Dresden, Germany*

Christian Schröter • *Max Planck Institute of Molecular Cell Biology and Genetics (MPI-CBG), Dresden, Germany*

Daniele Soroldoni • *Max Planck Institute of Molecular Cell Biology and Genetics (MPI-CBG), Dresden, Germany*

Marcel Tawk • *Department of Anatomy and Developmental Biology, University College London, London, UK*

Andrew J. Trotter • *Colon Molecular and Cell Biology Laboratory, Ludwig Institute for Cancer Research, Royal Melbourne Hospital, Parkville, VIC, Australia*

Akihiro Urasaki • *Division of Molecular and Developmental Biology, National Institute of Genetics, Mishima, Japan*

Zoltán M. Varga • *Zebrafish International Resource Center, University of Oregon, Eugene, OR*

Monte Westerfield • *Institute of Neuroscience, University of Oregon, Eugene, OR*

Zhongan Yang • *Department of Molecular, Cell, and Developmental Biology, University of California Los Angeles, Los Angeles, CA*

Part I

Mutagenesis and Mutants

Chapter 1

Highly Efficient ENU Mutagenesis in Zebrafish

Ewart de Bruijn, Edwin Cuppen, and Harma Feitsma

Summary

ENU (*N*-ethyl-*N*-nitrosourea) mutagenesis is a widely accepted and proven method to introduce random point mutations in the genome. Because there are no targeted knockout strategies available for zebrafish so far, random mutagenesis is currently the preferred method in both forward and reverse genetic approaches. To obtain high-density mutagenized zebrafish, six consecutive ENU treatments are applied at weekly intervals to adult male zebrafish by bathing them in ENU solution. With this procedure an average germ line mutation load of one mutation every 1.0×10^5–1.5×10^5 basepairs is reached routinely in our lab.

Key words: Zebrafish, ENU, Mutagenesis, TILLING, Method, Mutant, Knockout.

1. Introduction

N-Ethyl-*N*-nitrosourea (ENU) is an alkylating agent and one of the most powerful mutagens known. It exerts its mutagenic activity by ethylating oxygen or nitrogen atoms in DNA bases. These DNA adducts are not mutations themselves, but they induce error-prone replication which can result in a full mutation after two rounds of replication *(1)*. Adult male animals are traditionally subjected to repeated treatments with ENU in order to introduce point mutations in premeiotic germ cells. After one full round of spermatogenesis, these males are mated with wild-type females in order to generate a set of progeny that possess many unique heterozygous point mutations. In zebrafish, ENU mutagenesis may be accomplished with relative ease by bathing the fish in ENU solution *(2, 3)*.

ENU mutagenesis in zebrafish has been used extensively in forward mutagenesis screens in the search for early developmental

Graham J. Lieschke et al. (eds.), *Zebrafish,* Methods in Molecular Biology, vol. 546
DOI: 10.1007/978-1-60327-977-2_1

phenotypes *(4)*. Today, almost any scoreable phenotype has been the subject of a forward screen—including cancer, maternal effects and epilepsy *(5–8)*. For recessive mutations, multiple crosses of the carrier are necessary to generate homozygotes that will reveal the phenotype. However, zebrafish are unique in that they allow screening of haploid progeny or homozygous diploids created by pressure treatment *(9)*. Positional cloning of mutations with phenotypes of interest is used to identify the underlying gene.

Targeted approaches such as homologous recombination—the method of choice for generating mouse knockouts—are currently not available in the zebrafish because of the lack of pluripotent embryonic stem cells. However, the high efficiency of ENU mutagenesis has stimulated researchers to develop a procedure that uses random mutagenesis to inactivate genes of interest. Target-selected mutagenesis enables identification of mutations of interest through screening of the coding parts in the gene of interest in a DNA library of the mutant fish. *(10, 11)*. Because ENU induces random mutations and can cause almost all possible basepair changes, a range of alleles is usually obtained, which includes nonsense mutations, mutations that affect transcript splicing and missense mutations. This procedure has been successfully used to generate a series of knockout zebrafish *(12–16)*. Resequencing data reveal that the ENU mutagenesis procedure introduces approximately one mutation every

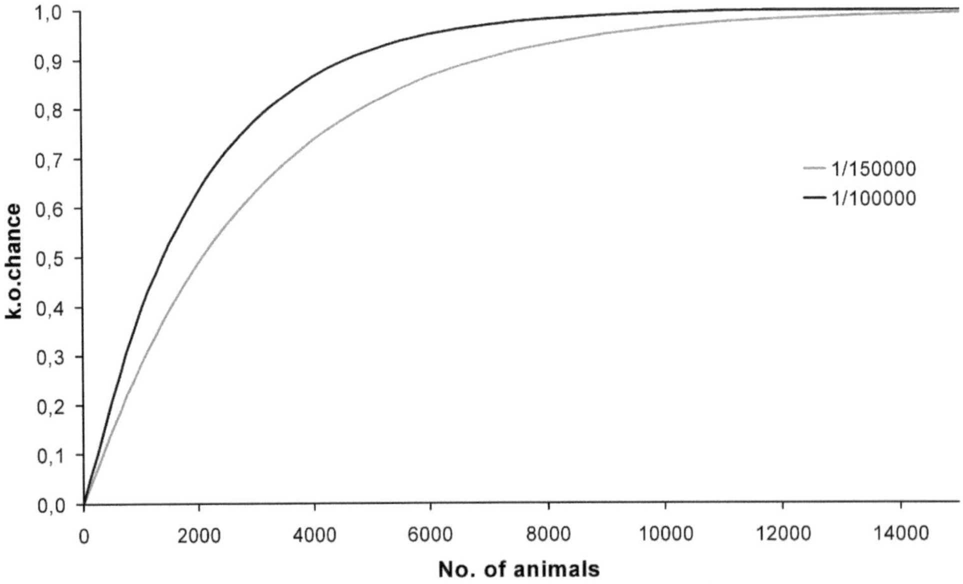

Fig. 1. Relationship between the chance of obtaining a nonsense mutant for a gene of interest (knockout chance) and the number of mutant animals screened. Graph shows the relationship for mutation frequencies of 1/100,000 basepairs (*black line*) and 1/150,000 basepairs (*grey line*). This is calculated for an average gene for which 2 amplicons of 500 basepairs each are screened.

100,000–150,000 basepairs (unpublished data), which indicates that a mutant library of a few thousand zebrafish would theoretically be sufficient to obtain a knockout for any individual gene (**Fig. 1**).

The method described in this chapter provides a detailed step-by-step protocol for highly efficient and reproducible ENU mutagenesis in zebrafish, suited for both reverse and forward genetic approaches.

2. Materials

2.1. Preparation of Mutagenesis Workspace

1. Dark, silent room with fume hood (*see* **Note 1**).
2. One-way paper to absorb spilled ENU.
3. 3-L plastic beaker.
4. 1M Sodium hydroxide (NaOH) (*see* **Note 2**).
5. 100-mL Measuring cylinder.
6. 250-mL Plastic beaker.
7. 10 mM Acetic acid (HAc). Store at 4°C. Warm to room temperature prior to treatment (*see* **Note 3**).
8. Three 1.5-mL screwcap Eppendorf tubes per treatment.
9. 20 L of 10 mM phosphate (sodium) buffer (pH 6.6) per 50 mutagenized fish. Prepared fresh for each treatment from 1 M phosphate stock solution (pH 6.6). Readjust pH to 6.6 with HCl (*see* **Note 4**). The phosphate stock solution is an appropriate mix of ~62.5 mL of 1 M sodium dihydrogen phosphate (NaH_2PO_4) and ~37.5 mL of 1 M disodium hydrogen phosphate (Na_2HPO_4), which should be stored at room temperature.
10. 50 mL tube.
11. 0.1M Sodium hydroxide (NaOH) (*see* **Note 2**).
12. Large fish tanks and lids for housing fish overnight.
13. 8 L of eggwater *(2)* per 50 mutagenized fish.
14. Tricaine methanesulphonate (MS-222) (ethyl 3-aminobenzoate methanesulphonate) ($C_{10}H_{15}NO_5S$) (Sigma A-5040). A stock solution contains 400 mg tricaine powder in 97.9 mL MilliQ water with approximately 2.1 mL 1 M Tris–HCl (pH 9.0). Adjust pH to 7.0 and store in refrigerator. Dilute stock 25 times with fish water for anaesthetic solution *(2)*.
15. Pots (approximately 0.5 L) for phosphate buffer and ENU solution during mutagenesis. One pot can hold up to six fish.
16. Sieves that fit in pots.

2.2. Preparation of ENU Stock Solution

1. Safety gear for protection during handling ENU include lab coat, mouth mask, protective eyewear, and two layers of gloves (*see* **Note 5**).
2. *N*-Ethyl-*N*-nitrosourea (ENU) ($C_3H_6N_3O_2$) in 1 g isopac (Sigma N 3385). One isopac is sufficient for treatment of 50 fish. Use a new isopac for each treatment (*see* **Note 6**), preferably from the same production batch for one mutagenesis experiment (*see* **Note 7**). Isopacs are stored at −20°C. **WARNING**: ENU is extremely hazardous (*see* **Subheading 3**).
3. Scissors and blunt tweezers.
4. Small needle.
5. 50 mL syringe with appropriate needle.
6. Two pieces of Parafilm per treatment.
7. Standard laboratory pipets p20, p200 and p1000 with tips.
8. 100 μL quartz cuvette.

2.3. Mutagenesis Treatment

1. 10 mL pipet with pipeting balloon.

2.4. Cleaning and ENU Waste Disposal

1. Alkaline waste container for ENU solutions.
2. Solid waste container for ENU materials.
3. 25% NH_3 solution (ammonia) (*see* **Note 2**).

3. Methods

The mutagenesis procedure in zebrafish uses six consecutive treatments at weekly intervals, which means that **Subheadings 3.2–3.5** are repeated six times, including preparation of some of the chemicals. Because ENU is extremely hazardous to researchers and the environment, special attention must be paid to aspects of safety and waste disposal throughout the protocol.

3.1. Selection and Preparation of Fish

1. Male fish of TL (or Tubingen Longfin) strain (*see* **Note 8**) between 3 and 8 months of age are selected based on health and then tested for fertility prior to mutagenesis. The selected fish should be maintained in optimal condition (*see* **Note 9**).
2. Feed the fish well the morning prior to treatment.
3. A well-laying batch of female AB fish (*see* **Note 10**) should be ready for outcrossing with males during and after (*see* **Note 11**) mutagenesis treatments.

3.2. Preparation of Mutagenesis Workspace

1. The mutagenesis room should be made as stress-free as possible by dimming lights, placing blinds on windows, and reducing external noise. (*see* **Note 1**).
2. Prepare fume hood by covering hood and floor with one-way paper.
3. Was and dry pots, sieves, and fish tanks. Fill the 3 L beaker with 2 L 1 M NaOH as an inactivation solution.
4. 94.9 mL 10 mM HAc is put into the 250 mL beaker.
5. Prepare three 1.5 mL Eppendorf tubes with 290 μL, 990 μL and 1,500 μL phosphate buffer, respectively, for OD measurement. The first two tubes are for making a precise 3,000-fold dilution of the ENU solution. The third tube will serve as reference.
6. Fill a 50 mL tube with approximately 20 mL of 0.1 M NaOH for OD measurement disposal.
7. Pieces of parafilm, scissors and tweezers should be available in the fume hood; andsyringes and needles should be unpacked. Pipets and tips are put nearby.
8. Prepare the large (4 L) fish tanks for overnight recovery with an equal mix of phosphate buffer (*see* **Note 12**), egg water, fish water (O/N fish water), and 10 mL of MS-222 (*see* **Note 13**). Prepare near the fume hood.
9. Pots, sieves, and consumables should be set up as shown in **Fig. 2**. Fill pots inside the hood with 290 mL phosphate buffer for the ENU solution. The remaining buffer is used to fill the pots for preincubation and post-mutagenesis wash.

Divide fish into groups of 5 or 6 per sieve in the pots with phosphate buffer. Leave fish for 30 min to acclimate to the lower pH.

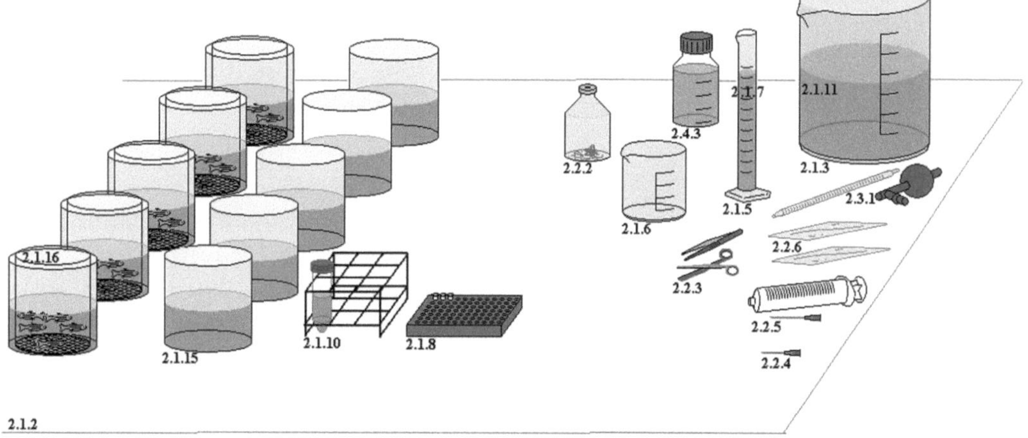

Fig. 2. Organization of mutagenesis workspace. All the materials and equipment mentioned in **Subheading 2** are indicated.

3.3. Preparation of ENU Stock Solution

1. Put on safety wear.
2. Unpack ENU bottle and remove the complete metal ring from the top with tweezers. Push the small needle into the rubber top for depressurization.
3. Fill the syringe with 94.9 mL 10 mM HAc twice and inject into the ENU bottle with the needle. Place the syringe and needle into the inactivation solution.
4. Remove the small needle from the bottle and save for use.
5. Fold one piece of parafilm around the top of the bottle to prevent leakage.
6. Shake vigorously for approximately 10 min or until ENU is completely dissolved (*see* **Note 14**). Place finger on top of bottle during shaking for additional prevention of leakage.
7. Depressurize bottle by reinserting the small needle. The small needle, parafilm, and rubber top can then be removed and put into the inactivation solution.
8. Change gloves after this action to prevent contamination. Dispose used gloves in the solid waste container (*see* **Note 15**).
9. Pipet the 10 μL ENU solution into the Eppendorf tube containing 290 μL phosphate buffer and mix.
10. Transfer 10 μL of this dilution into the Eppendorf tube containing 990 μL (OD dilutionMix).
11. The active ENU concentration is measured by OD in a spectrophotometer (*see* **Note 16**). Because this apparatus is usually not in the mutagenesis room, safety wear is taken off and Eppendorf tube containing OD dilution, the Eppendorf containing phosphate buffer reference, p200 pipet and tips, 50 mL tube with 0.1 M NaOH, and a quartz cuvette are taken to the spectrophotometer.
12. Measure the ENU concentration at a wavelength of 238 nm (*see* **Note 17**) (extinction coefficient = 5,830 M^{-1} cm^{-1}) with phosphate buffer as a reference *(2)*. Dispose of tips and OD dilution in the NaOH tube and rinse the cuvette multiple times with phosphate buffer and water.
13. Once back in the mutagenesis room, all contaminated disposables and solutions are inactivated.
14. The ENU concentration of the measurement dilution can be calculated with the formula [ENU] = (0.1715A) mM. Because the OD dilution is diluted 3000 times, absorbance (A) should be around 0.1749 for 1 isopac of ENU. The volume of ENU stock solution needed for 300 mL 3.3 mM ENU mutagenesis solution can be calculated via 1.92/A mL.

3.4. Mutagenesis Treatment

1. With a 10-mL pipet, add the calculated volume of ENU solution to the mutagenesis pots to achieve an end concentration of 3.3 mM ENU.
2. Transfer the sieves with the fish into the ENU working solution. Shake once to mix.
3. Incubate for 1 h in complete silence (*see* **Note 1**).
4. Inactivate leftover ENU solution in the bottle immediately by adding inactivation solution and then placing the entire bottle in the inactivation solution beaker.
5. After 1 h, place fish back in phosphate buffer and bathe for 10 min (*see* **Note 18**) to wash away ENU.
6. Transfer fish into overnight fish water tanks covered with lids (*see* **Note 19**).
7. Add 10–20 mL inactivation solution to the ENU solution pots for inactivation.
8. Postpone further cleaning until the next day to avoid disturbing the fish during recovery.
9. Perform the next mutagenesis treatment in 1 wk. In total, six treatments are applied.

3.5. Cleaning and ENU Waste Disposal

1. After recovery, the fish may be set back into the aquarium system with a little bit of food.
2. Clean the workspace. It should be noted that inactivated ENU is still toxic and that safety gear must still be worn.
3. Transfer disposables from the inactivation beaker to the solid waste container.
4. Empty the pots and the fish tanks in the sink while diluting with tap water.
5. Discard all other liquids in the alkaline waste container.
6. Was and dry all pots, sieves, fish tanks, and beakers. Clean non-disposable labware with a towel dampened with ammonia. Inactivate potential droplets and aerosols in the hood by sprinkling ammonia on the one-way paper. Dispose of paper in the solid waste container.
7. Clean fume hood with ammonia.
8. Safely store all materials for the next treatment.

3.6. Generation of F1 Mutants and Determination of Mutagenesis Efficiency

1. When aiming for mutagenized pre-meiotic germ cells, progeny from crosses 3 wk after the last mutagenesis treatment (*see* **Note 20**) may be used for setting up a library or screen (*see* **Note 21**). We routinely mutagenize 50 males to generate about 5,000 F1 progeny, which is sufficient to raise a library of about 3,000 adult F1 fish for screening.

2. The achieved mutation frequency may be tested at the single locus level by crossing males to albino females and then scoring for albino progeny. However, we found this method to be imprecise because of the low number of albino fish that are routinely retrieved. Therefore, we set up a screening protocol at the single basepair level by directly resequencing DNA from part of the progeny.
3. A mutant library can be kept alive and further crossed for recessive forward genetic screens. Alternatively, sperm may be stored frozen (*see* **Note 22**).

4. Notes

1. Stress is the leading cause of death in mutagenized fish. Prepare the mutagenesis room accordingly and tell your colleagues that you are performing an experiment and ask for silence. Mark the room with warning signs that an experiment is being performed. Precautions may be taken by placing towels between the doors and posts. The room should be at an appropriate temperature.
2. ENU is highly unstable at high pH levels with a half-life of less than 1 min at pH > 8.5. Therefore, the use of basic solutions (0.1–1 M NaOH, ammonia) is the preferred way to inactivate spills and contaminated equipment and tools.
3. Maintaing the solution and ENU isopac at room temperature facilitates dissolving of ENU.
4. pH should be checked carefully, as it highly influences ENU activity and stability.
5. It is recommended that two persons undertake the treatments for concentration, handling and safety reasons.
6. It is possible to store the ENU solution in 10 mM HAc in the freezer, but activity is affected and it poses the risk of bottle breakage. Therefore, we strongly recommend to use a new isopac for every treatment and not to store any ENU.
7. We have seen variation between different isopacs. This is reduced when taking all of them from the same production batch (i.e. LOT number).
8. We recommend using TL as wild-type strain for mutagenesis. Internal studies showed that TLs are the most effective in relation to mutation frequency and survival.
9. The male fish should be looked after very carefully and water quality should be kept optimal. Feed only and sufficiently artemia during the period of mutagenesis.

10. Outcrossing against another background strengthens the offspring. We routinely use AB as the strain for outcrossing.
11. Inter-treatment matings are supposed to activate spermatogenesis and therefore increase the efficiency of multiple treatments, but there is no experimental evidence for this.
12. This is to reduce the pH shock after treatment.
13. Slight sedation of fish during recovery may reduce anxiety-induced death. The working concentration of MS-222 is 1/9th of the usual anesthetizing dose.
14. Dissolving of ENU takes time. Be patient, as residual lumps of ENU powder will affect concentration measurement and mutagenesis. If particles do not dissolve, remove them in order to obtain a homogenous solution.
15. Change gloves anytime you think is necessary to assure safety.
16. The ENU concentration should always be measured by OD because of variation between isopacs andloss of activity during preparation.
17. Concentration measurements at 395 nm are routinely used in mouse mutagenesis experiments but showed large deviations in our hands, probably because of the different solvent that is used.
18. This time may be extended. Transfer a first sieve to the overnight fish tanks to see how the fish react. If they have difficulties, they may be left in the wash solution for a longer period of time.
19. Take care that during transfer no ENU solution is spilled, and if so, inactivate and clean up directly.
20. This avoids mosaics.
21. We use a maximum of 800 siblings (F1s) per mutagenized male to prevent clonal effects, but there is no experimental evidence for this.
22. Check the fertility of the males prior to freezing as F1 progeny derived from ENU-mutagenized fish tend to be more fragile and have lower fertility rates as compared with wild-type fish.

Acknowledgments

The authors would like to thank Freek van Eeden for details provided on the mutagenesis protocol. This work was supported by funds from the European Union-funded FP6 Integrated Project ZF-MODELS.

References

1. Noveroske, J. K., Weber, J. S., and Justice, M. J. (2000). The mutagenic action of N-ethyl-N-nitrosourea in the mouse. *Mamm. Genome* **11**, 478–483.
2. Mullins, M. C., Hammerschmidt, M., Haffter, P., and Nusslein-Volhard, C. (1994). Large-scale mutagenesis in the zebrafish: in search of genes controlling development in a vertebrate. *Curr. Biol.* **4**, 189–202.
3. Solnica-Krezel, L., Schier, A. F., and Driever, W. (1994). Efficient recovery of ENU-induced mutations from the zebrafish germline. *Genetics* **136**, 1401–1420.
4. Haffter, P., Granato, M., Brand, M., Mullins, M. C., Hammerschmidt, M., Kane, D. A., Odenthal, J., van Eeden, F. J., Jiang, Y. J., Heisenberg, C. P., Kelsh, R. N., Furutani-Seiki, M., Vogelsang, E., Beuchle, D., Schach, U., Fabian, C., and Nusslein-Volhard, C. (1996). The identification of genes with unique and essential functions in the development of the zebrafish, Danio rerio. *Development* **123**, 1–36.
5. Baraban, S. C., Dinday, M. T., Castro, P. A., Chege, S., Guyenet, S., and Taylor, M. R. (2007). A large-scale mutagenesis screen to identify seizure-resistant zebrafish. *Epilepsia* **48**, 1151–1157.
6. Moore, J. L., Rush, L. M., Breneman, C., Mohideen, M. A., and Cheng, K. C. (2006). Zebrafish genomic instability mutants and cancer susceptibility. *Genetics* **174**, 585–600.
7. Dosch, R., Wagner, D. S., Mintzer, K. A., Runke, G., Wiemelt, A. P., and Mullins, M. C. (2004). Maternal control of vertebrate development before the midblastula transition: mutants from the zebrafish I. *Dev. Cell* **6**, 771–780.
8. Wagner, D. S., Dosch, R., Mintzer, K. A., Wiemelt, A. P., and Mullins, M. C. (2004). Maternal control of development at the mid-blastula transition and beyond: mutants from the zebrafish II. *Dev. Cell* **6**, 781–790.
9. Cheng, K. C., and Moore, J. L. (1997). Genetic dissection of vertebrate processes in the zebrafish: a comparison of uniparental and two-generation screens. *Biochem. Cell Biol.* **75**, 525–533.
10. Wienholds, E., and Plasterk, R. H. (2004). Target-selected gene inactivation in zebrafish. *Methods Cell Biol.* 77, 69–90.
11. Wienholds, E., van Eeden, F., Kosters, M., Mudde, J., Plasterk, R. H., and Cuppen, E. (2003). Efficient target-selected mutagenesis in zebrafish. *Genome Res.* **13**, 2700–2707.
12. Berghmans, S., Murphey, R. D., Wienholds, E., Neuberg, D., Kutok, J. L., Fletcher, C. D., Morris, J. P., Liu, T. X., Schulte-Merker, S., Kanki, J. P., Plasterk, R., Zon, L. I., and Look, A. T. (2005) tp53 mutant zebrafish develop malignant peripheral nerve sheath tumors. *Proc. Natl Acad. Sci. U S A* **102**, 407–412.
13. Feitsma, H., Leal, M. C., Moens, P. B., Cuppen, E., and Schulz, R. W. (2007). Mlh1 deficiency in zebrafish results in male sterility and aneuploid as well as triploid progeny in females. *Genetics* **175**, 1561–1569.
14. Hurlstone, A. F., Haramis, A. P., Wienholds, E., Begthel, H., Korving, J., Van Eeden, F., Cuppen, E., Zivkovic, D., Plasterk, R. H., and Clevers, H. (2003). The Wnt/beta-catenin pathway regulates cardiac valve formation. *Nature* **425**, 633–637.
15. Wienholds, E., Koudijs, M. J., van Eeden, F. J., Cuppen, E., and Plasterk, R. H. (2003). The microRNA-producing enzyme Dicer1 is essential for zebrafish development. *Nat. Genet.* **35**, 217–218.
16. Wienholds, E., Schulte-Merker, S., Walderich, B., and Plasterk, R. H. (2002). Target-selected inactivation of the zebrafish rag1 gene. *Science* **297**, 99–102.

Chapter 2

Production of Pseudotyped Retrovirus and the Generation of Proviral Transgenic Zebrafish

Li-En Jao and Shawn M. Burgess

Summary

This chapter describes a method for generation of the high-titer pseudotyped Moloney murine leukemia virus (MLV) that efficiently infects zebrafish embryos (i.e., more than 25 retroviral copies per cell). Injection techniques are also described for production of the retrovirus-infected mosaic "founder" fish. We describe a quantitative PCR (qPCR)-based assay as a quick way to assess the infectivity after each round of viral production and injection. Most of the required equipment is commercially available and commonly present in most research laboratories.

Key words: Zebrafish, Retrovirus, Pseudotyped Moloney murine leukemia virus, Insertional mutagenesis, Injection.

1. Introduction

The use of zebrafish as a vertebrate model organism in research has increased substantially in the past two decades, following the demonstration that zebrafish are amenable to large-scale forward mutagenesis screens *(1, 2)*. Typically, forward genetic screens were generally limited to studies using invertebrates such as flies, worms, and yeast. Because of its small size, fecundity, and fast embryonic development *ex utero*, zebrafish made it possible to apply large-scale genetic screens to the study of vertebrate-specific processes that affect development and disease. Two large-scale forward genetic screens used the chemical ethylnitrosourea (ENU) as a mutagen and generated more than 6,600 observable embryonic mutations. However, because ENU mainly produces point

Graham J. Lieschke et al. (eds.), *Zebrafish,* Methods in Molecular Biology, vol. 546
DOI: 10.1007/978-1-60327-977-2_2

mutations, laborious positional cloning becomes necessary for identifying most of the mutated genes. To date, only about 160 of the genes responsible for the corresponding phenotypes identified in these two large-scale ENU-based screens have been cloned.

Insertional mutagenesis is a method complimentary to chemical mutagenesis, where foreign DNA is inserted into the genome, disrupting gene expression. The major advantage of insertional mutagenesis is the ease of identifying the mutated genes. The inserted foreign DNA (e.g., transposons or retroviruses) can act as a molecular landmark that allows for rapid cloning of the adjacent flanking genomic sequences. Moloney murine leukemia virus (MLV) pseudotyped with the envelope glycoprotein from vesicular stomatitis virus (VSV-G) has been developed as an insertional agent in zebrafish *(3)*. Pseudotyping renders MLV the ability to infect a broader range of hosts, including zebrafish cells. It also increases the stability of viral particles, allowing for the increase of the viral titers 1000-fold through ultracentrifugation *(4, 5)*. A large-scale forward insertional mutagenesis screen based on this pseudotyped MLV system was performed successfully, identifying approximately 500 observable embryonic recessive mutations. These mutants represent about 385 different genes; 335 of which have been identified *(6–10)*.

The pseudotyped MLV system can also be used as a transgenesis tool for the purpose of gene delivery. For example, a large-scale enhancer detection screen has used this MLV system to deliver an "enhancer-trap" vector into zebrafish and generated more than 1,000 transgenic lines expressing the reporter yellow fluorescent protein (YFP) in various tissues and cells *(11)*.

The pseudotyped MLV system provides the advantage that it is the most efficient insertional agent in vertebrates to date. Using a high-quality preparation of virus, almost all injected founders will transmit integrations through the germline with an average of ten copies per cell in the F1 progeny *(12, 13)*. In this chapter, we describe protocols to generate highly infective pseudotyped MLV particles and techniques to inject zebrafish embryos with the pseudotyped MLV. A quick quantitative polymerase chain reaction (PCR) assay is also described for the early assessment of infectivity after injection.

2. Materials

2.1. Cell Culture

1. 600-mL cell culture flasks (Nalge Nunc International, Rochester, NY).
2. Poly-L-lysine, 0.01% *(w/v)* (Sigma, St. Louis, MO).

3. 1X phosphate-buffered saline (PBS) (Invitrogen, Carlsbad, CA, Carlsbad, CA).
4. 0.25% Trypsin-ethylenediamine tetraacetic acid (EDTA) solution (Invitrogen).
5. DMEM growth medium: A 500-mL bottle of Dulbecco's modified Eagle' medium (DMEM, Invitrogen) is supplemented with fetal bovine serum (FBS; Hyclone, Logan, UT) to 10% *(v/v)* and 1X penicillin-streptomycin (Invitrogen).
6. Lipofectamine™ transfection reagent (Invitrogen).
7. pHCMV-G plasmid *(5)*.
8. Opti-MEM I medium (Invitrogen).
9. Bottle top filters, 0.2 μm pore size (Nalge Nunc International).
10. HEK293-based viral packaging cell line, for best results with stable MLV gag-pol expression and stable proviral mRNA expression.

2.2. Virus Concentration

1. Ultracentrifuge (XL-90, Beckman Coulter, Fullerton, CA).
2. SW28 (or equivalent ultracentrifuge rotor) (Beckman Coulter).
3. Ultracentrifuge tubes, 40 mL (Beckman Coulter).

2.3. Embryo Preparation

1. One 2-L Erlenmeyer flask (for preparing 1X Holtfreiter's).
2. One 1-L Erlenmeyer flask (for preparing 1X Holtfreiter's solution containing polybrene).
3. Three to four 100-mL beakers (for collecting embryos).
4. 25-mL pipets (for washing embryos).
5. 1 M HEPES solution (Sigma), stored at 4°C.
6. 80 mg/mL polybrene (10,000X stock): This stock solution is made by adding appropriate amount of water to the lyophilized powder of Sequa-brene (Sigma). Store stock solution at –20°C.
7. 10X modified Holtfreiter's solution: For 2 L of 10X stock solution, mix 70 g NaCl, 2.6 g $CaCl_2 \cdot (2H_2O)$, and 1 g KCl, filter-sterilized or autoclaved. This 10X stock is stable for months at room temperature. For preparing 1X Holtfreiter's working solution, dilute the 10X stock into 1X and buffer it with HEPES to a final concentration of 5 mM at pH 7.0. This 1X Holtfreiter's solution should be made freshly each time as HEPES is not stable at room temperature. Two to three liters of 1X Holtfreiter's solution is usually required for preparing 3–4 clutches of embryos. Set aside 0.5–1 L of 1X Holtfreiter's and add polybrene to a final concentration of 8 μg/mL.

8. 10 mg/mL Pronase: This stock is prepared by adding an appropriate amount of water to the pronase powder (Roche Applied Science) and incubating the solution at 37°C for 1 h (self-digestion step). The solution is then aliquoted (e.g. 500 µL per tube) and stored at –20°C.

2.4. Pre-injection Preparation

1. Needle puller (Sutter Model P-2000, Sutter Instrument, Novato, CA).
2. Quartz or glass capillaries (Sutter Instrument).
3. 10-cm Petri dishes.
4. Microscope slides (75 X 25 mm, 1 mm thick).
5. 2% *(w/v)* agarose made in 1X Holtfreiter's solution (for making injection ramps).
6. Six-well tissue culture dishes (Corning, Lowell, MA).

2.5. Virus Injection

1. Injection apparatus (*see* **Subheading 3**).
2. Scalpel blade (for cutting off the tip of the injection needle).
3. Injection hood with dissecting microscope.
4. 160 µg/mL polybrene (20X stock).
5. 1% *(w/v)* phenol red in 1X PBS.
6. 5.75-in. Wide bore pasteur pipet (cat. no. 13–678–30, Fisher Scientific, Pittsburgh, PA).
7. 0.22-µm-filtered system water.
8. 10-cm Petri dishes (for raising the injected embryos during the first 5 d).

2.6. Virus Evaluation

1. Proteinase K lysis buffer: 10 mM Tris–HCl (pH 8.0), 100 mM NaCl, 10 mM EDTA, 0.4% sodium dodecyl sulfate [SDS], and 100 µg/mL proteinase K (freshly added from a 200X stock, Invitrogen). This solution can be prepared beforehand with all the components added except proteinase K. Lysis buffer without proteinase K can be stored at room temperature for months.
2. Isopropanol.
3. 70% ethanol.
4. Heating blocks set at 50°C and 37°C.
5. Viral (SFG) probe: 5′-FAM-CTGCTGACCACCCCCAC-CGC-TAMRA-3′, stored at –20°C and protected from light.
6. RAG1probe:5′-HEX-GCGCAACGGCGGCGCTC-TAMRA-3′, stored at –20°C and protected from light.
7. SFG primers:
   5′-CGCTGGAAAGGACCTTACACA-3′
   5′-TGCGATGCCGTCTACTTTGA-3′.

8. RAG1 primers:
   5′-ATTGGAGAAGTCTACCAGAAGCCTAA-3′
   5′-CTTAGTTGCTTGTCCAGGGTTGA-3′.
9. Platinum® *Taq* DNA polymerase and buffers (Invitrogen) or equivalent PCR reagents: 10X PCR buffer, 50 mM $MgCl_2$, and Platinum® *Taq* polymerase (5 U/μL).
10. 10 mM dNTP Mix (Invitrogen).
11. iCycler (Bio-Rad, Hercules, CA) or equivalent real-time PCR machine
12. 96-well PCR plates for iCycler (Bio-Rad).
13. Reference fish DNA samples with known numbers of proviral copies (e.g., reference DNA with one and four proviral copies per cell).

3. Methods

3.1. Cell Culture

Pseudotyped retroviruses are typically generated in a packaging cell line while the cells are transiently transfected with vectors expressing the pseudotyped envelope proteins. We used a cell line, termed GT186 *(12)*, for packaging the pseudotyped MLV. The GT186 cell line was derived from 293 human embryonic kidney cells that stably express the *gag* and *pol* genes of MLV under the control of the human cytomegalovirus (CMV) promoter *(14)*. There is also a stable chromosomal integration of the proviral genome whose RNA transcript is packaged into the MLV active viral core particles. The proviral genome contains full MLV long terminal repeats (LTRs), the packaging signal (ψ), and a reporter, *lac*Z gene, used for titering purposes. Plasmid pHCMV-G (the VSV-G protein driven by the CMV promoter) is transiently transfected into GT186 cells. The expressed VSV-G proteins encapsulate the viral core particles at the plasma membrane and the active pseudotyped retroviruses are released into the tissue culture medium and collected. It is critical to express the VSV-G proteins only transiently because they are cytotoxic when expressed at the levels necessary to generate high-titer pseudotyped viruses. A detailed protocol for generating a large batch of virus sufficient for a 1-wk schedule of injections follows.

3.1.1. Day 1: Grow the Cells on Poly-Lysine-Coated Tissue Culture Flasks

1. Coat the 600-mL flasks with poly-L-lysine. Draw 25 mL of 0.01% *(w/v)* poly-L-lysine solution to cover the surface of each flask for few seconds and then transfer the solution to the next flask. Repeat the step until all flasks are "rinsed." Leave the "rinsed" flasks horizontally in the hood for 5 min and then

stand all flasks up to let the excess solution flow down to the bottom of the flasks (*see* **Note 1**).

2. Use a sterile aspirating pipet connected to a vacuum source to dry the surface of flasks until no liquid can be seen.
3. To have the appropriate density of cells at the time of transfection, typically we grow GT186 cells in five 600 mL flasks (surface area 185 cm^2, without the poly-L-lysine coating) until about 80% confluence and then split the cells into 10 poly-L-lysine-coated 600-mL flasks the day before transfection (*see* **Note 2**).
4. To trypsinize the cells, the cells in each flask are washed with 20 mL of 1X PBS once. 3 mL of trypsin–EDTA solution is then added slowly to cover the entire cell monolayer. Sit the flasks under the hood for about 3–5 min to ensure that all cells detach from the surface and from each other (HEK293 cells actually adhere to each other better than to the flask).
5. While waiting for trypsinization to complete, aliquot 20 mL of DMEM growth medium to the poly-L-lysine-coated flasks. Stop trypsinization by adding 7 mL of DMEM growth medium into each flask. Resuspend thoroughly, and pool all cells into one flask (the total volume should be around 50 mL). Mix thoroughly and aliquot 5 mL of the cell suspension into each poly-L-lysine-coated flask containing 20 mL of DMEM growth medium (*see* **Note 3**).

3.1.2. Day 2: Transfection Using Lipofectamine™ (LF)

1. We found that the optimal cell density just prior to transfection is 70–80% confluence. The optimal LF:DNA ratio is 15:1 (μL:μg). For each 600 mL flask, 8 μg of pHCMV-G plasmid is used (and thus 120 μL of LF is needed). In total 80 μg of pHCMV-G plasmid and 1200 μL of LF are needed for ten flasks.
2. To prepare the DNA/LF complex, in one 15-mL conical tube, mix 80 μg of pHCMV-G plasmid with 5 mL of Opti-MEM I medium without serum and antibiotics. In another 15-mL conical tube, mix 1,200 μL of LF with 4 mL of Opti-MEM I medium without serum and antibiotics.
3. Combine the solutions from these two 15-mL conical tubes and mix by gently pipeting up and down three times.
4. Incubate the DNA/LF solution for 15–20 min at room temperature.
5. During this incubation, rinse the cells with 20 mL of Opti-MEM I medium without serum or antibiotics and replenish the cells with 19 mL of Opti-MEM I medium without serum or antibiotics in each flask (*see* **Note 4**).
6. Aliquot 1 mL of DNA/LF mix to each flask (for a final total volume of 20 mL). Mix gently and return the cells to the 37°C incubator.

7. Incubate at 37°C for 8–12 h.
8. Discard the DNA/LF mix and replenish the cells with 20 mL of DMEM growth medium (containing 10% FBS and antibiotics). Return to the 37°C incubator.

3.1.3. Day 3: First Collection and Change Medium

1. Between 21 and 24 h post-transfection, collect the medium and filter through a 0.2-μm filter, and replenish the cells with fresh 20 mL of DMEM growth medium. Return to the 37°C incubator. The filtered, collected medium is stored at 4°C (first collection).

3.1.4. Day 4: Second Collection

1. Between 44 and 48 h post-transfection, collect the medium again through a 0.2-μm filter. Combine the second collection with the first. The medium is then subjected to ultracentrifugation (*see* **Note 5**).

3.2. Virus Concentration

The volume of collected medium (~400 mL) requires two rounds of centrifugation in an SW28 rotor. To reduce the sample loss, we use the same centrifuge tubes (five tubes) for both rounds of centrifugation.

1. 38 mL of collected medium is added into each SW28 ultracentrifuge tube. The tubes need to be balanced to within 0.1 g of each other. The medium is centrifuged in a SW28 rotor at 27,000 rpm for 1.5 h at 4°C (131,453 × *g*).
2. The supernatant is gently poured into a beaker containing bleach and the tube is inverted briefly on a piece of paper towel to allow liquid to collect near the top rim of the tube.
3. Aspirate the rim of the tubes briefly and return the tubes to the tube adapters.
4. Load the second half of the medium to the same five centrifuge tubes and repeat the centrifugation step.
5. After the second centrifugation, a Pasteur pipet is used to aspirate away the excess liquid. It is important to ensure that all the excess liquid is removed from the side of the tube; small amounts of medium running back into the pellet can significantly dilute the virus, but also be careful not to over-dry the sample (*see* **Note 6**).
6. For each tube, gently resuspend the virus pellet in 30 μL of 1X PBS by pipeting up and down several times gently; avoid introducing bubbles (*see* **Note 7**).
7. Place a small piece of Parafilm over the tube and leave the sample at 4°C for 4 h to overnight.
8. Pool all the samples into a 1.5-mL microtube. Typically about 100 μL of virus suspension can be collected from five centrifuge tubes.

3.3. Embryo Preparation

The embryos are injected at the blastula stage, approximately 3 h after fertilization. Because of this small window of time for injection and the many embryos to be injected, it is essential to generate clutches of embryos in "waves" by setting up crosses at different times throughout the morning. We typically set up three to four tanks of adult fish for crosses the evening before injection. The next morning female and male fish are crossed to generate embryos at 45- to 60-min intervals. Before injection, the chorions of embryos need to be removed. The detailed protocol for embryo preparation follows.

1. The evening before injection, 6–8 females and 3–4 males are kept separate in a large breeding box. The most common large breeding box is made by cutting out the bottom of a "double-width" mouse cage and gluing a piece of wire screen over the hole. The cut-out mouse cage is stacked into another mouse cage, which is filled with system water. Place the males in the lower section by putting them in before putting in the insert. Place the females in the "upper" chamber. The belief is that keeping the fish together, but unable to breed, will maintain the "interest" to breed longer than if they are kept completely separate. It is unclear whether this is actually true. Typically three to four breeding boxes are set up for a day's injections.
2. At "dawn" when the lights come on in the fish facility, mix the females and males from the first breeding box in the top portion of the breeding box. The fish are allowed to mate for no more than 15 min after the first embryos are released.
3. After the fish are removed, the embryos are collected by pouring the water containing the embryos through a small tea strainer. The embryos are rinsed with sterile methylene-blue-containing system water briefly, and rinsed into a 100-mL beaker.
4. Replace the methylene-blue-containing system water with 20 mL of freshly made 1X Holtfreiter's solution.
5. Add 150 μL of 10 mg/mL pronase solution to the embryos and swirl gently. The pronase will begin digesting away the chorions of the embryos. Periodically swirl the beaker gently to help separate the chorions from the embryos.
6. It is critical to stop the pronase digestion promptly so that the embryos do not suffer unwanted damage. Usually we stop the pronase treatment when about 15–20 embryos appear to be dechorionated. To estimate the number of dechorionated embryos, gently swirl the beaker and the embryos that have lost their chorions will collect in the middle of the beaker. Over-digestion will result in fragile embryos and poor survival.
7. To stop the pronase digestion, slowly add 60 mL of 1X Holtfreiter's solution to the beaker and gently pour off. Do not pour off all the buffer as surface tension will damage the

embryos. Repeat the wash steps 6–8 times until most chorions are washed away. However, it is normal to see a small portion of embryos with chorions still attached (in fact, when all embryos have lost their chorions, it usually indicates that the embryos have suffered over-digestion by pronase). For the final wash, use 1X Holtfreiter's solution containing 8 μg/mL of polybrene (*see* **Note 8**).

8. Incubate the embryos in 1X Holtfreiter's solution containing 8 μg/mL of polybrene in a 28°C incubator for approximately 2.5–3 h.
9. To inject as many embryos as possible on a single day, space crosses 45 min to 1 h apart for the remaining breeding boxes. If all three or four rounds of crosses give significant numbers of embryos (>300 embryos per cross), this will allow the injector 3–4 h of injection time in the afternoon with appropriately timed embryos available throughout the period of injection; 1000–2000 embryos can be injected per person in an afternoon.

3.4. Pre-injection Preparation

3.4.1. Needle Preparation

Microinjection needles may be pulled in any commercial needle puller. We use the Sutter Model P-2000 (Sutter Instrument). The needles we use are pulled from quartz capillaries with 1.0 mm outer and 0.7 mm inner diameter. The pulled needle should be long and thin as the injections are made into the intercellular space instead of into the cells directly. Thus, a long, thin needle can slip in between cells with a minimal amount of damage to the integrity of the cell cap.

3.4.2. Injection Ramps

Agarose ramps are used to secure and position embryos during injection. To make an injection ramp, place a regular microscope slide in the bottom of a 10-cm Petri dish and then pour about 15 mL of 2% *(w/v)* agarose (made with 1X Holtfreiter's solution) to just cover the slide. A second slide is then placed on the top of the first slide and rested on the rim of the Petri dish at an angle (**Fig. 1**). After the agarose hardens, carefully remove the second slide to create a ramp and a groove in the bottom of the dish. Fill the dish with 1X Holtfreiter's solution containing 8 μg/mL of polybrene and incubate at 32°C for at least 1 h before injection.

3.4.3. Recovery Dishes

After injection, embryos will be transferred from the ramp into the recovery dishes, which are regular six-well tissue culture dishes filled with 1X Holtfreiter's solution containing 8 μg/mL of polybrene. Prepare several of those buffer-filled dishes and place them in a 37°C incubator at least 1 h before the injections begin.

3.5. Virus Injection

3.5.1. Injection Apparatus

The injection apparatus we use is a simple device. It consists of a 20-mL syringe with a 20-gauge syringe needle, connected to a needle holder by polypropylene tubing (**Fig. 2**). The front of

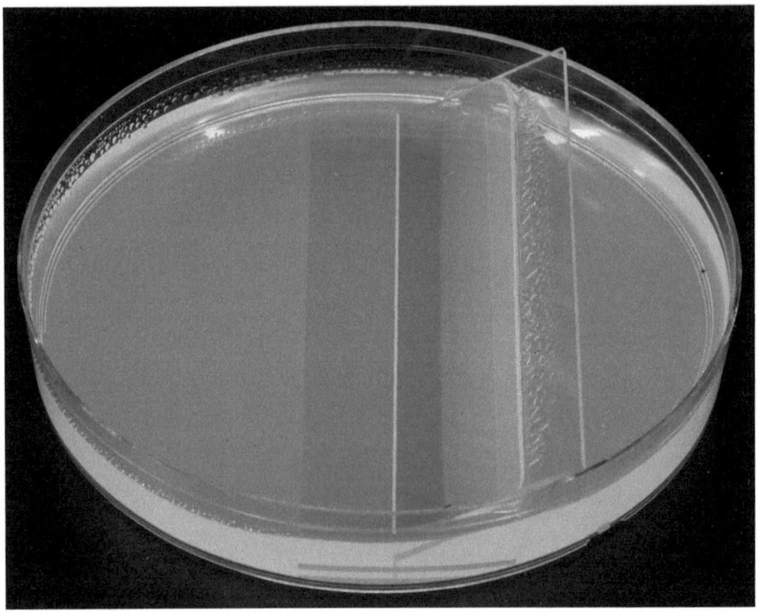

Fig. 1. Injection ramp.

the needle holder has a small silicon gasket, which is used as an airtight seal when the needle is inserted. The end of the pulled needle is sealed and needs to be removed with a scalpel blade. This cut is made under a dissecting microscope by lowering the blade straight down onto the needle. The location of the cut should be close to the point as we want the needle hole to be very small. However, because the viral preparations are slightly particulate, the hole needs to be large enough so that it does not get clogged with debris. The needle is then put into the silicon gasket and held in place by a screwed cap that fits over the end of the needle holder.

3.5.2. Loading Virus into the Needle

Just before injection, add polybrene to a final concentration of 8 μg/mL from a 20X polybrene stock and a trace amount of 1% phenol red (2 μL every 100 μL of concentrated virus) to the virus suspension prepared in **step 8 of Subheading 3.2** to help visualize the virus during injection. The virus stock now is ready for use in injections (*see* **Note 9**). To load the virus into the needle, on a clean microscope slide, add a drop (~15 μL) of the virus stock under the dissecting microscope, which is mounted in an injection hood (*see* **Note 10**). The virus is then drawn up into the needle by applying a vacuum with the 20-mL syringe. The concentrated virus often contains a significant amount of small debris, which inevitably will clog the needle. When the needle is clogged, apply the pressure on syringe while lifting the needle out of the surface near the edge of the virus drop. This helps

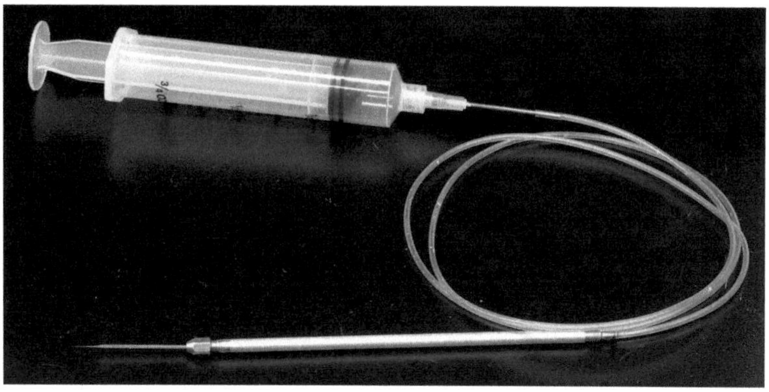

Fig. 2. Injection apparatus.

clear up the clog and also leaves the debris on the edge of the drop so that the same debris will not be sucked into the needle again. By repeating the "drawing and pushing" steps, the needle should be filled up within 5–10 min. Be careful not to introduce air bubbles into the needle by always keeping the needle under the liquid surface while applying a vacuum. We find it easier to use a hand-held needle holder rather than a micromanipulator for both aspiration of the virus as well as for injection (*see* **Note 11**). The main reason the virus is aspirated into the needle instead of being loaded with a "gel-loading" pipet tip from the back end, is to prevent clogging of the needle during injections.

3.5.3. Injection Procedure

Once the needle is filled with virus, take out the agarose ramp from the 32°C incubator and place the embryos in a single row at the bottom edge of the ramp using a 5.75-in. wide-bore Pasteur pipet. Caution should be taken while transferring the embryos because they are very fragile without the protection of the chorions (*see* **Note 12**). The embryos placed on the ramp should be between the 512- and 2000-cell stages. This is the optimal time frame for infecting the germ cells. At this stage there are four primordial germ cells, which will divide into 20–30 cells in the next few hours *(15)*, and the cell cap provides a "space" where the virus can be injected and retained long enough for infection. The injection ramp is then placed on a dissecting microscope in an injection hood. To start the injection, set the plunger of the 20-mL syringe half way along the syringe barrel and gently push the syringe to initiate the virus slowly flowing out of the needle point. The needle point is then inserted into the cell mass (**Fig. 3**). Only a very gentle "touch" is needed to get the point into the cell mass. Avoid penetrating the yolk with the point, for this is frequently fatal to the embryo. Each embryo is injected 5–6 times at different locations around the cell cap. The injector should

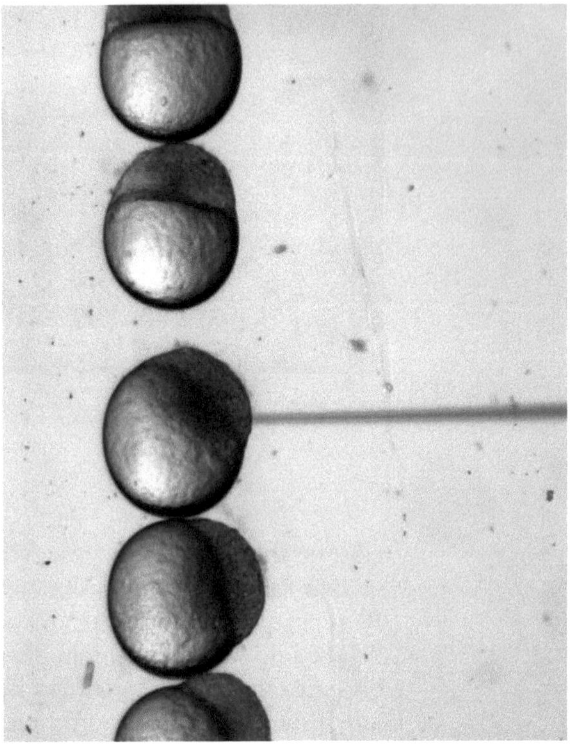

Fig. 3. Injection procedure. Dechorionated embryos (512- to 2,000-cell stages) are placed at the bottom edge of the agarose ramp. Viruses are injected into each embryo at different locations around the cell cap.

try to distribute the phenol red color evenly over the entire cell mass. The flow rate of the virus out of the needle tip should be kept as constant as possible to ensure that every embryo receives a similar amount of virus. It usually takes 15–20 min to inject one row of embryos (~100 embryos in an agarose ramp). After every embryo on the ramp has been injected, the injector returns to the beginning of the ramp and repeats the injection process. But this time only 2–3 "pokes" are given to each embryo. Tests in our laboratory have shown that there is generally an increase in overall virus infection when the injections are done twice to each embryo (*see* **Note 13**).

Once the injections are complete, the embryos are carefully transferred from the injection ramp to the pre-warmed six-well dishes. Each well of the six-well dish is loaded with a similar number of embryos. A single six-well dish usually holds embryos from one or two ramps of injected embryos (i.e., ~100–200 embryos). The six-well dish is then placed in a 37°C incubator for a 90-min heat shock period. After heat shock, the dish is transferred to a 32°C incubator and left for about 20 min. The dish is then taken out of the incubator again; the broken, unfertilized embryos, and their debris are removed from each well under the dissecting microscope. This cleanup process is critical to ensure the proper development of the remaining healthy-looking embryos. After the cleanup, the dish is returned to a 32°C incubator and left overnight. The next morning embryos that appear to have developed normally are transferred into filtered system water in a 10-cm Petri dish and raised in the normal fashion at 28°C.

3.6. Virus Evaluation

Because the production of pseudotyped virus depends on transient transfection of the packaging cells with the plasmid encoding VSV-G, the titer of the virus can vary significantly between different viral preparations. The most direct assessment of the efficacy of the virus from different viral preparations is to determine the level of infectivity (i.e., average proviral copies per cell) in the injected founders. Although this value does not indicate the actual number of proviral copies that will be transmitted to the next generation, it is a fairly good predicator of overall transgenicity, which correlates with germline transmission. To have a quick assessment on the efficacy of the virus, we randomly select two to three embryos from each batch of injected embryos, isolate their genomic DNA, and determine the average proviral copies in the isolated DNA by using a multiplex qPCR-based assay (designated as Embryo Assay) *(7)*. The number of proviral insertions per cell is computed by measuring the amplification rate of the *SFG* locus, which is specific to proviral DNA, and comparing the ratios of threshold values between founder embryo DNA and DNA with known copy numbers of proviral insertions; the results are normalized to the control *RAG1* locus, which is simultaneously measured. A detailed procedure of a typical Embryo Assay is as follows.

1. Select 2–3 healthy-looking 2- to 3-d-old injected embryos and put them into a 1.5-mL microtube (*see* **Note 14**).
2. Remove the residual water and add 100 μL of lysis buffer (with freshly added 100 μg/mL proteinase K) to the embryos.
3. Incubate the microtube in a heat block at 50°C. Resuspend the solution every 10 min to help the lysis process until the solution becomes homogenous without any visible tissue. This step usually takes 30–40 min.
4. Add 100 μL of isopropanol to the solution, vortex, and centrifuge in a benchtop centrifuge at the maximum speed for 5 min at room temperature to precipitate the genomic DNA. A black pellet should be seen in the bottom of the tube after centrifugation.
5. Discard the supernatant and wash the pellet with 200–400 μL of 70% ethanol. Repeat the centrifugation step.
6. Discard the supernatant and dry the pellet by applying a gentle suction over the pellet using a vacuum attached to a Pasteur pipet with the tip covered by a 200-μL pipet tip.

7. A 50 μL *Bam*HI restriction enzyme digestion reaction is setup to digest the genomic DNA in the pellet by adding the reaction mix directly into the tube with the dried pellet.
8. Incubate at 37°C for 30 min with occasional resuspension by pipeting up and down to break up the pellet.
9. During the 30 min restriction enzyme digestion, setup the qPCR "master mix" without the DNA in a volume that after the addition of 3 μL of the *Bam*HI-digested genomic DNA into the master mix, the final concentrations of the reaction components will be as follows and the total reaction volume per well will be 25 μL: 1X PCR buffer containing 5 mM $MgCl_2$, 0.4 mM dNTP, 0.2 μM of each *SFG* and *RAG1* primer, 0.5 μM *RAG1* probe, 0.25 μM *SFG* probe, 2.5 U of Platinum® *Taq* polymerase. We usually setup the reaction in duplicates for each DNA sample tested in a 96-well PCR plate. Each run also contains wells of a reference control from a fish with known copy number of inserts.
10. The PCR is carried out with an iCycler (Bio-Rad) using HEX-530 and FAM-490 channels for *RAG1* and *SFG* probes, respectively. The cycle profile is 2 min at 95°C, followed by 40 cycles of (15 s at 95°C, 30 s at 60°C).
11. At the end of the run, the *RAG1* and *SFG* threshold cycles (Cts; the cycle at which the amount of product passed a certain threshold in the linear amplification range) are calculated for each sample. A ΔCt value is defined by subtracting the *SFG* Ct from the *RAG1* Ct. The larger the ΔCt value, the greater the number of proviral copies for any given sample is. By subtracting the reference's ΔCt from each sample's ΔCt (the reference contains known *N* proviral copies per cell), we calculate the ΔΔCt, which can then be used in the following formula to estimate the number of average proviral copies per cell in any given fish: the average proviral copies per cell in a given fish = $N \times 1.9^{\Delta\Delta Ct}$.

4. Notes

1. Coating the cell-growing surface with the positively charged poly-L-lysine molecules helps the GT186 cells remain firmly attached to the surface, making it ready for the subsequent transfection procedure. 600-mL flasks can be replaced with the 15-cm diameter cell culture dishes for cell culture. Using flasks makes the handling easier during the collection of medium, which contains the viral particles. However, flasks are significantly more expensive than dishes.

2. GT186 cells will gradually lose the efficacy of packaging viral particles, possibly because of the instability of the integrated viral genome and/or *gag-pol* genes after prolonged propagation. It is thus critical to use GT186 cells with a limited number of passages. After a batch of cells has been propagated for a while (~3–4 months) and the infectivity starts to drop, we thaw a new vial of early-passage cells from the frozen stock.
3. 293 cells attach to each other more strongly than they do to the bottom of the flask. To ensure even plating of the cells, you must actively dissociate the cells from each other. To get the cells into a single-cell suspension, we rest the pipet tip against the bottom of the flask while pipeting up and down the cell suspension. In this way the solution is passing through a narrow space, resulting in a greater shearing force that can separate the cell aggregates more efficiently.
4. Opti-MEM I medium can be replaced with serum-free DMEM medium from **steps 2** to **5 of Subheading 3.1.2**. The washing step (**Subheading 3.1.2**, **step 5**) may also be carried out using 20 mL of 1X PBS.
5. At 48-h post-transfection, you should see many cell-to-cell fusions in the VSV-G-transfected GT186 cells. This usually indicates a good production of infectious MLV particles as the expressed VSV-G proteins tend to induce cell fusion.
6. It is helpful to prevent the liquid running down the sides of the centrifuge tube, by scratching the inner side of the tube with the Pasteur pipet to make a spiral path while aspirating the excess liquid. Any residual liquid will thus be trapped by the scratched grooves instead of running back to the virus pellet when the tube is upright and the pellet is suspended in PBS.
7. Usually the whiter the suspension at this point, the higher the viral titer will be.
8. Promptly stopping the pronase digestion appears to be one of the most critical steps to have good quality embryos for injection. Most chorions will come off of the embryos during the subsequent washing steps, not during the digestion period. Thus, do not hesitate to stop digestion even when only few embryos have lost their chorions. It takes some practice to get the timing right, but generally when about 20 embryos have lost their chorions, the washing step should be begun. During the washing step, it is convenient to use a 25-mL pipet. After drawing the solution into the pipet, usually we detach the pipet from the pipettor and use the thumb to hold and release the solution slowly into the beaker in an effort trying not to overly disturb the embryos, minimizing the potential damage to the delicate embryos.

9. The virus stock can be stored at 4°C for up to 5 d without significantly losing titer. We do not recommend freezing down the virus since the viral titer will decrease by approximately 50% after freezing.
10. VSV-G pseudotyped MLV can also infect humans. Proper precautions should be used in handling this type of virus.
11. Loading the needle with virus can be one of the most difficult steps in the whole procedure because the needle tip will inevitably be clogged by the small debris in the viral preparation. We have found the best way to load the needle is to move the syringe plunger almost, but not all the way down to the bottom of the syringe. This allows a strong vacuum to be applied to the needle, but leaves a small space to apply pressure in the other direction if the needle becomes clogged. We also found that it is helpful not to start applying the pressure to clear the clog until the needle is completely clogged or the virus just drawn into the needle will almost be completely pushed out during the clearing process, resulting in almost no gain in the loading process. It is also important not to draw in any air bubbles because the small bubbles also clog the needle tip. Once the needle is full, it is better to disconnect the hose again (it is also a good idea to learn how to do the unplugging just using one hand) and move the plunger to about half way up the syringe. This gives more control over applying pressure during injection.
12. It is important to use the Pasteur pipet with an opening wide enough (e.g. at least 2 mm in diameter) to transfer embryos to prevent them from crushing each other as they pass through the opening. To place the embryos on the ramp as a neat line, slowly draw the embryos into the pipet from the beaker (it is important to fully release the bulb before retrieving the pipet tip out of the surface so that the pipet is completely filled with the solution without trapping any air in the very front of the pipet tip). Immerse the pipet tip just under the surface of the solution in the ramp. Let gravity draw the embryos down and out of the pipet without squeezing the bulb. While the embryos are dropping out of the pipet, move the pipet along the ramp to have the embryos drop as a single line.
13. The injection of virus into embryos is a learned skill. It takes time for an injector to get acquainted with the technique. It is a delicate balance between giving the embryo enough virus to achieve high infectivity, and fatally damaging the embryo with excessive pokes and virus. Several pointers below should help a novice shorten the learning process: (1) Because the virus is injected between cells, the needle needs to be very thin to reduce the chance of poking into the cells directly.

However, using a fine needle also means an increased difficulty of loading the virus. The injector should resist the urge to cut a larger needle hole, which will not only easily disrupt the embryo but also cause needless waste of virus. (2) Try to do several small injections, evenly throughout the entire cell mass. The tip of the needle can be used to orient the embryos. A very gentle "touch" should be enough to penetrate the needle into the cell mass. Always avoid damaging the yolk. The hand holding the injection apparatus should rest firmly on the stage of the microscope to increase stability. While doing the second round of injections, pay extra attention to those embryos with lighter dye tracer. One or two extra injections can be given to those lighter-colored embryos. (3) Pay attention to the virus flow out of the needle tip. Use the hand on the syringe plunger to control the flow rate by gently pushing or pulling the plunger when needed during injection. If the needle hole is fine enough, the virus flow should stay constant for quite a while after a gentle push. (4) Micromanipulators are not recommended. It is difficult at first to inject "freehand," but eventually it is much faster than the micromanipulator, and high injection numbers are desirable to offset reduced survival.

14. You should wait until the embryos are at least 2 d-old before performing the embryo infection assay to prevent measuring the "unintegrated" viral DNA, which is still present in the early embryos. This will result in artificially high estimates of infection. It is also important to avoid selecting the unhealthy looking embryos to assay because the measured values from those fish do not represent the infectivity of the healthy population (again, they tend to be inflated values).

Acknowledgements

This research was supported by the Intramural Research Program of the National Human Genome Research Institute, National Institutes of Health.

References

1. Driever, W., Solnica-Krezel, L., Schier, A. F., Neuhauss, S. C., Malicki, J., Stemple, D. L., Stainier, D. Y., Zwartkruis, F., Abdelilah, S., Rangini, Z., Belak, J., and Boggs, C. (1996). A genetic screen for mutations affecting embryogenesis in zebrafish. *Development* **123**, 37–46.

2. Haffter, P., Granato, M., Brand, M., Mullins, M. C., Hammerschmidt, M., Kane, D. A., Odenthal, J., van Eeden, F. J., Jiang, Y. J., Heisenberg, C. P., Kelsh, R. N., Furutani-Seiki, M., Vogelsang, E., Beuchle, D., Schach, U., Fabian, C., and Nusslein-Volhard, C. (1996). The identification

of genes with unique and essential functions in the development of the zebrafish, Danio rerio. *Development* **123**, 1–36.

3. Lin, S., Gaiano, N., Culp, P., Burns, J. C., Friedmann, T., Yee, J. K., and Hopkins, N. (1994). Integration and germ-line transmission of a pseudotyped retroviral vector in zebrafish. *Science* **265**, 666–9.
4. Burns, J. C., Friedmann, T., Driever, W., Burrascano, M., and Yee, J. K. (1993). Vesicular stomatitis virus G glycoprotein pseudotyped retroviral vectors: concentration to very high titer and efficient gene transfer into mammalian and nonmammalian cells. *Proc Natl Acad Sci U S A* **90**, 8033–7.
5. Yee, J. K., Miyanohara, A., LaPorte, P., Bouic, K., Burns, J. C., and Friedmann, T. (1994). A general method for the generation of high-titer, pantropic retroviral vectors: highly efficient infection of primary hepatocytes. *Proc Natl Acad Sci U S A* **91**, 9564–8.
6. Amsterdam, A., and Hopkins, N. (2006). Mutagenesis strategies in zebrafish for identifying genes involved in development and disease. *Trends Genet* **22**, 473–8.
7. Amsterdam, A., Burgess, S., Golling, G., Chen, W., Sun, Z., Townsend, K., Farrington, S., Haldi, M., and Hopkins, N. (1999). A large-scale insertional mutagenesis screen in zebrafish. *Genes Dev* **13**, 2713–24.
8. Amsterdam, A., Nissen, R. M., Sun, Z., Swindell, E. C., Farrington, S., and Hopkins, N. (2004). Identification of 315 genes essential for early zebrafish development. *Proc Natl Acad Sci U S A* **101**, 12792–7.
9. Gaiano, N., Amsterdam, A., Kawakami, K., Allende, M., Becker, T., and Hopkins, N. (1996). Insertional mutagenesis and rapid cloning of essential genes in zebrafish. *Nature* **383**, 829–32.
10. Golling, G., Amsterdam, A., Sun, Z., Antonelli, M., Maldonado, E., Chen, W., Burgess, S., Haldi, M., Artzt, K., Farrington, S., Lin, S. Y., Nissen, R. M., and Hopkins, N. (2002). Insertional mutagenesis in zebrafish rapidly identifies genes essential for early vertebrate development. *Nat Genet* **31**, 135–40.
11. Ellingsen, S., Laplante, M. A., Konig, M., Kikuta, H., Furmanek, T., Hoivik, E. A., and Becker, T. S. (2005). Large-scale enhancer detection in the zebrafish genome. *Development* **132**, 3799–811.
12. Chen, W., Burgess, S., Golling, G., Amsterdam, A., and Hopkins, N. (2002). High-throughput selection of retrovirus producer cell lines leads to markedly improved efficiency of germ line-transmissible insertions in zebra fish. *J Virol* **76**, 2192–8.
13. Wang, D., Jao, L. E., Zheng, N., Dolan, K., Ivey, J., Zonies, S., Wu, X., Wu, K., Yang, H., Meng, Q., Zhu, Z., Zhang, B., Lin, S., and Burgess, S. M. (2007). Efficient genome-wide mutagenesis of zebrafish genes by retroviral insertions. *Proc Natl Acad Sci U S A* **104**, 12428–33.
14. Miyoshi, H., Takahashi, M., Gage, F. H., and Verma, I. M. (1997). Stable and efficient gene transfer into the retina using an HIV-based lentiviral vector. *Proc Natl Acad Sci U S A* **94**, 10319–23.
15. Yoon, C., Kawakami, K., and Hopkins, N. (1997). Zebrafish vasa homologue RNA is localized to the cleavage planes of 2- and 4-cell-stage embryos and is expressed in the primordial germ cells. *Development* **124**, 3157–65.

Chapter 3

Undertaking a Successful Gynogenetic Haploid Screen in Zebrafish

Judith E. Layton

Summary

Chemical mutagenesis using *N*-ethyl-*N*-nitrosourea is the current method of choice for dense mutagenesis in zebrafish. Methods are available for both pre-meiotic and post-meiotic sperm mutagenesis; in this chapter, pre-meiotic mutagenesis is described. Mutated males are crossed with untreated females to create an F1 generation that is heterozygous for the mutations. The F1 females can be screened directly by making haploid embryos using in vitro fertilization (IVF) with ultraviolet (UV)-irradiated sperm. This approach requires substantially fewer fish and less aquarium space than the classical F2 generation screen and is feasible for a small research group. Production of haploid embryos is described in detail.

Key words: Mutagenesis, ENU, Gynogenetic, Haploid, Screen Zebrafish.

1. Introduction

Chemical mutagenesis is established as an unbiased approach to achieving dense mutagenesis of a genome *(1–3)*. In zebrafish, the chemical of choice is *N*-ethyl-*N*-nitrosourea (ENU) and it is usually used to mutate male fish. ENU is an alkylating agent causing transfer of an ethyl group to DNA (and carbamoylation of proteins). During subsequent rounds of DNA replication, the DNA adducts most commonly give rise to AT to TA transversions and AT to GC transitions *(1–3)*. Less frequently, small deletions occur *(4)*. These mutations will most commonly cause recessive phenotypes.

Graham J. Lieschke et al. (eds.), *Zebrafish*, Methods in Molecular Biology, vol. 546
DOI: 10.1007/978-1-60327-977-2_3

During spermatogenesis, spermatogonia (sperm stem cells) undergo several rounds of mitosis before meiosis followed by sperm maturation. Post-meiotic mutagenesis results in sperm that carry the mutations on one DNA strand, requiring further DNA replication to fix the mutation. Fertilization and subsequent rounds of DNA replication result in mosaic progeny that may not carry the mutation in their germ-line. Thus, recovery of these mutations is variable. It has also been reported that deletions are more frequent than point mutations in post-meiotic mutagenesis *(5)*. Mutation of pre-meiotic spermatogonia, on the other hand, results in mutations that are present on both DNA strands in the sperm and progeny that are non-mosaic and heterozygous for the induced mutations. Most screens have been undertaken with protocols designed to cause spermatogonial mutation, however, sperm mutation protocols have also been used *(5–7)*. This chapter describes a protocol causing spermatogonial mutation. The treatment conditions giving the best results have been established by others *(8, 9)* and used successfully in two large scale screens *(10, 11)*.

After mutagenesis, the treated males are crossed with untreated females of the same strain and F1 families raised that are heterozygous for the induced mutations. An advantage of using zebrafish is that the F1 females can be tested for recessive mutations directly by producing haploid embryos as described below *(12)*. This approach requires far fewer fish to be raised and tested than the classical approach of Haldane in which F2 families are screened by pair-wise matings *(3, 9)* and is feasible for a small research group. Although screening occurs one generation earlier than in a classical screen, ultimately the putative F1 mutants must be recovered in the F2 generation, so the number of generations required is the same as in the Haldane method, but the number of F2 fish required is substantially reduced.

The haploid screen saves time, aquarium space and labor but the use of haploids has some disadvantages. Haploid embryos develop reasonably normally for the first day or two, but become progressively more abnormal with age and die around day 4 or 5. They have a characteristic appearance, being shorter than normal, with shorter, often irregular somites and circulation defects. Therefore the chosen screening characteristic should be tested in haploid embryos to establish the feasibility of a haploid screen. It is difficult to stage haploid embryos precisely because of their abnormalities; their rate of development is slower than diploid embryos and is more variable, both between and within clutches, thus ideally the screening method should not require precise staging. The fish stock to be used in the screen should be tested for proportion of good quality haploids before ENU treatment (**Fig. 1**). There should be at least 50% good haploid embryos or scoring for the phenotype of interest will be too difficult *(13)*.

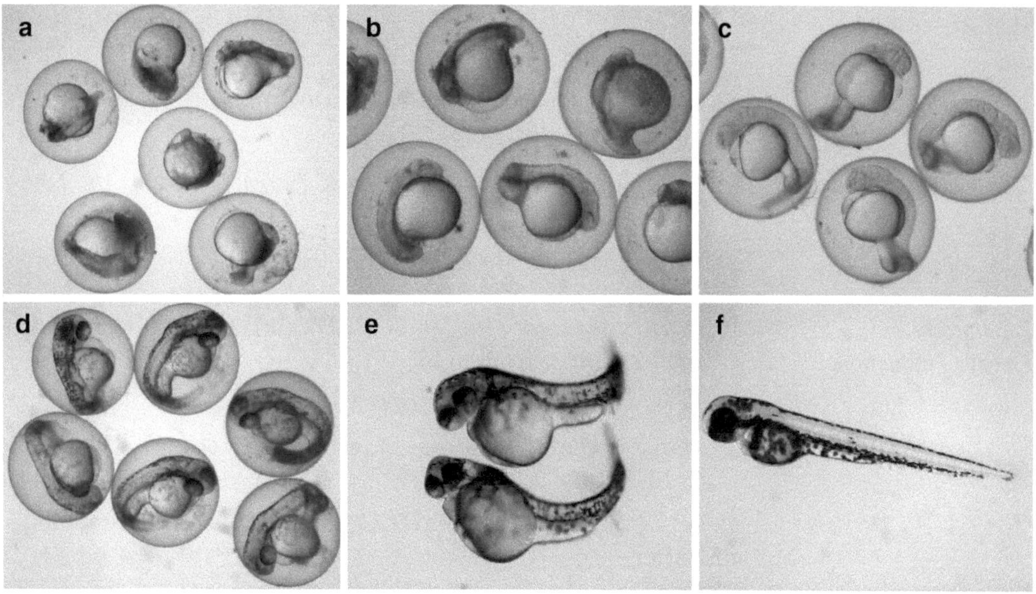

Fig. 1. Appearance of haploid embryos. Typical appearance of haploid embryos (**a–e**), compared with a diploid embryo (**f**). (**a**)–(**c**), 24–27 hpf; (**d**)–(**f**), 52 hpf. (**a**) poor quality haploids (discard); (**b**), intermediate quality; (**c**)–(**e**), good quality. (**c**), (**d**) *golden* eggs, (**a**), (**b**), (**e**), (**f**), St. Kilda eggs.

The methods for ENU treatment of fish and production of haploid embryos are described in detail below.

2. Materials

2.1. ENU Treatment (for 2 × 6 fish)

1. Eight 600-mL or 1 L plastic beakers and two beaker covers (e.g., large Petri dish).
2. Two beaker inserts for transferring fish (can be made from a cut up 2 L plastic cylinder of diameter smaller than that of the beakers, with nylon mesh attached with silicone to cover one end).
3. Sieve or small net in case of escapees or dead fish.
4. Insulated box large enough to contain at least two beakers.
5. 50 mL syringe and large and small gauge needles.
6. One pair large blunt forceps.
7. Isopac bottle of 1 g ENU (Sigma N3385-1G).
8. 0.5 M Na_2HPO_4.
9. 0.5 M NaH_2PO_4.
10. Glacial acetic acid.

11. Aquarium system water.
12. Bench paper.
13. Cytotoxic waste disposal bags/containers.
14. Disposable gloves.
15. ENU inactivating solution: 20% *(w/v)* $Na_2S_2O_3$, 1% *(w/v)* NaOH.
16. Large container for inactivation bath.

2.2. Haploid Embryo Production

1. *Benzocaine stock (100X).* 10 g/100 mL ethanol. Cover with foil to protect from light.
2. Beakers for benzocaine (600 mL ×2).
3. Hank's premix solution (aliquoted and kept at −20°C) (*see* **Subheading 2.3**).
4. 0.174 g aliquot of $NaHCO_3$ in 5 mL tube (for Hank's buffer).
5. Plastic spoon.
6. Forceps, scissors, scalpel blades.
7. Gauze, tissues and paper towels.
8. Microtube pestles.
9. 1.5 mL microtubes.
10. Small spatula.
11. Marker pen.
12. Timer.
13. Sterile pipet tips.
14. Pipets (10, 200 and 1,000 μL).
15. Dissecting microscope.
16. 90-mm Petri dishes.
17. 500 mL tissue culture flasks and beakers for fish containing system water.
18. Two watch glasses in shallow plastic box.
19. Cling wrap.
20. Bottles of water and egg water.
21. UV light box (e.g., Stratagene UV lamp, 254 nm, 15 W, mounted under roof, 45 cm from base of box).

2.3. Hank's Buffer Premix

1. *Stock #1.* 8.0 g NaCl, 0.4 g KCl, 100 mL purified water.
2. *Stock #2.* 0.358 g Na_2HPO_4 (anhydrous), 0.60 g KH_2PO_4, 100 ml purified water.
3. *Stock #4.* 0.72 g $CaCl_2$ 50 mL purified water.
4. *Stock #5.* 1.23 g $MgSO_4 \cdot 7H_2O$, 50 mL purified water.

Combine in order: 10.0 mL stock #1, 1.0 mL stock #2, 1.0 mL stock #4, 86.0 mL purified water and 1.0 mL stock #5. Store for several weeks at 4°C or longer at –20°C.

3. Methods

3.1. ENU Treatment of Male Zebrafish

WARNING: ENU is carcinogenic as well as mutagenic and must be used with great care to avoid exposure of laboratory personnel. Protective clothing (ie, gown, gloves, glasses) must be worn. ENU should only be used in a cytotoxic hood and all waste should be disposed of appropriately. The ENU treatment is toxic to fish and it is likely there will be some deaths.

To minimize stress, fish should be kept quiet and in the dark during treatment. If the hood is in a shared laboratory, ask other lab members to be as quiet as possible. Treatment is carried out at a lower than normal temperature, also to reduce stress. Our experience with ENU treatment is summarized in **Table 1**.

1. Bring the fish to be treated (*see* **Note 1**) over from the aquarium the day before (or at least a few hours beforehand) to allow equilibration to 20–22°C.

Table 1
Effect of Fish Strain and Dose of ENU on Outcome

Strain	ENU treatment (mM)	No. treated	Died during treatment	Infertile	Gave offspring	Mutants recovered[a]
St. Kilda[b]	2.5	18	1[c]	0	15	3
	3.0	13	3	2	8	12
St. Kilda-total		31	4	2	23	15
WIK	2.5	19	0	5	14	1
	3.0	6	2	2	2	0
WIK-total		25	2	7	16	1
AB*	2.5	7	0	0	7	4
All strains	2.5	44	1	5	36	8
	3	19	5	4	10	12

[a]For details of the type of mutants, *see (14)*
[b]Local pet shop strain
[c]Two escaped during treatment

2. Turn on cytotoxic hood. Cover hood and floor with bench paper. Make sure that cytotoxic waste containers are available and put a cytotoxic waste bag in the hood. Always wear two pairs of gloves and change gloves frequently, especially if contamination is suspected. Remove and discard gloves when leaving the cytotoxic hood. Use disposable equipment where possible, otherwise use equipment (e.g., beakers) that is dedicated for ENU use only.
3. Make up stock ENU solution in the cytotoxic hood. Remove the ENU isopac from the can and equilibrate to room temp. Remove metal collar with blunt forceps and insert small needle to allow air flow. Inject 2 × 50 mL acetic acid solution into the isopac bottle containing 1 g ENU. The acetic acid concentration is calculated to give a final 10 mM concentration, taking into account the amount in the ENU (see label on bottle).

For example: For an ENU batch containing 1.2% acetic acid

- 0.012 g in 1 g ENU.
- MW acetic acid = 60.02, final volume = 100 mL.
- Molarity = (0.012/60.02) × 10 = 2 mM.
- Therefore make up 8 mM acetic acid.
- Glacial acetic acid = 17.5 M. Hence 8 mM requires 0.457 mL/L.

Shake vigorously intermittently over several hours until most of the ENU is dissolved (*see* **Note 2**). The stock will be 85 mM. Using a syringe and needle or disposable pipet, transfer the solution to 15 mL tubes. The unused solution can be stored in the tubes in a sealed container at −20°C for a few weeks and used for subsequent treatments.

- o Make up 10 mM sodium phosphate buffer, pH 6.6, from stock 0.5 M solutions of Na_2HPO_4 and NaH_2PO_4. (28 mL NaH_2PO_4 and 12 mL Na_2HPO_4 in 2 L is about right). Check the pH because the activity of ENU is pH-dependant. Dilute 14.6 mL ENU solution (for 2.5 mM) or 17.7 mL (for 3 mM ENU) into 500 mL phosphate buffer. Use a disposable pipet to measure the amount of ENU required. Treat up to six fish in 500 mL.
- o For each beaker of ENU, prepare two more beakers of about 600 mL of the phosphate buffer to rinse the fish in after treatment and a beaker of aquarium system water for recovery. All solutions and fish need to be at 20–22°C to minimize fish stress. Use ice if necessary for cooling.
- o Transfer the fish to the beaker insert (cylinder with mesh over one end) and place in the beaker of ENU. Cover cylinder to prevent fish jumping out (e.g.,with Petri dish). From now

until the end of recovery, fish must be kept quiet and dark to minimize stress (e.g., inside a thick polystyrene foam box in the cytotoxic hood). Turn off the hood, shut the front cover and leave fish in ENU for 1 h.

- After 1 h, open up and turn on the hood and transfer fish (in the insert) sequentially into the two beakers of phosphate buffer for a few minutes each to wash off ENU solution.
- Transfer fish to a beaker of system water (in the insert) and leave them to recover for at least 3 h in the quiet and dark (can be overnight). Fish can then be returned to the aquarium in fresh system water. Their tank should be labeled toxic and aquarium staff notified not to touch them for 24 h. Any fish that die in the first 24 h should be disposed of in a cytotoxic waste container.
- All solutions and non-disposable equipment must be inactivated in a bath of 20% sodium thiosulphate ($Na_2S_2O_3$) plus 1% NaOH, preferably separate baths for the ENU solutions and the equipment. The volume should be at least twice the volume of liquid to be inactivated. A 10 L plastic box is convenient for this and can also be used for storage of ENU equipment. For 5 L of solution, add 1 kg $Na_2S_2O_3$ and 50 g NaOH. Put the container(s) of inactivating solution in the cytotoxic hood, pour in all solutions and immerse any non-disposable equipment used. Leave overnight to inactivate and dispose of the next day. Solid waste is disposed of in cytotoxic waste containers.
- Repeat the ENU treatment twice, at 3–4 day intervals (*see* **Note 3**).
- Cross the ENU-treated males with untreated females, preferably as individual pairs to monitor their fertility. For about the first 2–4 wk, the males will produce heavily mutated sperm, resulting in abnormal embryos that will mostly not survive. Any survivors will be mosaic for the ENU-induced mutations, which will not necessarily be carried in the germ-line, therefore the mutations are often not recoverable. After this period, sperm will be derived from mutated spermatogonia, giving rise to non-mosaic embryos, heterozygous for the ENU-induced mutations.

3.2. Testing the Mutation Rate

It is advisable to test the effectiveness of the ENU treatment by determining the mutation rate at an indicator locus in a non-complementation test. The recessive pigmentation mutants (*gol, alb, spa, bra*) are useful for this. Cross the ENU-treated males with homozygous mutant indicator females and determine the frequency of phenotypically mutant embryos produced. The target locus specific inactivation rate frequency should be about

1/1,000, and this is approximately what was achieved overall at the *gol* locus with the regimens described in **Table 1**. For a haploid screen, a higher mutation rate is undesirable because it may give too many abnormal embryos.

3.3. Raising Offspring

Set up each mutated male (F0) with one or two wild-type females of the same strain to produce embryos that will be the F1 generation. Number each male and use a labeling system for the offspring so that you can identify which male they were derived from. This may be useful when positionally cloning mutations, because one can determine whether a sequence variation that is a potential mutation was ENU-induced by testing whether it was present in the original ENU-treated male ancestor or not. Freeze the male to provide DNA later when enough embryos have been raised.

The number of fish required will be determined by the number of genomes to be screened. For a gynogenetic haploid screen, only females will be screened and it is unlikely that good eggs will be obtained from all females, therefore about a fourfold excess of the number of fish intended to be screened should be raised.

Another factor to consider is that a male has about 500–1000 spermatogonia, so it is advisable to screen no more than 1000 genomes derived from one ENU-treated male to reduce the possibility of detecting the same mutation twice. In reality, we did not come close to this limit in any of our F1 families.

3.4. Screening

3.4.1. Production of Haploid Embryos

Haploid embryos are produced by in vitro fertilization of eggs with UV-irradiated sperm. The sperm are prepared in Hank's buffer, which prevents their activation until water is added.

1. Prepare Hank's Buffer:
   - (a) Thaw out an aliquot of Hank's premix and dispense 990 μL volumes into 1.5 mL microtubes.
   - (b) Dissolve 0.174 g $NaHCO_3$ in 5 mL water (must be made fresh).
   - (c) Add 10 μL $NaHCO_3$ solution to each 990 μL Hanks premix aliquot.
   - (d) Put 800 μL aliquots of complete Hank's into Eppendorf tubes on ice.
   - (e) Refreeze unused Hank's premix (can be kept at 4°C for a short time).
2. Chill two watch glasses:
   - (a) Place ice in shallow plastic box and cover with cling film to prevent ice getting onto the watch glasses. Put watch glasses on the covered ice.
   - (b) Wipe condensation from watch glass just before use as water activates sperm.

3. Prepare anesthetic: 100 μL of Benzocaine stock into 100 mL aquarium system water. (Other suitable anesthetics can be used).
4. Male fish dissection for sperm collection:
   (a) Place up to four males in benzocaine (no need to do singly here as they are to be killed) (*see* **Note 4**).
   (b) Remove fish and blot dry, cut off head and make incision along belly.
   (c) Place fish belly-up on a gauze pad and open out abdomen.
   (d) Pull gut out gently and discard.
   (e) Remove testes (white, opaque organs, one sitting on each side of swim bladder, **Fig. 2**). It is easier to use a dissecting microscope for this step.
   (f) Put testes in Hank's buffer on ice (two fish into 300 μL, 4–6 fish into 800 μL, *see* **Note 5**)
   (g) Homogenise testes gently with a microtube pestle or use the plunger from a 1 mL syringe.
   (h) Allow debris to settle out on ice (sperm is in supernatant).
5. UV-irradiated sperm:
   (a) Check UV light in light box is working (use safety glasses).

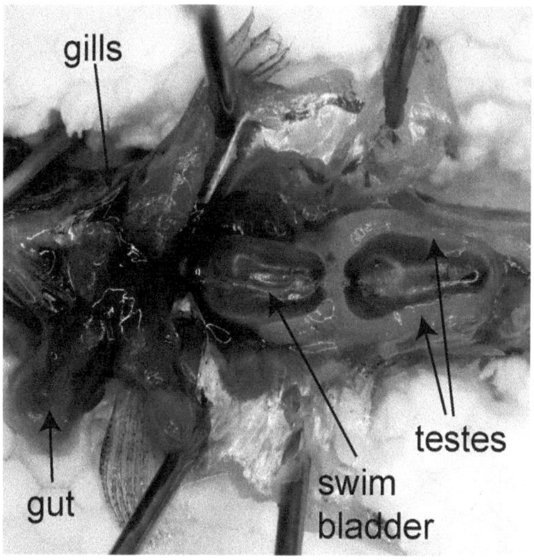

Fig. 2. Ventral view of a dissected male zebrafish showing the position of the testes. The abdomen of the fish has been opened and the testes can be seen lying on either side of the swim bladder. The displaced gut and the gills are indicated. Rostral end to the left, caudal end to the right.

   (b) Put sperm suspension (excluding debris) onto watch glass on ice, first making sure there is no condensation on the glass, and spread out the suspension to make as thin a layer as possible. Discard used pipet tip and tube to prevent inadvertent mixing of irradiated and non-irradiated sperm.
   (c) Put in UV light box for 1 min, open light box.
   (d) Swirl gently to mix and irradiate in light box for additional 1 min 15 s (2 min 15 s total, *see* **Note 6**).
   (e) Transfer with a clean pipet tip into a clean tube.
   (f) Further dilute with 200 μL/tube Hank's if required (*see* **Note 5**).

6. Extract eggs from female fish (*see* **Note 7**):
   (a) Best egg production is obtained in the first 1–2 h after the lights come on in the aquarium (*see* **Note 8**).
   (b) Anesthetize fish as lightly as possible, one at a time.
   (c) Place fish on a damp paper towel and dab dry with a tissue.
   (d) Place fish in a Petri dish, tail to left, belly away from squeezer (for right-handed person).
   (e) Hold fish's tail with middle finger, and support fish's back with pointer finger (left hand).
   (f) Dampen fingers, shake off excess water (right hand).
   (g) Gently stroke belly (from head towards tail) with a little pressure. Mature eggs will be expelled easily. Don't press too hard or the fish may be injured (*see* **Note 9**).
   (h) If eggs result (ripe eggs are yellowish and not white or watery), remove eggs from the female's fins, side and underneath with a dry spatula.
   (i) Place fish in a recovery beaker containing system water (if no eggs) or individual flask (if eggs obtained).

7. Fertilize eggs:
   (a) Add 50 μL UV irradiated sperm to a good clutch of eggs and mix with the pipet tip.
   (b) Leave 30 s.
   (c) Add 700 μL purified water.
   (d) Leave 1 min (label Petri dish and flask while waiting with ID number).
   (e) Dilute in egg water, incubate at 28°C.
   (f) After several hours, remove unfertilized eggs and change egg water if necessary (*see* **Note 10**).

3.4.2. Method of Detection of Mutations of Interest

The method for detection of mutations in the screen (such as microscopic examination of live embryos, whole mount *in situ* hybridization, or immunohistochemistry) should be well established and demonstrably reliable before starting the screen. It is helpful to become familiar with the variation of the screening characteristic in normal haploid embryo populations, in order to distinguish mutations from normal variation. A true mutation should be present in 50% of the embryos and have a reasonably consistent phenotype. When scoring embryos, first discard those that are quite abnormal (in our experience about 25%, with various abnormalities, **Fig. 1a**) and examine the remainder.

3.5. Mutant Recovery

When an interesting phenotype is detected in the haploid embryos, the corresponding F1 female should be crossed as soon as possible with a wild-type male of the same strain to propagate the mutation. If possible, also cross the F1 female with a male of a different strain to create a mapping cross. The F2 progeny are pair-wise mated to test for the presence of the mutation. One half of the F2 progeny will be carrying the putative mutation, thus one quarter of F2 pairs should produce embryos with one quarter having the mutant phenotype.

How many pairs should be tested before deciding that a mutation was not recovered? To attain 99% certainty that nothing was missed, 16 independent pairs (with at least 16 embryos from each) should be tested. Note that the phenotype in diploid embryos may be slightly different from the haploid phenotype. In addition, there may be more than one phenotype in a family (*see* **Note 11**).

4. Notes

1. Males for ENU treatment must be young and healthy for best survival and maximum fertility. Initially, we pre-tested males for fertility, but found this was unnecessary because none were infertile.
2. ENU did not ever completely dissolve. After about 2 h, most of it dissolved. It is possible to determine the concentration spectrophotometrically, but this is not recommended because of the possibility of contaminating the laboratory.
3. The ENU dose and number of treatments is a balance between achieving maximum mutation rate and minimum death. Our experience is summarized in **Table 1**. The lack of mutants recovered from the WIK lines suggests that a higher ENU dose may be required for this strain. We did not

determine the mutation rate at an indicator locus for WIK or AB*. Much more extensive data has been obtained by Solnica-Kretzel et al. *(9)*.

4. Males of any genetic background can be used because their DNA is inactivated, but they should be young and healthy, and well fed to give the best fertilization rates.
5. These sperm preparations can be diluted up to twofold with Hank's buffer after irradiation with no decrease in fertilization ability.
6. The time required for complete sperm DNA inactivation will need to be determined empirically. We used eggs from *gol* mutants fertilized with UV-treated wild-type sperm to detect any diploids, which would be normally pigmented. The UV lights lose strength with time so be alert for the appearance of diploids and change the lamp when this occurs. When one becomes familiar with the appearance of haploids, the appearance of diploids among them will be easily noticed.
7. To optimize the efficiency of the screen, it is worth trying to maximize egg production. We improved the proportion of females giving good quality eggs in a single in-vitro fertilization (IVF) attempt from 12 to 43%. Our final procedure was to test all potential tanks of F1 fish for spawning the week before undertaking IVF (this often required testing of about 90 4 L tanks). The best egg producers were selected and "mops" (artificial weed) were left in these tanks to encourage spawning until females were separated. The day before squeezing, the females were separated into tanks on the main system and given extra blood worms. The density of fish was kept below 30 per 4 L tank (for all screening fish) and 15 per 4 L tank for separated females. It is important to maintain optimum water quality and water flow rate. We were initially advised that females should be separated from males for 1-2 wk before squeezing. This does not make sense because the presence of males stimulates egg production and the egg production cycle is about 2 d *(15)*. We tested setting up pairs in laying boxes with dividers instead of separating the females, but this did not improve results and was time consuming. Fish can be re-squeezed after a recovery period of 4 wk. With the optimized regime, we obtained good eggs from 56% of females on first attempt, 38% on second attempt and 39% on third attempt. Females that gave sufficient eggs were not included in subsequent squeezing.
8. We tested production of haploid embryos from wild-type females in the morning, when the lights came on compared with the afternoon (with a fresh batch of UV-irradiated sperm). The survival at 24 h of fertilized eggs was 85%

(morning, 254 eggs, five clutches) vs 24% (afternoon, 344 eggs, six clutches), indicating that the egg quality was much better in the morning. The light cycle would have to be changed for IVF to be optimal later in the day.

9. Deaths should be rare and are most likely caused by squeezing too hard, using too much anesthetic or using unhealthy fish.
10. In our hands, fertilization rate in haploid embryos was usually at least 50%, but varied widely from 0 to 90%, even with one batch of sperm, reflecting egg quality. There was also some death of abnormal (fertilized) embryos during the first 24 h, also quite variable, but usually less than 50%.
11. In our screen, recovery in the F2 generation took over 12 mo beyond the end of the screen. We recovered 20 of 30 putative mutants.

Acknowledgments

The author thanks Dr. Graham Lieschke for the initial planning of the screen and for encouragement during the project; Drs. Len Zon and Nick Trede for lessons in making haploid embryos and providing their ENU treatment protocol; John Hayman, Luke Kapitany, Dora McPhee, Fiona Connell and Hannah Woolcott for assistance with the screen; and John Hayman for the photo in **Fig. 2**. During the work from which our experience is derived, Dr. Lieschke's laboratory was/is supported by a Wellcome Trust Senior Research Fellowship, NHMRC grants 234708, 461222, 461208, 516750, ARC grant DP0346823 and NIH grant HL079545.

References

1. Noveroske, J. K., Weber, J. S., and Justice, M. J. (2000). The mutagenic action of N-ethyl-N-nitrosourea in the mouse. *Mamm Genome* **11**, 478–483.
2. Balling, R. (2001). ENU mutagenesis: analyzing gene function in mice. *Annu Rev Genomics Hum Genet* **2**, 463–492.
3. Knapik, E. W. (2000). ENU mutagenesis in zebrafish – from genes to complex diseases. *Mamm Genome* **11**, 511–519.
4. Shibuya, T., and Morimoto, K. (1993). A review of the genotoxicity of 1-ethyl-1-nitrosourea. *Mutat Res* **297**, 3–38.
5. Imai, Y., Feldman, B., Schier, A. F., and Talbot, W. S. (2000). Analysis of chromosomal rearrangements induced by postmeiotic mutagenesis with ethylnitrosourea in zebrafish. *Genetics* **155**, 261–272.
6. Riley, B. B., and Grunwald, D. J. (1995). Efficient induction of point mutations allowing recovery of specific locus mutations in zebrafish. *Proc Natl Acad Sci U S A* **92**, 5997–6001.
7. Alexander, J., Stainier, D. Y., and Yelon, D. (1998). Screening mosaic F1 females for mutations affecting zebrafish heart induction and patterning. *Dev.Genet.* **22**, 288–299.

8. Mullins, M. C., Hammerschmidt, M., Haffter, P., and Nusslein-Volhard, C. (1994). Large-scale mutagenesis in the zebrafish: in search of genes controlling development in a vertebrate. *Curr Biol* **4**, 189–202.
9. Solnica-Krezel, L., Schier, A. F., and Driever, W. (1994). Efficient recovery of ENU-induced mutations from the zebrafish germline. *Genetics* **136**, 1401–1420.
10. Driever, W., Solnica-Krezel, L., Schier, A. F., Neuhauss, S. C., Malicki, J., Stemple, D. L., Stainier, D. Y., Zwartkruis, F., Abdelilah, S., Rangini, Z., Belak, J., and Boggs, C. (1996). A genetic screen for mutations affecting embryogenesis in zebrafish. *Development* **123**, 37–46.
11. Haffter, P., Granato, M., Brand, M., Mullins, M. C., Hammerschmidt, M., Kane, D. A., Odenthal, J., van Eeden, F. J., Jiang, Y. J., Heisenberg, C. P., Kelsh, R. N., Furutani-Seiki, M., Vogelsang, E., Beuchle, D., Schach, U., Fabian, C., and Nusslein-Volhard C. (1996). The identification of genes with unique and essential functions in the development of the zebrafish, Danio rerio. *Development* **123**, 1–36.
12. Kimmel, C. B. (1989). Genetics and early development of zebrafish. *Trends Genet* **5**, 283–288.
13. Walker, C. (1999). Haploid screens and gamma-ray mutagenesis. *Methods Cell Biol* **60**, 43–70.
14. Hogan, B. M., Layton, J. E., Pyati, U. J., Nutt, S. L., Hayman, J. W., Varma, S., Heath, J. K., Kimelman, D., and Lieschke, G. J. (2006). Specification of the primitive myeloid precursor pool requires signaling through Alk8 in zebrafish. *Curr Biol.* **16**, 506–511.
15. Nagel, V. (1986). Untersuchungen zur Eiproduktion beim Zebrabarbling (*Brachydanio rerio*). *J Appl Ichthyol* **4**, 173–181.

Chapter 4

Cryopreservation and In Vitro Fertilization at the Zebrafish International Resource Center

Carrie Carmichael, Monte Westerfield, and Zoltán M. Varga

Summary

In recent decades, laboratories throughout the world generated several thousand mutant, transgenic, and wild-type zebrafish lines and more lines continue to be produced. At the same time, relatively little effort has been expended to develop reliable, high-throughput, standardized, long-term cryopreservation storage methods, even though laboratories and the research community as a whole struggle to maintain the large number of lines alive. Safe and reliable methods for maintaining these valuable genetic resources are vital for future biomedical research.

Cryopreservation is the most efficient method for large-scale, long-term storage of important genetic materials. It extends the time offspring can be produced from individual fish, reduces the need to maintain live populations, and can prevent catastrophic loss of irreplaceable research lines. Cryopreservation is also the most cost-effective alternative for maintaining genetic resources because it reduces costs for animal and facility maintenance, personnel, and space. In addition, it provides novel opportunities to develop new types of research using large numbers of lines. For example, several genetic strategies, such as TILLING—or enhancer and gene trapping—depend on the use of cryopreservation to bypass generations of live organisms until a strain is revived for research.

This chapter describes and discusses the current cryopreservation method used at the Zebrafish International Resource Center. This method is derived from the initial protocol developed for zebrafish over 20 years ago that has recently been refined.

Key words: Danio rerio, Cryoprotectant, Germplasm, Liquid nitrogen repository, Sperm, Sperm library, In-vitro fertilization.

1. Introduction

In recent years, large- and small-scale genetic screens around the world have generated thousands of novel zebrafish strains *(1, 2)*. In addition, a number of new technologies became available,

Graham J. Lieschke et al. (eds.), *Zebrafish,* Methods in Molecular Biology, vol. 546
DOI: 10.1007/978-1-60327-977-2_4

such as generation of transgenic reporter lines, targeted knockout of genes *(3, 4)*, TILLING (Targeting Induced Local Lesions In Genomes) *(5)*, viral and transposon based insertions *(1, 2, 6)*, and enhancer or gene trap strategies *(7, 8)*. Several mutagenesis programs and consortia will use these new technologies to generate, identify, and characterize thousands of genes in the near future (http://www.zf-models.org/). Thus, further increase in the number and variety of zebrafish strains is certain.

In spite of the significant cost to generate these lines, little effort has been made to safely preserve them for future generations of researchers. Typically, the capacity of laboratories to maintain strains is limited by personnel and facility space. Many lines are in danger of becoming extinct without a concerted effort to bank them.

Currently, cryopreservation is the preferred method for preserving samples from endangered species and for storing rare or important genetic materials to maintain genetic diversity. Cryopreservation is the best alternative to live stocks for several reasons *(1)* Facility space is efficiently used; *(2)* the effective reproductive time of males is extended; *(3)* valuable resources do not become extinct if a line cannot be maintained live; *(4)* reverse genetic mutageneses such as TILLING, are more feasible because live fish do not need to be maintained; *(5)* cryopreservation is highly cost-effective because it saves funds needed for electricity, water, filtration, tank cleaning, repair, food, and personnel.

However, the cryopreservation protocols currently used by the zebrafish research community can be relatively difficult to learn and laboratories cannot adapt them easily. Moreover, success rates and fertilization rates vary considerably among laboratories, most likely because a number of different protocols have been developed *(5, 9)* as modifications of the original procedure by Walker and Streisinger *(10)*, which was based on the original cryopreservation method of Harvey et al. *(11)*. These modifications have largely been tested empirically by determining whether or not they affect or improve overall post-thaw fertilization rates. Without a strong understanding of the biophysical principles that govern the physiology of frozen cells, it is unclear whether the community currently uses the most practical, reliable, and efficient method today.

Cryopreservation generates several conditions that can lead to severe cell damage *(12)*. These include solution damage, intracellular and extracellular ice formation, and dehydration. When tissues are cooled slowly, water moves out of cells and ice forms in the extracellular space. The movement of water out of cells can lead to cellular dehydration, and too much extracellular ice can lead to mechanical damage by crushing. Although some organisms and tissues can tolerate extracellular ice, intracellular ice crystals almost always severely damage cells. Solutes that are excluded from the ice crystal structure during freezing will affect the balance of osmotic pressure. The stresses associated with dehydration and intracellular ice crystallization are particularly damaging to cells because intracellular structure, organelles, and molecules are severely disrupted *(12, 13)*.

The risks of cellular damage can be greatly reduced by adding a cryoprotectant to the freezing solution, which lowers the freezing temperature and increases the viscosity of the liquid. Instead of crystallizing, the solution forms amorphous ice (vitrification). Vitrification of water can also be achieved by extremely rapid cooling, in the absence of a cryoprotectant *(14, 15)*. Cryoprotectants and solutes in general increase the viscosity and decrease the freezing temperature. To achieve these effects inside the cell, cryoprotectants must readily penetrate the cell membrane. Unfortunately, common cryoprotectants such as methanol, glycerol, dimethylsulfoxide (DMSO), or *N*,*N*-dimethyl acetamide (DMA) are often toxic in high concentrations. Therefore, one of the challenges to develop an effective cryopreservation method is limiting the damage produced by the cryoprotectant itself.

Recently, three modifications of the original Harvey method *(11)* have been reported *(5, 9, 16)*. The first study tested ethylene glycol, DMSO, glycerol, and DMA as potential cryoprotective agents to replace methanol that has cytotoxic effects and is known to mutate DNA *(9)*. The second modification aimed at procedural optimization and led to a more streamlined protocol that can be more easily adapted to various laboratory conditions *(5)*. However, both methods produce only modest improvements on Harvey's protocol in terms of post-thaw fertilization rates, materials, and procedure (**Table 1**, *see* **Note 1**). Recently, another variation that omits the use of powdered milk as an anticoagulant of sperm tails has been developed in a study where four cryoprotectants: DMSO, DMA, methanol, and glycerol were tested side by side *(16)*.

These three methods may work sufficiently well for some research laboratories, however there is insufficient information or rigor to allow them to be used in a robust way for high-throughput applications. Moreover, some protocols still include the use of additives such as powdered milk, and storage containers such as glass capillary tubes and cryovials, that are inefficient and hamper necessary quality controls such as motility estimation and cell density measurements. In addition, the use of uncrushed, crushed, or powdered dry ice for freezing presents significant challenges for standardization of freezing rates within or among the different protocols.

A goal of the Zebrafish International Resource Center (ZIRC) is to serve as a repository for zebrafish lines and to preserve as many of these invaluable resources as possible for the research community. Therefore, ZIRC aims to develop robust cryopreservation

Table 1
Overview of Published Zebrafish Cryopreservation Methods

Factor	Study: Harvey et al. *(11)*	Morris et al. *(9)*	Draper et al. *(5)*	Yang et al. *(16)*
Cryoprotectant	10% methanol	10% DMA	8% methanol	8% methanol
Extender	Ginsburg	BSMIS	Ginsburg	HBSS
Collection	Squeezing	Dissection	Squeezing	Dissection
Container	Capillary	Capillary	Cryovial	Straw
Dilution	HBSS	BSMIS	HBSS	HBSS
Freezing method	Dry ice	Dry ice	Dry ice	Controlled-rate freezer
Freeze time	30 min	30 min	20 min	–
Freeze rate	16°C/min	?	?	10°C/min
Motility	43 ± 12%	12 ± 6%	?	78 ± 10%
Thawing	Air	37°C	33°C, 8 s	40°C, 5 s
Fertility	51 ± 36%[a]	14 ± 10%	28 ± 18%	33 ± 20%

A cross-comparison of key steps and agents used in zebrafish cryopreservation methods

DMA, *N,N*-dimethyl acetamide; *HBSS*, Hanks' balanced salt solution; *BSMIS*, buffered sperm motility-inhibiting solution

[a]Reported as % hatching, which may differ from % fertility. *See* **Note 1** for discussion of the issues in comparing different protocols and their outcomes

protocols for its own purposes and for laboratories in the research community to ensure that resources can be stored for future generations of researchers. Here, we present and discuss the cryopreservation protocol developed by Draper and Moens *(5)*, which was adapted from the cryopreservation method by Harvey et al. *(11)*. The ZIRC currently uses this method because it has the highest throughput capability and therefore fits ZIRC's need to cryopreserve large numbers of zebrafish strains.

2. Materials

2.1. Sperm Freezing Solutions

1. *Tricaine anesthetic stock solution.* 400 mg Tricaine powder, 97.9 mL ddH_2O, add 2.1 mL 1 M Tris–HCl (pH 9.0). Mix components in an amber glass bottle with a screw cap and

adjust to pH 7.0 with 1 M Tris-HCl (pH 9.0) (*see* **Note 2**). Store refrigerated. To anesthetize fish, add 4.2 mL Tricaine stock solution to 100 mL fish water in a crystallizing dish (*see* **Note 3**).

2. *10X Ginsburg Fish Ringer's (the salt mixture used in the cryopreservation solution) (17).* To 400 mL ddH_2O add in order, 32.5 g NaCl, 1.25 g KCl, 1.75 g $CaCl_2{\cdot}2H_2O$, adjust with ddH_2O to 500 mL. To avoid precipitation of its components, it is important to add each reagent in this order and to allow each reagent to dissolve completely before adding the next. Autoclave and refrigerate. 10X Ginsberg stock solution can be stored in the refrigerator and used as needed.
3. *10X Sodium bicarbonate ($NaHCO_3$) (the buffering component in the cryopreservation solution).* 50 mL ddH_2O, 0.10 g $NaHCO_3$. Must be prepared fresh each day of cryopreservation.
4. *1X Ginsberg Fish Ringer's mix.* (To make 25 mL) 2.5 mL 10X Ginsberg Fish Ringer's, 2.5 mL 10X $NaHCO_3$, 20 mL ddH_2O. The 1X Ginsberg Fish Ringer's is the final working solution that is prepared before each freeze event.
5. *Cryopreservation solution without methanol (A).* 10 mL 1X Ginsburg Fish Ringer's (at room temp.), 1.5 g Powdered Skim Milk (Carnation® instant milk; *see* **Note 4**).
6. *Cryopreservation solution with methanol (B).* 9 mL 1X Ginsburg Fish Ringer's (at room temp.), 1 mL methanol (Acetone free, absolute), 1.5 g powdered skim milk (Carnation® instant milk). To prevent precipitation of components, it is important to mix Ginsberg Fish Ringer's with methanol before adding powdered skim milk. For comment on the choice of methanol (*see* **Note 5**). Clearly label cryopreservation media *with* and *without* methanol (*see* **Note 6**).

2.2. Sperm Freezing Materials

1. *Powdered dry ice from liquid CO_2.* To obtain more reproducible freezing rates, we maximize contact between dry ice and the freezing vessels (**Subheading 2.2**, **items 12** and **13**, and **Subheading 3.1.3**, **steps 9**, **10**). To this end, we use powdered, not crushed dry ice. To produce powdered dry ice, attach a fire extinguisher cone to the liquid output (**Fig. 1a**) of a CO_2 tank that has a liquid siphon. It is useful to angle the cone 90° downward. Position a Styrofoam box or cooler directly under the cone to catch the dry ice. For safety purposes, wear a face shield, ear protection, and cryogloves. Open the liquid CO_2 output valve quickly so that it does not surge and clog the output valve. Compact the powdered dry ice in the cooler as it is produced (*see* **Note 7**). When the box is filled with powdered dry ice (a depth of 20 cm or more), turn off the liquid CO_2 output valve.

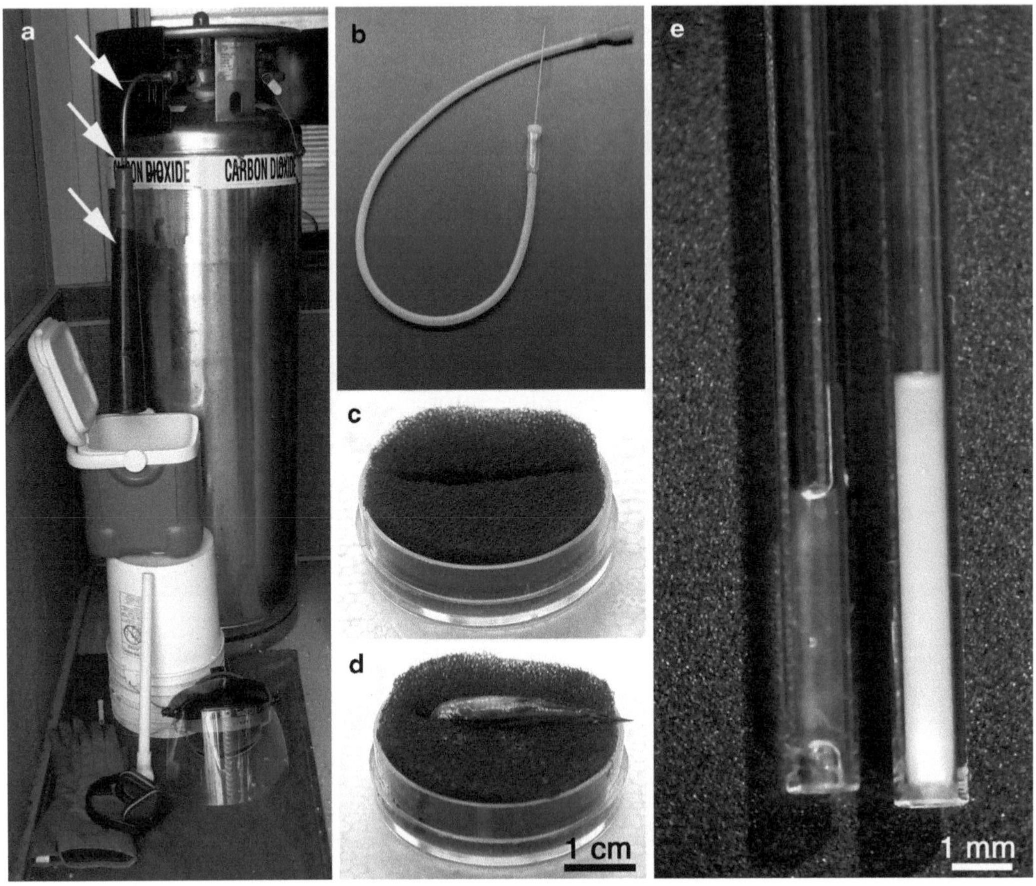

Fig. 1. Zebrafish sperm cryopreservation tools. (**a**) Tools to make powdered dry ice from liquid CO_2. A fire extinguisher cone is attached to the liquid output of a liquid CO_2 tank (*arrows*). At the bottom: cooler, safety gloves, goggles, ear protection, isolation pad, and PVC tubing to compact dry ice powder. (**b**) The aspirator tube assembly. The aspirator is a soft rubber tube adaptor that has a mouthpiece on one end and a capillary holder on the other. (**c**) Sponge fish holder. A 35 × 10 mm Petri dish with a sponge that is cut to hold males. (**d**) The sponge needs to be moistened before fish are placed in it. (**e**) Examples of empirically determined sperm quality. On the *right*, milky, opaque milt indicates high density of sperm cells whereas lower quality sperm (*left*) is watery and more transparent. The observed difference in opacity correlates only loosely with sperm cell density counts using a hemocytometer and does not correlate well with post-thaw fertilization rates. *Scale bars*: (**c, d**) 1 cm; (**e**) 1 mm.

2. 10 μL disposable capillary micropipets (Drummond Scientific Cat. No. 2-000-010; includes one aspirator tube assembly).
3. Aspirator tube assembly (**Fig. 1b**; Sigma-Aldrich cat. no. A-5177).
4. Plastic spoon to remove fish from MESAB (*see* **Note 8**).
5. Sponge fish holder (to hold male while squeezing; **Fig. 1c, d**).
6. 2 glass crystallizing dishes (one for tricaine anesthetic, the other for rinse and recovery; **Fig. 2a**, 6/7).

7. 0.5 mL microcentrifuge tubes for cryopreservation solution (mark tubes whether they will be with or without methanol) in 0.5 mL microcentrifuge rack. (Prepare the correct number of tubes according to the number of fish lines used. Divide the tubes into one row without methanol and one with methanol; **Fig. 2b**, 17/18).
8. Dissecting microscope with incident stage lighting.
9. Forceps (Millipore cat. no. XX6200006).
10. Watch glasses 3″ diameter (Corning cat. no. 9985-75).

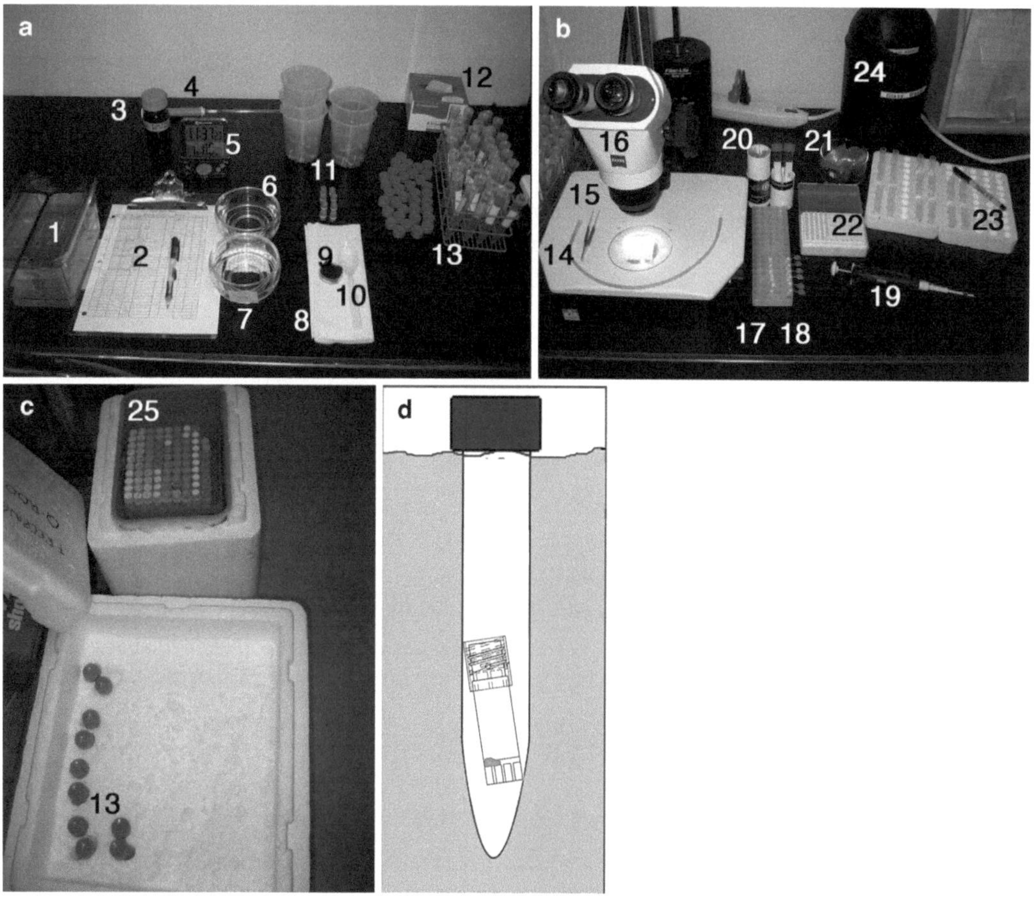

Fig. 2. Preparation of the cryopreservation workspace. (**a**) Assistant's side, (**b**) squeezer's side, (**c**) box with powdered dry ice and Falcon tubes (*foreground*), Styrofoam container with liquid nitrogen fiberglass tray insert and color-coded screwcap cryovials. (**a**) On the assistant's side 1. fish tank; 2. data records sheet; 3. Tricaine stock solution; 4. Pasteur Pipet; 5. timer; 6. recovery dish; 7. anesthetic dish; 8. Stack of paper towels; 9. Sponge fish holder (*see also* **Fig. 1 c, d**); 10. spoon; 11. screwcaps for cryovials (in cups and on table); 12. paper wipes; 13. rack with 15 mL Falcon tubes and lids. (**b**) On the squeezer's side: 14. aspirator tube assembly; 15. forceps; 16. dissection scope; 17. row of tubes containing cryoprotectant solution without methanol; 18. row of tubes containing cryoprotectant solution with Methanol; 19. micropipettor; 20. 10 μL glass capillaries, prelabeled at 1.67 cm; 21. watchglasses; 22. micropipet tips; 23. pre-labeled empty cryovials; 24. waste container for glass and sharp objects. (**c**) 13. Container with powdered dry ice and inserted Falcon Tubes on the *left*; 25. liquid nitrogen and immersed color-coded cryovials. (**d**) Profile of 15 mL Falcon Tube containing a 2 mL cryovial with sample, submerged in dry ice.

11. 10 × 10 cryoboxes (Taylor-Wharton cat. no.R24K-9C44).
12. 15 mL conical tubes (Falcon cat. no. 352099).
13. 2 mL cryogenic vials (Corning cat. no. 430488).
14. Cryogenic vial color-coded cap inserts (Corning cat. no. 430499).
15. Nalgene scissor-type forceps (NNI cat. no. 6320 0010).
16. 10 μL single channel pipetter and tips (preset to 10 μL).
17. Cryogloves.
18. Black marker (water insoluble).
19. Timer.
20. Paper towels.
21. Liquid nitrogen or vapor freezer.
22. 4-L Liquid nitrogen Dewar flask.
23. Styrofoam cooler retrofitted with fiberglass tray insert (US Plastic Corp. cat. no. 49267, **Fig. 2c**, 25).

2.3. Solutions for Sperm Thawing

1. *Hank's Stock Solution 1.* 8.0 g NaCl, 0.4 g KCl in 100 mL ddH_2O. Store refrigerated.
2. *Hank's stock solution 2.* 0.358 g Na_2HPO_4 anhydrous, 0.60 g KH_2PO_4 in 100 mL ddH_2O. Store refrigerated.
3. *Hank's stock solution 4.* 0.72 g $CaCl_2$ in 50 mL ddH_2O. Store refrigerated.
4. *Hanks' stock solution 5.* 0.601 g $MgSO_4$ in 50 mL ddH_2O. Store refrigerated.
5. *Hank's stock solution 6.* 0.35 g $NaHCO_3$ in 10.0 mL ddH_2O. Must be prepared fresh for each cryopreservation session.
6. *Hank's premix.* 10.0 mL Hank's stock solution 1, 1.0 mL Hank's stock solution 2, 1.0 mL Hank's stock solution 4, 86.0 mL ddH_2O, 1.0 mL Hanks' sock solution 5. Must be combined in this order to avoid salt precipitation. Store refrigerated.
7. *Hank's final.* 9.9 mL Hank's premix, 0.1 mL Hank's stock solution 6.

2.4. Materials for Sperm Thawing

1. Assembled water bath.
2. 100 μL single channel pipetter and tips (set on 70 μL).
3. 35 × 10-mm Petri dishes.
4. Paper towels.
5. 2 Metal spatulas (Fisherbrand cat. #14-373-25A).
6. 2 Crystallizing dishes (for Tricaine and rinsing).
7. Plastic spoon.

8. Black marker.
9. Scissor-type forceps.
10. 2 L glass flask filled with fish water.
11. Glass measuring pipets (1 and 5 mL).
12. 1 Gallon recovery tank.
13. Pre-sorted healthy females the afternoon before squeezing (selected by looking for opaque area around urogenital opening). Use no more than 15 fish per 1 gallon tank; do not feed again until after squeezing.
14. List of samples to be thawed and their locations.

3. Methods

3.1. Sperm Cryopreservation

This procedure is generally carried out in teams of two but can be done alone under certain circumstances (*see* **Note 9**). An ideal workspace includes a ca. 1.5 × 1 m countertop (**Fig. 2a, b**) and two chairs with wheels.

3.1.1. Preparations in Advance

1. Prepare solutions (**Subheading 2.1, items 1** and **2**) and collect materials (**Subheading 2.2, items 2–22**).
2. Mark capillary micropipets at 1.67 cm (3.33 μL) to normalize sperm volume.
3. With black marker, pre-label 2 mL Corning vials with numbers and appropriate stock information. Place vials in order on the vial storage rack (supplied with vials). Prepare two vials for each male because samples will be split in half and aliquoted into two cryovials. Insert color-coded caps into vial lids, alternate colors between fish stocks. Remove the lids and sort into plastic cups, set aside.
4. It is convenient to prepare a sheet to record the freezing information. The records sheet can be filled in ahead of time with stock and vial information and which vial cap color represents each stock. Later, locations of each vial can be recorded here and time can be kept for when samples are to be moved in and out of dry ice. This information will help maintain organization during freezing and can later be transferred to a database that tracks freezing data and sample locations.
5. Sort males into tanks the afternoon before they are to be squeezed (use no more than 20 fish per 1 gallon tank, do not feed until after squeezing).

3.1.2. Preparations on the Morning of Cryopreservation

1. Make fresh sperm freezing solutions (**Subheading 2.1, items 3–6**) and place on orbital mixer for 20 min.
2. Mix 4.2 mL Tricaine solution with 100 mL fish water in a crystallizing dish.
3. Add fish water to a second crystallizing dish to rinse fish before squeezing and to hold fish recovering from anesthesia and squeezing.
4. Place fish tanks with males and nets on a cart near workspace.
5. Fill styrofoam container with approximately 20 cm of dry ice (**Fig. 2c**).
6. Fill 4-L Dewar flask with liquid nitrogen.
7. Set aside lid and place cryobox in an insulated fiberglass tray. Fill with liquid nitrogen until the level is three-fourth way up the cryobox. Keep lid on cooler. Refill as needed while freezing.
8. Fill 0.5 mL centrifuge tubes on left with cryopreservation solution **without** methanol and label clearly (**Fig. 2b**, 17).
9. Fill 0.5 mL centrifuge tubes on right with cryopreservation solution **with** methanol and label clearly (**Fig. 2b**, 18).
10. Dampen sponge in fish holder with fish water.

3.1.3. Cryopreservation Procedure

1. Assemble the aspirator tube. Place the marked glass capillary in the rubber adapter on one end of the aspirator tube assembly (**Fig. 1b**).
2. Anesthetize the male fish. Place 1–2 males in a crystallizing dish that contains Tricaine diluted in fish water. Once gill movement has slowed, remove fish and rinse briefly in second crystallizing dish containing fish water (*see* **Notes 2** and **3**).
3. Dry the fish. Place fish on stack of paper towels and use a spoon to roll and blot the fish. Pay special attention to dry the region around the urogenital opening between the anal fins (*see* **Note 10**). Place fish belly up into foam fish holder (**Fig. 1c, d**)
4. Position the fish. Place male under the dissecting scope and expose the urogenital opening. Gently spread the anal fins apart with the end of the capillary tube.
5. Collect the sperm. Expel sperm by gently stroking the sides of the fish from posterior to anterior with smooth forceps (Millipore). Place capillary tube near the urogenital opening and collect sperm with gentle suction as it is expelled. Note the sample quantity and quality on the record sheet (*see* **Fig. 1e**, **Notes 11** and **12**). Avoid feces that might be expelled along with the sperm. Return fish to crystallizing dish without

Tricaine for recovery. An advantage of this method is that it permits pooling sperm from different males (*see* **Notes 13** and **14**).

6. Normalize the cryoprotectant concentration. If sperm volume reaches or exceeds the normalizing mark on the capillary tube (1.67 cm or greater), proceed directly to **step** 7. If the sperm volume does **not** reach the normalizing mark, add Cryopreservation solution **without** methanol up to the mark to normalize the volume.
7. Add the cryoprotectant to the sperm. Pull up Cryopreservation solution **with** methanol to the orange mark on capillary (total volume is 20 μL). Expel sperm and cryoprotectant mixture onto a clean watch glass (avoid bubbles). Use pipettor (set to 10 μL) to mix by stirring and pipeting up and down 1–2 times. Be quick; **steps 7–10** should be completed in 30 s (*see* **Note 15**).
8. Place the sperm into the cryovials. Pipet 10 μL of sperm/cryoprotectant medium into the bottom of a labeled 2 mL cryovial. Pipet the remaining 10 μL into a second cryovial, labeled with the same information (*see* **Note 16**).
9. Transfer the cryovials into Falcon tubes. Close cryovials with preselected color cap and drop each one into the bottom of a 15 mL Falcon tube (at room temperature). Cap Falcon tubes.
10. Freeze the sperm. Insert the cryovial-containing Falcon tubes immediately into the powdered dry ice. The tubes in the dry ice should be deep enough so that only the cap protrudes (**Fig. 2c, d**). Start a timer that runs continuously and record on the record sheet the time each pair of samples is placed in the dry ice. Organize the tubes in dry ice so that you keep track of the samples while continuing to freeze and add more. Keep each sample on dry ice for 20 min (*see* **Note 17**).
11. Place the cryovials into liquid nitrogen. After the sperm samples have been frozen on dry ice for 20 min, transfer them two at a time to a cryobox in an insulated fiberglass tray that is partially filled with liquid nitrogen. Keep track of the location of each sample in the box using the record sheet. Store the sperm in the fiberglass tray until all of the samples can be transferred to the liquid nitrogen freezer at once. Refill the tray as needed with liquid nitrogen from the Dewer flask. Wear cryogloves and protective glasses when handling liquid nitrogen and when placing the cryobox in the liquid nitrogen freezer for long-term storage. *See* **Note 18** for comments about sperm quality control and its impact on variability of post-thaw fertilization rates.

3.2. *In Vitro* Fertilization with Cryopreserved Sperm

We find that this works best with teams of two. One person squeezes eggs from females while the other person thaws sperm and fertilizes eggs.

3.2.1. Preparations in Advance

1. Prepare solutions (**Subheading 2.3**, **items 1–4** and **6**) and collect materials (**Subheading 2.4**, **items 1–14**).

3.2.2. Preparations on the Morning of Thawing

Start working as soon as the lights are turned on in the fish room.

1. Set a water bath to 33°C.
2. Mix 4.2 mL tricaine solution and 100 mL fish water in a crystallizing dish.
3. Fill second crystallizing dish with fish water for rinsing.
4. Fill 1-gallon recovery tank with fish water.
5. Make Hank's final and place on ice (prepare fresh solutions (**Subheading 2.3**, **items 5** and 7))
6. Put females and nets on cart and position near workspace.
7. Fill 4-L Dewar flask with liquid nitrogen.
8. Fill fiberglass tray (insulated in Styrofoam cooler) with liquid nitrogen.
9. Identify vials by location in freezer and put in tray with liquid nitrogen.

3.2.3. In Vitro Fertilization with Cryopreserved Sperm

1. Anesthetize females. Place females in a finger bowl containing tricaine diluted in fish water. Once gill movement has slowed, remove fish and dip in fish water to rinse off excess tricaine.
2. Dry the fish. Blot female dry on a paper towel (excess water on eggs will activate sperm prematurely and decrease or prevent optimal fertilization rates).
3. Squeeze the females. Squeeze eggs from females into 35-mm plastic Petri dish. Dampen fingers in the recovery dish and blot on paper towel. Place two fingers on the dorsal side of the fish to support the back. With one finger of the other hand, express the eggs by gently pressing from anterior to posterior along the ventral side (**Fig. 3a**). Start just behind the pectoral fins and move slowly and gently toward the tail (*see* **Note 19**).
4. Collect the eggs. If "good" eggs are obtained (*see* **Note 20**), use the metal spatula to gently move eggs away from the fish body (**Fig. 3b**). For recovery, slide the fish out of the dish into a 1-gallon tank with clean fish water.
5. Combine the eggs from several females. If possible, try to obtain several clutches of eggs (*see* **Note 21**) and combine them in one dish using the metal spatula. Do not use "bad" or "mixed" eggs (Fig. 3d). If you cannot get three "good"

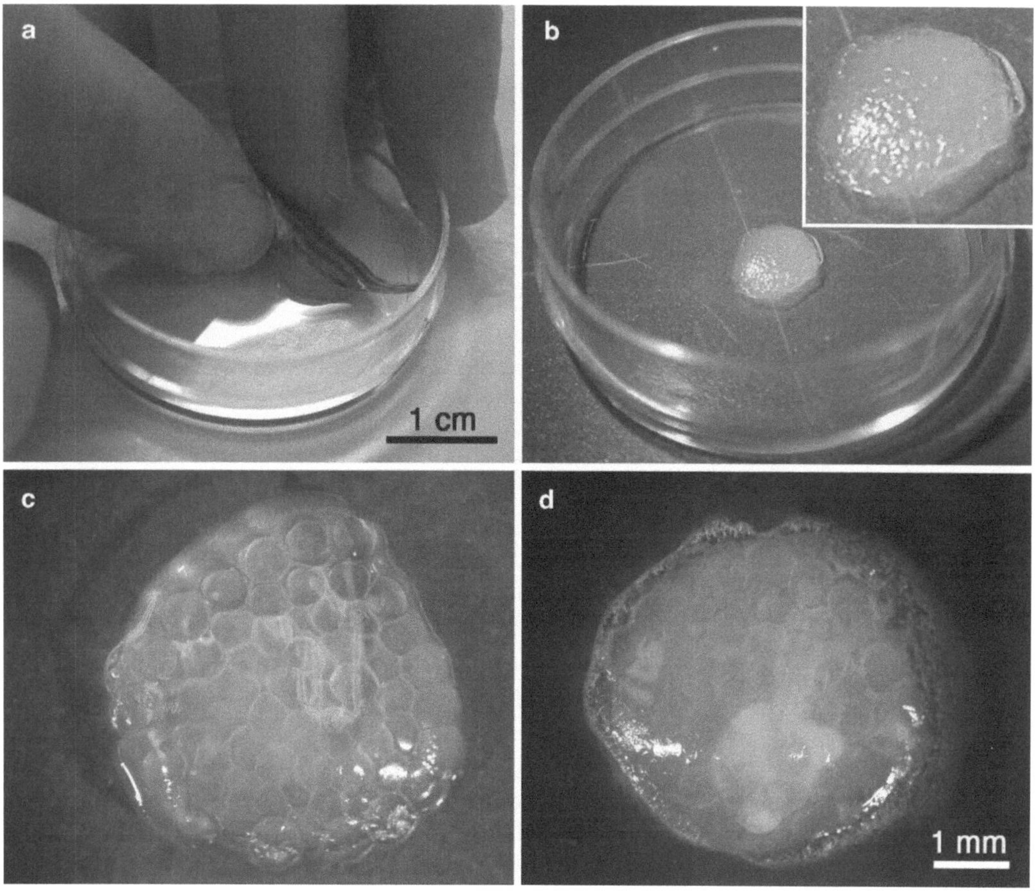

Fig. 3. Visual assessment of egg quality. (**a**) Anesthetized female is squeezed gently to release eggs into 35 mm Petri dish. (**b**, **c**) "Good" quality batches of eggs. Eggs are yellowish in incident light and appear translucent with transmitted light. (**d**) Lower quality batch of eggs. Note the opaque, white eggs intermingled with normal looking ones. Scale bars: (**a**) 1 cm; (**b–d**) 1 mm (*see Color Plates*).

clutches within 4 minutes of your first clutch, proceed to **step 6** (one "good" clutch might be sufficient **Fig. 3c**). Keep dish covered in a humidity chamber while sperm are thawed (*see* **Note 22**).

6. Thaw the sperm. Remove sperm vial from liquid nitrogen and remove cap. Make sure no liquid nitrogen is inside the vial by tipping the vial upside down. Immerse vial about half way into a 33°C water bath for 8–10 s.
7. Add Hank's final solution. Quickly add 70 mL Hank's Final to vial and mix by pipeting up and down. Immediately add the mixture to the eggs and stir gently with pipet tip.
8. Activate the sperm for fertilization. Without delay, activate sperm and eggs by adding 750 μL fish water. Swirl the dish to

mix eggs and sperm. Incubate for about 5 min at room temperature.

9. After 5 min, fill the dish with fish water and maintain it at 28°C. Later, transfer fertilized embryos to 100 × 20 mm Petri dishes. After 2–3 h, count fertilized and unfertilized embryos. To determine the fertilization rate, infertile eggs need to be counted as well (*see* **Note 23**). Distribute fertilized embryos to 50 per dish (100 × 20 mm) before submitting them to the nursery.
10. Take good care of the larvae! For the first 5 d, change the water daily and remove dead embryos. *See* **Note 24** for further comments about variability in post-thawing fertilization rates.

4. Notes

1. Until the present, there has been no attempt at standardization of protocols and terms. Developing standards to compare protocols and results among laboratories is difficult because of inconsistent use of terminologies. In 2005, the ZIRC, in collaboration with the Smithsonian Institution and the Louisiana State University Agricultural Center, hosted a symposium on the cryopreservation of zebrafish lines at which several examples of these types of discrepancies came to light.

   One problem is that common cryobiological terms are used differently in reference to laboratory standards and therefore, can be misleading. For example, one laboratory may pool 300–1000 unfertilized eggs from several females to ensure the successful recovery of a line, whereas another lab may use 50–100 eggs. The first laboratory reports 100% success in recovering a line (fertilization rates may be as low at 5%), whereas the second lab reports only 30% success, even though their fertilization rates are higher when fertilization per egg is compared. In addition, fertilization rates are also difficult to compare in the absence of standardized concentrations of frozen sperm samples.

   Several other relatively poorly understood parameters may contribute to variability in post-thaw fertilization rates such as the quality of eggs, conditioning of males, the quantity and "turbidity" of sperm samples (**Fig. 1e**) at the time of freezing, and even the definition of successful fertilization—which can range anywhere from sperm–egg penetration to first cell division to development until 24 h, hatching, or inflation of the swim bladder.

In addition to terminology differences, a number of technical differences were identified within the cryopreservation method itself depending on the goals of individual labs and the extent to which they have modified the protocol by Walker and Streisinger *(9, 10, 18)*. For instance, different types of storage vessels and liquid nitrogen freezers are used, either of which could affect post-thaw success rates.

2. Tricaine (3-amino benzoic acid ethyl ester, also called ethyl m-aminobenzoate, MESAB, and MS-222) is available in powdered form from Sigma (cat. no. A-5040), or as Finquel (part no. C-FINQ-UE) from Argent Chemical Laboratories, Inc. Purchase the smallest amount possible and store the Tricaine powder at −20°C because the compound deteriorates over time.
3. Add more drops of Tricaine stock to the anesthetic solution when anesthesia of fish takes longer than 3–4 min. Tricaine is absorbed and diluted by water dripping from nets as more fish are anesthetized. Keep two males in anesthetic at all times to maintain a steady pace when performing the procedure. During anesthesia, fish may occasionally bleed from their gills. Because pH is buffered in the Tricaine solution, we exclude pH fluctuations as a cause for gill bleeding and speculate that stress and high blood pressure might be contributing factors. Fish that bleed from the gills do not typically recover from anesthesia.
4. Powdered skim milk was used in the original protocol to avoid clumping of sperm cell tails *(11)*. Unfortunately, powdered milk has disadvantages because important quality parameters such as cell motility and density cannot be assessed before the cells are frozen and because this particular brand of powdered milk is not readily available everywhere. We recently reported that powdered milk can be omitted *(16)*. However, these results were obtained from sperm collected from dissected testes that were frozen with a controlled rate freezer at cell concentrations ranging from 10^7 to 10^8 cells/mL.
5. Currently, few alternative cryoprotectants other than methanol have been tested for their suitability to cryopreserve zebrafish sperm. Alternative cryoprotectants that have been tested include DMA, ethylene glycol, DMSO, glycerol, and ethanol *(9, 16)*. Currently, methanol appears to be the most reliable cryoprotectant. Glycerol is least suitable for this purpose *(9, 16)*.

6. Label each solution clearly to distinguish between cryopreservation medium **with** and **without** methanol. After preparing the freezing solutions, mix the solutions well on a vortex to obtain a fluid consistency and to avoid clumping of powdered milk. Rock both solutions gently on an orbital mixer for 20 min. Before use, aliquot the solutions into labeled 0.5 mL centrifuge tubes while avoiding surface bubbles. Use one set of cryopreservation media tubes for each fish line to avoid cross contamination of samples.
7. To compact the dry ice, do not use metallic instruments or tools because of the risk of electrical shock. When dry ice particles exit the liquid CO_2 dewar, they crystallize and generate electric potential due to friction in the fire extinguisher cone. PVC tubing is suitable. It is also helpful to use an isolating pad (*see* **Fig. 1a**).
8. To drain fluids more easily from the plastic spoon, drill 3–4, 2 mm holes into the tip of the spoon.
9. We find that it is helpful to work in teams of two: one assisting and one squeezing. The assistant anesthetizes fish, dries them before squeezing, and ensures recovery after the sperm samples have been obtained (**steps 1**, **2**, **4** and recovery of fish in **step 5**). The assistant is also the record keeper.

   The other person squeezes males, mixes sperm with cryoprotectant, aliquots it into two cryovials, and hands the vials to the assistant to be capped and placed on dry ice (**steps 5–10**). The assistant then notes on the record sheet the time the vials are placed on dry ice and tracks the time that samples need to be moved from dry ice to liquid nitrogen. Either of the team members can carry out **step 11** (vials into liquid Nitrogen). However, it is usually the person assisting, who can keep better track of the time samples remain in dry ice.

   It is possible for one person to perform the procedure when, for example, males are used for preparation of genomic DNA samples and are not recovered from anesthesia. In this situation, it is not necessary to track fish recovery after squeezing and therefore assistance may not be required (*see* **Note 11**).
10. Sperm motility is completely blocked at 300 mOsmol/kg. One percent of sperm are motile around 288 mOsmol/kg. Decreasing the osmotic pressure of the solution further will activate more sperm *(16)*. Because water will dilute seminal fluid and activate sperm, it is important to dry the urogenital area as thoroughly as possible.

11. The preferred method for cryopreservation at the ZIRC is to expel sperm and revive the donor for future needs rather than dissecting testes. At ZIRC, a large amount of work goes toward identifying carriers for genetically distinct lines to provide as many lines as possible to the research community. After cryopreservation, revived males can be used again for additional freezing of germplasm, for in-house breeding of the line, to provide as adults, or to be used to generate embryos that are supplied to researchers. Because the ZIRC uses fish for multiple purposes, it is necessary to give males a sufficient recovery period after cryopreservation, typically 1 mo before they are anesthetized and squeezed again for cryopreservation. This helps to reduce stress and maximize a healthy reproductive lifespan. If the males are not needed for anything other than cryopreservation, it is possible to give them a shorter resting period of 2 weeks.
12. The amount and quality of sperm varies from fish to fish. Good sperm is white and opaque. Poor sperm appears watery. We keep track of sperm sample quality and assign numbers for quantity (empirical: milky, opaque sample; **Fig. 1e**, *right*).
    In general, the minimal requirements to keep a sample are either 0.5 μL of good sperm (3–4 mm in capillary), or 1.5 μL of poor sperm (**Fig. 1e**, left; or 6–7 mm in capillary).
13. Several samples can be pooled when collecting sperm into a glass capillary. This can be achieved by adding sperm from several males up to the normalizing mark on the capillary. In this case, the cryoprotectant is then added directly, omitting **Subheading 3.1**, **step 6**.
14. A general advantage of this protocol is that it is relatively easy to pool several samples and thus, to increase the total volume of sperm in individual samples. Additionally, pooling of samples is beneficial for post-thaw fertilization rates (Carmichael and Varga, personal observation; Bruce Draper, personal communication) because sperm from a fertile male may compensate for sperm from less fertile males in the mix.
15. Speed is important! The time between adding the cryoprotectant and placing sperm vials in dry ice (**steps 7–10**) should not exceed 30 s. Methanol permeates sperm cells rapidly *(11)*.
16. The use of standard laboratory pipets and 2 mL cryovials makes it easy to learn and perform the Draper–Moens protocol. However, with the Draper–Moens protocol, the small sample size stored in cryovials represents a disadvantage. With this method, the storage space in the freezer is

not efficiently used compared with the Walker–Streisinger Method *(10)* for example, where up to 40 capillaries with frozen sperm samples can be stored in one 5 mL cryovial. In contrast, the advantage to storing individual samples in 2 mL vials is that each sample can be labeled and handled independently. The use of 2 mL screw-cap cryovials is also beneficial for maintaining good safety standards, such as preventing the spread of pathogens among frozen samples.

17. It is crucial to observe the 20-min time limit. If samples are kept for less than 20 min on dry ice, intracellular crystallization may damage the cells as a result of incomplete freezing by the time they are submerged into liquid nitrogen. In contrast, if they remain on dry ice for longer than 20 min, they risk osmotic pressure imbalance and cellular dehydration. Unfortunately, the exact freezing rate obtained with these methods has not been accurately determined. Preliminary measurements indicate an asymptotic freezing curve with a relatively abrupt change in freezing rate 2 min after the sample has been placed into dry ice. Initially, the freezing rate is approximately 30°C/min(during the first 1–2 min) followed by a much slower freezing rate of approximately 2–3°C/min (3–10 min; Carmichael and Varga, unpublished observations).

    Other methods will produce different freezing rates that need to be tested for post-thaw fertility rates. Good results have been obtained in a recent series of tests with a controlled rate freezer at a constant freezing rate of 10°C/min *(16)*.

18. The relatively small sample sizes and the current practice of adding powdered skim milk prevent assessment of important sperm quality control parameters such as cell motility, morphology, density, and total cell count before the sperm is frozen. In addition, the current protocol specifies that the time between adding cryoprotectant to sperm and placing the sperm on dry ice is limited to 30 s or less, which also reduces the opportunity to assess sample quality. More research is necessary, and our recent studies suggest that exposure time to the cryoprotectant may be extended up to 10 min without major impact on sperm motility and cell survival *(16)*. Without assessment of these important sperm quality control parameters, it is difficult to predict resultant post-thaw fertilization rates. As a result, little is understood about the integrity of individual samples, and post-thaw fertilization rates continue to be highly variable. Yang and colleagues also found that a final dilution of ~10^8 cells/mL yielded the most optimal post-thaw fertilization rates, making the use of powdered milk unnecessary because coagulation of sperm

tails was less likely at this density. In the existing cryopreservation protocols, the cryoprotectant solutions are standardized at the expense of varying sperm densities. To maximize post fertilization success for individual samples, both cell density and final dilutions need to be calculated to obtain optimal results.

19. Only gentle pressure is necessary; if the female can produce eggs, they will be released readily. If gentle pressure fails to release eggs, do not continue to squeeze harder because this might injure the fish.
20. "Good" eggs have a yellowish, translucent color (**Fig. 3b**, **c**), whereas eggs that have remained in the female too long are white and watery (**Fig. 3d**). To maximize the likelihood of obtaining "good" eggs, collect them during the first 2 h after the light has been turned on in the fish facility. The characteristics of eggs are as follows: "good" – slightly yellowish, granular-looking eggs, translucent, not watery (**Fig. 3b, c**); "bad" – eggs already broken down, white-ish (**Fig. 3d**), watery, or with the consistency of baby cereal or cottage cheese (not shown); "mixed" – some potentially good eggs mixed with bad eggs (**Fig. 3d**).
21. Ideally, 3–4 clutches of eggs should be obtained. Additional clutches should be collected within 4 min of the first clutch to prevent the eggs from drying out (*see* **Note 22**). The number of eggs collected for post-thaw in vitro fertilization should take into account the typical survival rate in the nursery (if necessary) and the fertilization rates of previous test thaws (*see* **Note 23**). Combine the clutches using the metal spatula. Allow females to rest one month before they are anesthetized and squeezed again for in vitro fertilization.
22. To prevent the eggs from drying out, while obtaining additional eggs we prepare a humidity chamber for short-term storage by placing a moist paper tissue into a 100 × 20 mm Petri dish. Place the 35 × 10 mm dish containing the eggs upside down in the chamber.
23. Post thaw fertility rates may be highly variable because of differences in sperm quality and quantity among individual fish (*see also* **Note 3**). To test whether frozen samples can be recovered, we perform "test thaws" for each frozen line. We set a threshold of at least 5% post thaw fertility, which is calculated by counting all fertile embryos divided by the total number of infertile plus fertile embryos (no. fertile embryos/ no. total eggs). If the first sample does not pass the threshold, we thaw one or more additional samples, up to three samples per line. If all samples fail the test, new samples will be frozen and the process of test thawing will be repeated.

In the absence of quality controls such as cell counts and motility assessment before freezing, test thawing allows us to be confident that frozen lines can be recovered.

24. Several factors in addition to those already mentioned (*see* **Notes 18** and **23**) may also contribute to the variable outcome of post-thaw fertilization, including (but not limited to): skill level and proficiency of the individual researcher, materials used in different laboratories or between different freezing events, male variability, variations in sperm sample volume and quality, varying freezing rates, or thaw rates, and egg quality during in vitro fertilization. To predict sample reliability accurately and, therefore, the probability of recovering valuable research lines, current techniques including thawing and *in vitro* fertilization need to be further improved and standardized. Currently, the Draper–Moens method *(5)* on which this protocol is based is the most successful and efficient protocol and yields relatively consistent post-thaw fertility rates (averaging around 20% at the ZIRC; varying between 0 and 80%). With this method, 100% recovery of lines can be ensured by freezing a sufficient number of samples and by pooling samples for lines with low fertility rates.

Acknowledgments

We are grateful to Cecilia Moens and Bruce Draper for generously sharing information and teaching their cryopreservation method to the ZIRC staff. Additional information can be obtained from Jennifer Stout, Rafael Hernandez, and Cecilia Moens, HHMI, Fred Hutchinson Cancer Research Center, Seattle, WA, and Bruce Draper, now at the University of California, Davis. Direct questions to Bruce Draper (bwdraper@ucdavis.edu).

References

1. Driever, W., Solnica-Krezel, L., Schier, A. F., Neuhauss, S. C. F., Malicki, J., Stemple, D. L., Stainier, D. Y. R., Zwartkruis, F., Abdelilah, S., Rangini, Z., Belak, J., and Boggs, C. (1996). A genetic screen for mutations affecting embryogenesis in zebrafish. *Development* **123**, 37–46.
2. Haffter, P., Granato, M., Brand, M., Mullins, M. C., Hammerschmidt, M., Kane, D. A., Odenthal, J., van-Eeden, F. J. M., Jiang, Y.-J., Heisenberg, C.-P., Kelsh, R. N., Furutani-Seiki, M., Warga, R. M., Vogelsang, E., Beuchle, D., Schach, U., Fabian, C., and Nüsslein-Volhard, C. (1996). The identification of genes with unique and essential functions in the development of the zebrafish, *Danio rerio*, *Development* **123**, 1–36.
3. Meng, X., Noyes, M. B., Zhu, L. J., Lawson, N. D., and Wolfe, S. A. (2008). Targeted gene inactivation in zebrafish using engineered

zinc-finger nucleases, *Nat Biotechnol.* **26**, 695–701. Epub 2008 May 2025.

4. Doyon, Y., McCammon, J. M., Miller, J. C., Faraji, F., Ngo, C., Katibah, G. E., Amora, R., Hocking, T. D., Zhang, L., Rebar, E. J., Gregory, P. D., Urnov, F. D., and Amacher, S. L. (2008). Heritable targeted gene disruption in zebrafish using designed zinc-finger nucleases, *Nat Biotechnol.* **26**, 702–708. Epub 2008 May 2025.
5. Draper, B. W., McCallum, C. M., Stout, J. L., Slade, A. J., and Moens, C. B. (2004). A High-Throughput method for identifying ENU-induced point mutations in zebrafish, *Method Cell Biol.* 77, 91–112.
6. Amsterdam, A., Burgess, S., Golling, G., Chen, W. B., Sun, Z. X., Townsend, K., Farrington, S., Haldi, M., and Hopkins, N. (1999). A large-scale insertional mutagenesis screen in zebrafish, *Genes Dev* **13**, 2713–2724.
7. Ellingsen, S., Laplante, M. A., Konig, M., Kikuta, H., Furmanek, T., Hoivik, E. A., and Becker, T. S. (2005). Large-scale enhancer detection in the zebrafish genome, *Development* **132**, 3799–3811. Epub 2005 Jul 3727.
8. Kawakami, K. (2005). Transposon tools and methods in zebrafish, *Dev Dyn* **234**, 244–254.
9. Morris, J. P. IV., Berghmans, S., Zahrieh, D., Neuberg, D. S., Kanki, J. P., and Look, A. T. (2003). Zebrafish sperm cryopreservation with N,N-dimcthylacctamide, *Biotechniques* **35**, 956–958, 960, 962 passim.
10. Westerfield, M. (2007). *The Zebrafish Book; A Guide for the Laboratory Use of Zebrafish (Danio rerio)*, 3 ed., University of Oregon Press, Eugene.
11. Harvey, B., Kelley, R. N., and Ashwood-Smith, M. J. (1982). Cryopreservation of zebra fish spermatozoa using methanol, *Can. J. Zool.* **60**, 1867–1870.
12. Mazur, P. (2004) Principles of Cryobiology, in *Life in the Frozen State* (Fuller, B. J., Benson, E. E., and Lane, N., Eds.), p 672, Boca Raton, FL, CRC Press.
13. Tiersch, T. R., and Mazik, P. M. (2000). *Cryopreservation in Aquatic Species*, Baton Rouge, LA, The World Aquaculture Society.
14. Dattena, M., Accardo, C., Pilichi, S., Isachenko, V., Mara, L., Chessa, B., and Cappai, P. (2004). Comparison of different vitrification protocols on viability after transfer of ovine blastocysts in vitro produced and in vivo derived. *Theriogenology.* **62**, 481–493.
15. Isachenko, V., Isachenko, E., Montag, M., Zaeva, V., Krivokharchenko, I., Nawroth, F., Dessole, S., Katkov, II, and van der Ven, H. (2005). Clean technique for cryoprotectant-free vitrification of human spermatozoa, *Reprod Biomed Online* **10**, 350–354.
16. Yang, H., Carmichael, C., Varga, Z. M., and Tiersch, T. R. (2007). Development of a simplified and standardized protocol with potential for high-throughput for sperm cryopreservation in zebrafish Danio rerio, *Theriogenology.* **68**, 128–136. Epub 2007 June 2001.
17. Ginsburg, A. S. (1963). Sperm–egg association and its relationship to the activation of the egg in salmonid fishes, *J Embryol Exp Morphol.* 11, 13–33.
18. Harvey, B., Kelley, R. N., and Ashwood-Smith, M. J. (1982). Cryopreservation of zebrafish (*Brachydanio rerio*) spermatozoa using methanol, *Can. J. Zool* **60**, 1867–1870.

Part II

Transgenesis

Chapter 5

Transient and Stable Transgenesis Using *Tol2* Transposon Vectors

Hiroshi Kikuta and Koichi Kawakami

Summary

Transgenesis is an important methodology for studying the function of genes and genomes in model plants and animals. For the model vertebrate zebrafish, methods using the *Tol2* transposable element have been developed for this purpose. With these methods, the function of the transgene can be analyzed in both transient and stable transgenic fish. Recently, *cis*-sequences necessary for transposition of the *Tol2* element were revealed. This enabled development of transposon vectors containing minimal DNA sequences that are easily manipulated. More recently, several transposon vectors containing the Gateway sequence were created and reported. These are useful because any foreign sequences can be cloned into a transposon vector fairly easily and rapidly. This chapter describes the features of these transposon vectors, and protocols to perform transgenesis in zebrafish.

Key words: Transposon, Transposase, Transposition, *Tol2*, Transient transgenesis, Stable transgenesis, Microinjection, Green fluorescent protein.

1. Introduction

Transgenesis is an important methodology for studying the function of genes and genomes in model animals. In a model vertebrate zebrafish, methods for transgenesis have been developed. Stuart et al. injected plasmid DNA into one-cell stage embryos *(1)*. The injected DNA was integrated in the genome and transmitted to the next generation. Because it is easy to obtain a large number of eggs and to inject DNA solution into the large cytoplasm of eggs, this method has been a standard for creating transgenic zebrafish. For instance, transgenic fish expressing the green fluorescent protein

Graham J. Lieschke et al. (eds.), *Zebrafish,* Methods in Molecular Biology, vol. 546
DOI: 10.1007/978-1-60327-977-2_5

(GFP) were created *(2)*. Further, tissue-specific promoters were connected to the GFP gene and transgenic fish expressing GFP in temporally and spatially restricted fashions were constructed *(3, 4)*. However, researchers have sometimes encountered the following problems. First, the injected DNA is often integrated as a multimer. This may potentially cause variegated expression or silencing of transgene expression during passage of generations. Second, integrations are often associated with rearrangement of the genomic DNA surrounding the transgenes. This can present an obstacle in the way of cloning and characterizing the integration loci. Third, the germline transmission frequency is low; only 2–5% of the injected fish can give rise to transgenic fish in the next generation.

To circumvent these problems and to facilitate transgenic studies in zebrafish, efforts to develop transposon technologies have been carried out *(5–8)*. Among these, the medaka fish *Tol2* transposable element showed the highest germline transmission frequency *(9)* and is now widely used as a genetic tool in zebrafish *(10)*. Recently, *cis*-sequences required for transposition were characterized. Such analyses allowed construction of *Tol2* cloning vectors of smaller sizes, which are easily manipulatable *(11, 12)*. Furthermore, to facilitate cloning of a foreign DNA into *Tol2*, vectors that contain sequences necessary for application of the Gateway technology were developed *(13, 14)*. These transposon vectors can be used for different purposes, for instance, the construction of stable transgenic fish *(9)* or analyses of activities of a promoter, an enhancer or a gene of interest in transient expression assay *(15)*. In this chapter, we describe how the *Tol2* transposon vectors can be built and used for transient and stable transgenesis in zebrafish.

2. Materials

2.1. Cloning with Minimal Tol2 Vectors

1. T2AL200R150G *(11)*.
2. mini*Tol2* *(12)*.
3. Restriction enzymes and T4 DNA ligase.
4. QIAprep spin miniprep kit (QIAGEN).
5. BigDye terminator v3.1 cycle sequencing kit (Applied Biosystems).
6. DNA sequencer (ABI PRISM 3130x1, Applied Biosystems).

2.2. Cloning DNA into Tol2 Vectors by Using the Gateway Technology

2.2.1. Tol2 Vectors with the Gateway Cassette

1. pDest*Tol2*pA2 *(13)*.
2. pDest*Tol2*CG2 *(13)*.
3. p*Tol*Dest *(14)*.
4. p*Tol*EGFPDest *(14)*.
5. p*Tol*CherryDest *(14)*.
6. p*Tol*MTDest *(14)*.

2.2.2. Cloning DNA into an Entry Vector

1. pDONR221 (Invitrogen).
2. pCR8/GW/TOPO TA cloning kit (Invitrogen).
3. Competent DH5α High *Escherichia coli* (ToYoBo, Japan).

2.2.3. Construction of a Destination Vector

1. Gateway vector conversion system (Invitrogen).
2. T4 DNA ligase.
3. Calf intestinal alkaline phosphatase.
4. T4 DNA polymerase (if necessary).

2.2.4. Cloning DNA into a Destination Vector

1. Gateway Clonase enzyme mix: LR Clonase Buffer, LR Clonase, and proteinase K (Invitrogen).

2.3. Cloning a Zebrafish Genomic DNA Fragment into a Tol2 Enhancer Assay Vector with the Gateway Cassette

1. pGW-cfosGFP *(16)*.
2. pT2H-GW-hspEGFP (H. Kikuta, manuscript in preparation).
3. Expand High Fidelity PCR System (Roche).
4. PCR Buffer IV plus 25 mM $MgCl_2$ (ABgene).
5. dNTP mix (2.5 mM each).
6. Primers: fgf8DCR-f (5′-TCA CCG GTA CCA CCT ATT GA-3′); fgf8DCR-r (5′-TGA TTT CAC TCC TGC GCT AA-3′).

2.4. Microinjection of a Transposon-Donor Plasmid and Transposase mRNA

2.4.1. In Vitro Transcription of Transposase mRNA

1. pCS-TP *(9)*.
2. *Not*I.
3. mMESSAGE mMACHINE high yield capped RNA transcription SP6 Kit (Ambion).
4. Quick spin columns for radiolabeled RNA purification (Roche).
5. Phenol-chloroform, 100% ethanol, 2-propanol.

2.4.2. Microinjection

1. QIAfilter Plasmid Maxi kit (QIAGEN).
2. Puller (Narishige).
3. Glass capillary (Narishige).
4. Micropipet holder (Leica).

5. Teflon tube (inside diameter; 0.9 mm, outside diameter; 2.0 mm) (Narishige).
6. 20-mL Syringe.
7. Phenol red solution, 0.5% in DPBS (Sigma).
8. Agarose (Iwai chemicals).
9. Petri dish, 60 × 15 mm style (Falcon).

2.5. Excision Assay: Detection of Excision of Tol2 from a Donor Plasmid

1. Embryo lysis buffer: 10 mM Tris–HCl, 10 mM EDTA, 200 μg/mL proteinase K (add 1/100 vol of 200 mg/mL proteinase K stock solution just before use).
2. Primers: exL (5′-ACC CTC ACT AAA GGG AAC AAA AG-3′); exR (5′-CAA GGC GAT TAA GTT GGG TAA C-3′).

2.6. Transient and Stable Transgenesis

1. Fluorescence stereomicroscope (e.g., MZ 16FA, Leica).
2. 9-well glass depression plate (Corning).
3. Methylcellulose (Sigma).
4. 3-aminobenzoic acid ethyl ester (Sigma).

3. Methods

3.1. Cloning with Minimal Tol2 Vectors

The structures of T2AL200R150G and mini*Tol2* are shown in **Fig. 1**. T2AL200R150G contains 200-bp DNA and 150-bp DNA from the left and right terminals of the full-length *Tol2* respectively *(11)*. Any DNA fragments can be cloned between these *cis*-sequences, namely between the *Bgl*II and *Xho*I sites. mini*Tol2* contains 261-bp DNA and 202-bp DNA from the left and right terminals and a multi-cloning site (MCS) between them. A foreign DNA can be cloned at the MCS *(12)*. *Tol2* has a large cargo capacity. It has been shown that DNA of approximately 10 kb can be cloned without reducing the transposition activity *(11, 12)*.

1. Digest pT2AL200R150G or pmini*Tol2* with appropriate restriction enzymes. Insert a foreign DNA fragment at the cleavage site with T4 DNA ligase, transform *E. coli*, and pick antibiotic-resistant colonies.
2. Prepare plasmid DNA with QIAprep Spin miniprep kit and perform sequencing using BigDye terminator v3.1 cycle sequencing kit according to the manufacturer's instructions to confirm the structure of the plasmid. Analyze the sequence by using a DNA sequencer.

3.2. Cloning DNA into Tol2 Vectors by Using the Gateway Technology

To facilitate cloning of a foreign DNA fragment into *Tol2*, vectors containing sequences necessary for the Gateway technology were constructed *(13, 14)*. These vectors are listed in **Fig. 1**.

3.2.1. Tol2 Vectors with the Gateway Cassette

1. pDest*Tol2*pA2 and p*Tol*Dest contain an attR4-attR3 multisite Gateway cassette upstream of the polyA signal. Therefore, any DNA fragments cloned in the entry vector can be placed upstream of the polyA.

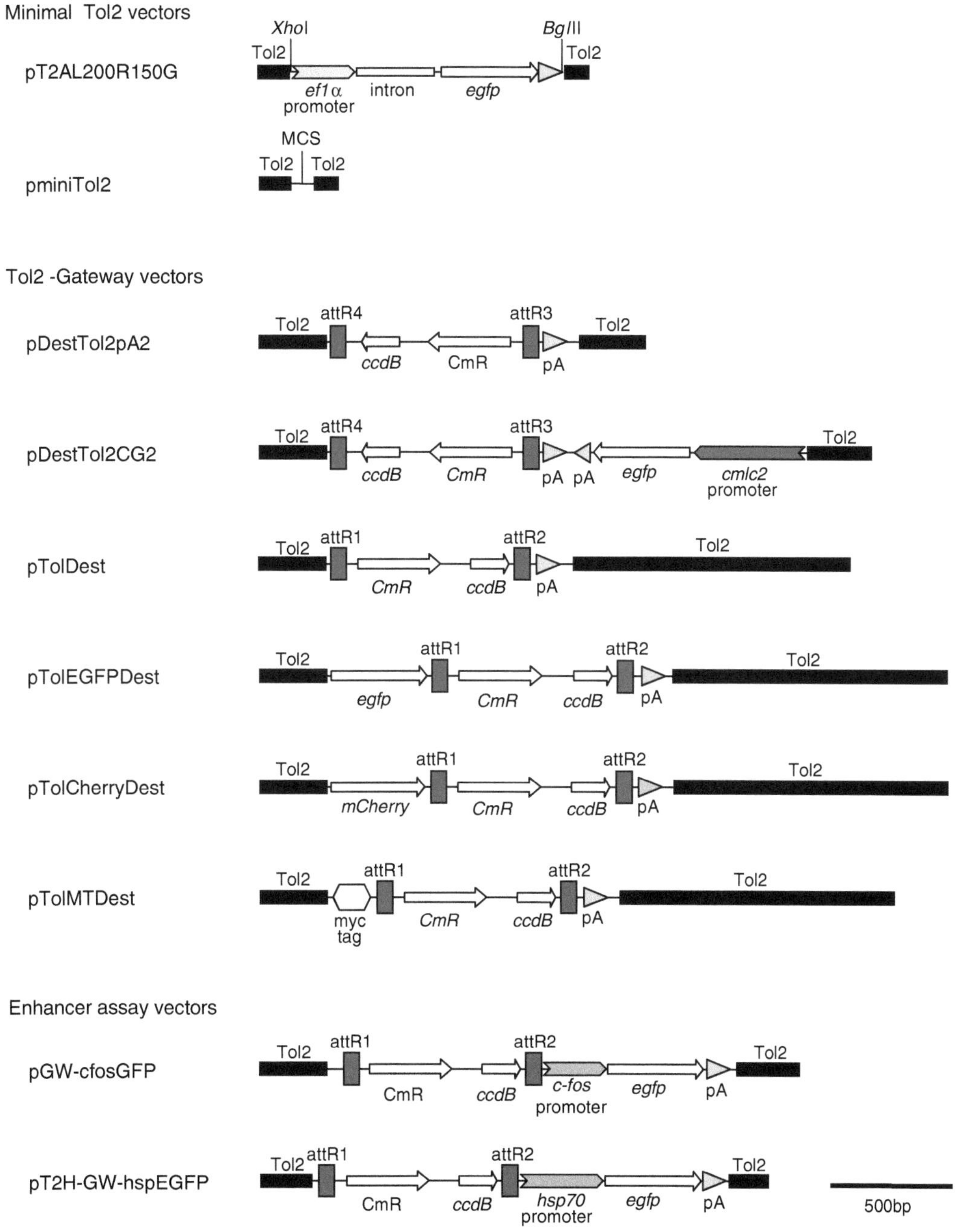

Fig. 1. *Tol2* vectors. The structures of *Tol2* vectors described in the text are shown.

2. pDest*Tol2*CG2 is essentially the same as pDest*Tol2*pA2 but contains the *cmlc2* promoter-GFP cassette to visualize a transgene as GFP expression in the heart.

3. p*Tol*EGFPDest, p*Tol*CherryDest and p*Tol*MTDest are designed to clone DNA downstream of the EGFP, mCherry, and *myc*-tag sequences. These vectors are used to synthesize fusion proteins with these tags. To express the fusion protein in desired regions, an appropriate promoter sequence should be cloned at the *Eco*RV site (**Fig. 2b**).

3.2.2. Cloning DNA into an Entry Vector

The major advantage of the Gateway technology is that researchers do not need to consider restriction enzyme sites.

1. The first step is to clone DNA into an entry vector that will result in appropriate attL and/or attR sites flanking the DNA fragment.
   a. Amplify a DNA fragment (either genomic DNA or cDNA) by PCR using primers that contain the attB1 or attB2 sequence at the end. Integrate the amplified DNA into the pDONR221 vector that contains the attP1 and attP2 sequence using the BP recombinase activity according to the manufacturer's instructions. The resulting entry plasmid contains attL1 and attL2 at each end of the cloned DNA.
   b. Alternatively, amplify a DNA fragment by PCR. Clone the amplified DNA into the pCR8/GW/TOPO entry vector by TA cloning according to the manufacturer's instructions. The resulting entry plasmid contains attL1 and attL2 at each end of the cloned DNA.
2. Multiple DNA fragments can be cloned into pDest*Tol2*pA2 and pDest*Tol2*CG2 by using the MultiSite gateway system. For this purpose, DNA fragments are cloned into different Types of entry vectors *(13)* that exploit a series of different attL and attR recombinase recognition sites to result in ordered integration of the multiple DNA fragments (**Fig. 2a**).
3. Prepare plasmid DNA and confirm the structure by sequencing.

3.2.3. Construction of a Destination Vector

A desired destination vector can be constructed by cloning the Gateway cassette that contains attR1, the *ccdB* gene, the chloramphenicol resistant gene, and attR2.

1. Linearize the vector of interest with an appropriate restriction enzyme that creates blunt ends. When cohesive ends are created, make blunt ends by treatment with T4 DNA polymerase.
2. Dephosphorylate the linearized vector with calf intestinal alkaline phosophatase. Ligate the vector with the gateway conversion cassette.

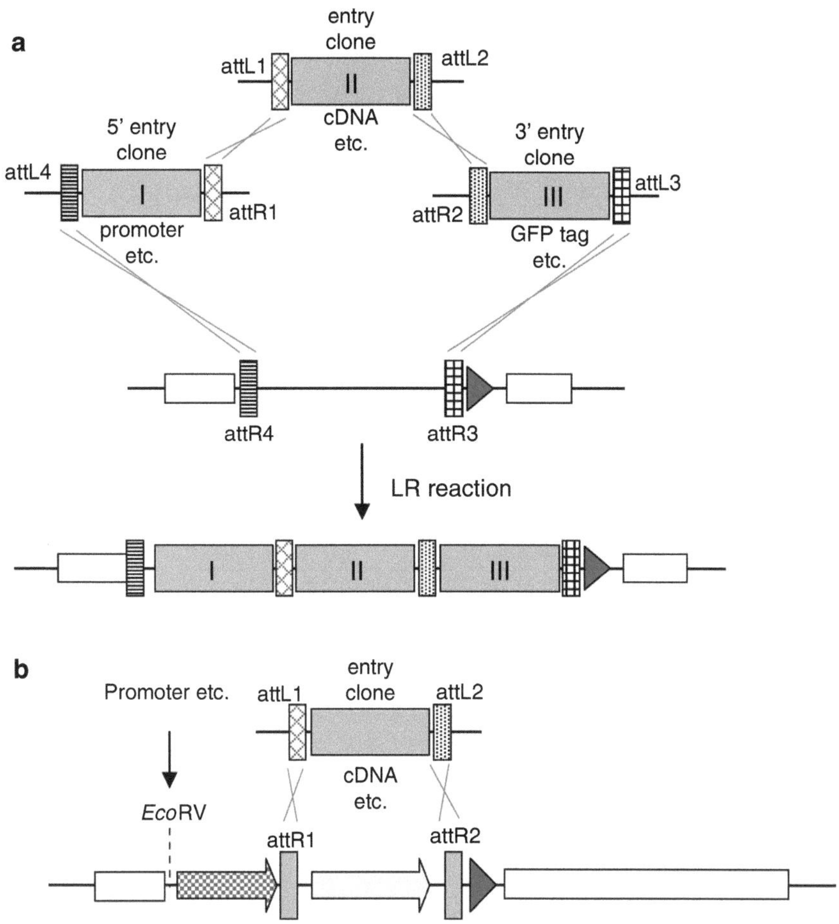

Fig. 2. Cloning DNA into *Tol2* destination vectors by the Gateway technology. (**a**) Cloning multiple DNA fragments into a destination vector (pDest*Tol2*pA2 or pDest*Tol2*CG2) using the MultiSite Gateway system. DNA fragments in the 5′ entry clone (e.g., promoter), in the (*middle*) entry clone (e.g., cDNA) and in the 3′ entry clone (e.g., a tag sequence at the COOH-terminus) can be combined in the exact order and orientation into a destination vector. (**b**) Cloning a DNA fragment into a destination vector (p*Tol*Dest, p*Tol*EGFPDest, p*Tol*CherryDest or p*Tol*MTDest) using the Gateway system. A DNA fragment in an entry clone (e.g., cDNA) can be cloned into a destination vector. A promoter DNA can be cloned at the *Eco*RV site upstream of the cDNA.

3. Mix 1 μL of the ligation product with 50 μL of One Shot ccdB Survival T1 Phage-Resistant Cells (supplied with the kit) according to the manufacturer's instructions. Pick chloramphenicol resistant colonies. Prepare plasmid DNA from the *E. coli* with QIAprep spin miniprep kit and confirm the structure by sequencing.

3.2.4. Cloning DNA into a Destination Vector

DNA fragments cloned in the entry vector are transferred into a destination vector through the LR reaction.

1. pDest*Tol2*pA2 and pDest*Tol2*CG2 contain attR4 and attR3. Multiple DNA fragments can be cloned together into these vectors. For instance, a promoter fragment, cDNA or a tag sequence cloned on different entry vectors that have attL4 and attR1, attL1, and attL2, or attR2 and attL3 at each end of the DNA fragments can be combined by MultiSite Gateway method (**Fig. 2a**).
2. p*Tol*Dest, p*Tol*EGFPDest, p*Tol*CherryDest, and p*Tol*MTDest contain attR1 and attR2. These are designed to clone DNA downstream of the EGFP, mCherry, and *myc*-tag sequences to make fusion proteins. In order to express the fusion proteins in desired regions, clone an appropriate promoter sequence at the unique *Eco*RV site (**Fig. 2b**).
3. Prepare plasmid DNA and confirm the structure by sequencing.

3.3. Cloning a Zebrafish Genomic DNA Fragment into a Tol2 Enhancer Assay Vector with the Gateway Cassette

pGW-cfosGFP and pT2H-GW-hspEGFP are designed to clone a genomic DNA fragment by using the Gateway technology and to analyze enhancer activities of the cloned DNA (**Fig. 1**). pGW-cfosGFP contains the Gateway cassette, the *c-fos* promoter, and the EGFP gene *(16)*. pT2H-GW-hspGFP contains the Gateway cassette, the *hsp70* promoter, and the EGFP gene. We will describe analysis of an enhancer of the *fgf8* gene by using pT2H-GW-hspEGFP (**Fig. 3**).

1. Identify genomic DNA sequences conserved between vertebrate species by comparing the zebrafish genomic sequence with human and/or mouse translated BLAT on the Ensemble database (**Fig. 3a**).
2. Design PCR primers (*see* **Note 1**) to amplify genomic DNA of interest. To clone a conserved region located close to the *fgf8* gene, fgf8DCR-f, and fgf8DCR-r are designed and used (**Fig. 3b**).
3. Perform PCR of 94°C for 2 min; 35 cycles of at 94°C for 30 s, at 56°C for 30 s, and at 72°C for 1 min; 72°C for 7 min in 50 μL 1X PCR buffer containing 2.5 mM $MgCl_2$, dNTP (0.25-mM each), 3.5 units of Expand High Fidelity Taq polymerase and 1 μM of primers. Analyze the PCR products by agarose gel electrophoresis.
4. Mix 4 μL of the PCR product, 1 μL of salt solution, and 1 μL of pCR8/GW/TOPO vector mix according to the manufacturer's instructions. Incubate at room temperature for 5 min.
5. Perform transformation by adding the 6 μL mixture to 50 μL of DH5α competent cells. Incubate the cells on ice for 30 min, heat at 42°C for 30 s, and then incubate on ice for 2 min.
6. Add 450 μL of SOC medium to the sample, Incubate at 37°C for 1 h, and plate 100 μL of the transformation mixture onto a LB medium plate containing ampicillin.

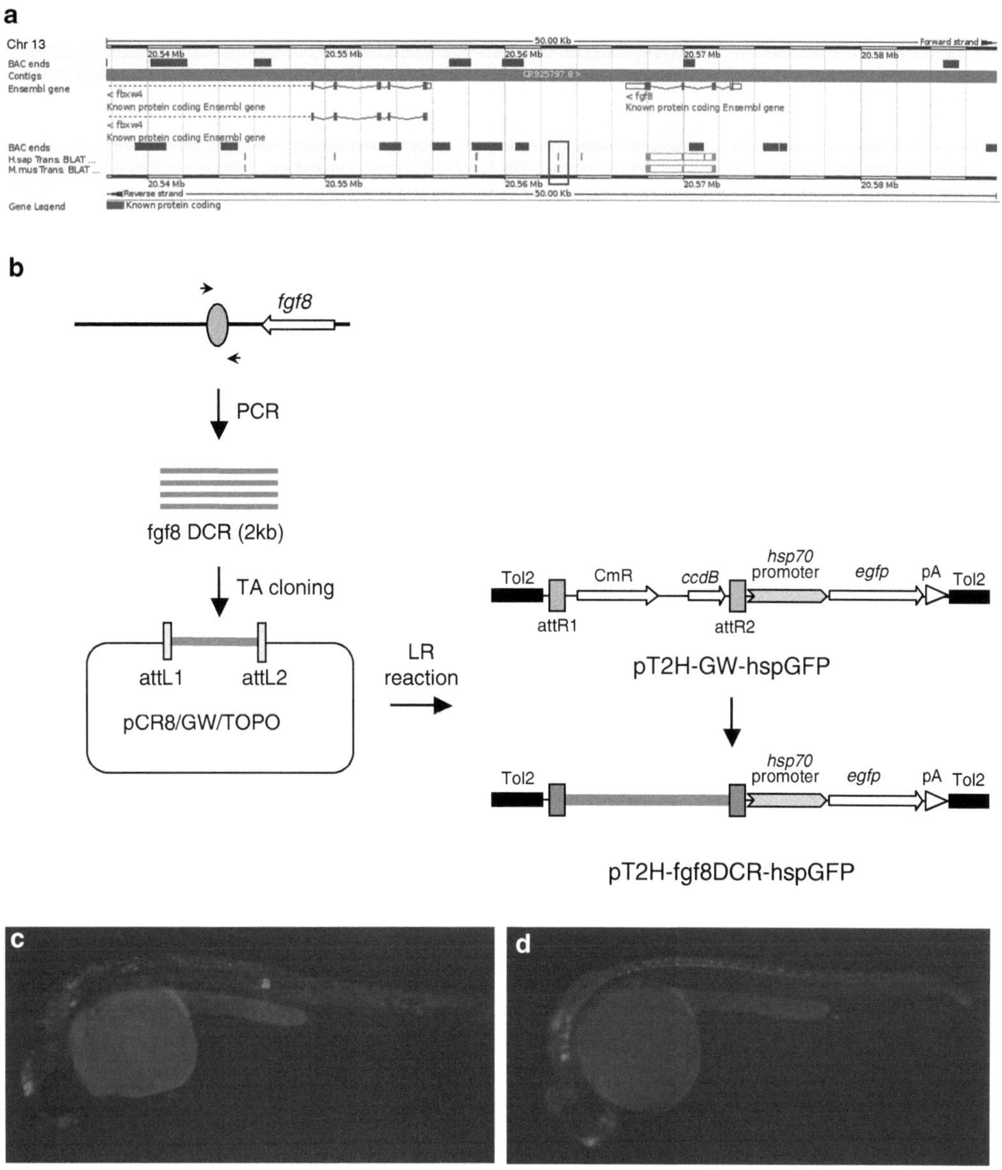

Fig. 3. Analysis of a genomic enhancer activity with a *Tol2*-Gateway enhancer assay vector: an example of analysis of a genomic enhancer of the *fgf8* locus. (**a**) Comparison of the zebrafish *fgf8* locus on the chromosome 13 with human and mouse translated BLAT. A red box indicates DNA sequence conserved among these species downstream of the *fgf8* gene. (**b**) The fgf8DCR is amplified by PCR from genomic DNA and cloned into the pCR8/GW/TOPO plasmid. Then, the fgf8DCR fragment is transferred to the enhancer assay vector pT2H-GW-hspGFP through the LR reaction, resulting in pT2H-fgf8DCR-hspGFP. (**c**) Transient transgenesis using pT2H-fgf8DCR-hspGFP. A side view of a 24 hpf-embryo injected with the transposase mRNA and pT2H-fgf8DCR-hspGFP. (**d**) Stable transgenesis using pT2H-fgf8DCR-hspGFP. Fish injected with the transposase mRNA and pT2H-fgf8DCR-hspGFP at the one cell stage were raised to adulthood and crossed with wild-type fish. A side view of the GFP-positive F1 embryo at 24 hpf. Both transient and stable transgenic embryos show essentially similar GFP expression patterns including expression in forebrain, midbrain-hindbrain boundary, and spinal cord (*see Color Plates*).

7. Pick ampicillin-resistant colonies. Prepare plasmid DNA with QIAprep spin miniprep kit and confirm the structure by sequencing.
8. Mix 100–300 ng of pCR8/GW/fgf8, 150 ng of pT2H-GW-hspEGFP, 2 μL of 5X LR Clonase Reaction Buffer, and TE buffer (pH 8.0) to final volume of 8 μL.
9. Add 2 μL of LR Clonase enzyme mix to the sample according to the manufacturer's instructions and mix well by flipping the tube. Incubate at 25°C for 1 h.
10. Add 1 μL of proteinase K solution and incubate the sample at 37°C for 10 min.
11. Transform 2 μL of the reaction into 50 μL of DH5α competent cells.
12. Prepare plasmid DNA with QIAprep spin miniprep kit and confirm the structure by sequencing. pT2H-fgf8DCR-hspGFP is thus created (**Fig. 3b**).

3.4. Microinjection Donor Plasmid and Transposase mRNA

3.4.1. In Vitro Synthesis of Transposase mRNA

1. Digest pCS-TP that contains transposase cDNA with *Not*I, purify once by phenol-chloroform extraction, precipitate with ethanol, rinse once with 70% ethanol, and dissolve in DEPC-treated water at the concentration of approximately 1 μg/μL.
2. By using 1 μg of the linearized DNA template and mMESSAGE mMACHINE SP6 kit, synthesize mRNA according to the manufacturer's instructions.
3. After incubation with DNase I, add DEPC-treated water to the sample up to 100 μL. Purify the sample with a spin column as follows. Centrifuge a column at 3,500 rpm (1,100 × g) for 2 min, add 100 μL of water and centrifuge again at 3500 rpm for 2 min. Then apply the sample to the column and spin at 3500 rpm for 4 min, and collect the elution.
4. Adjust the volume of the eluted sample to 135 μL by adding DEPC-treated water and add 15 μL ammonium acetate stop solution. Extract cDNA with an equal volume of phenol-chloroform, and then with an equal volume of chloroform. Transfer the aqueous phase to a new RNase-free tube.
5. Add an equal volume of isopropanol, mix well, chill the mixture for at least 15 min at –20°C. Centrifuge at 4°C for 15 min at the maximum speed. Carefully remove the supernatant solution and resuspend the RNA precipitate in 50 μL DEPC-treated water (~30 μg mRNA will be synthesized). Take an aliquot and adjust the concentration to 250 ng/μL for microinjection.

3.4.2. Microinjection

1. Prepare a transposon donor plasmid with QIAfilter plasmid maxi kit.
2. Purify DNA by phenol-chloroform extraction, precipitate in ethanol, rinse once with 70% ethanol, and suspend in nuclease-free H_2O at the concentration of 250 ng/μL.
3. Make the injection ramp of 1% agarose with a glass plate.
4. Prepare an injection needle from a glass capillary using a needle puller, and cut the tip with a sharp blade.
5. For preparation of the injected solution, mix the following components; 10 μL of 0.4 M KCl, 2 μL of Phenol Red solution, 2 μL of 250 ng/μL Transposase mRNA, 2 μL of 250 ng/μL the testing vector, 4 μL of water (final volume 20 μL). To remove unnecessary things that may clog the injection capillary, centrifuge the mixture at the maximum speed for 1 min, and transfer the upper 18 μL to a new tube.
6. Fill the microinjection mixture into the glass capillary and set the glass capillary to the injector (**Fig. 4a**).
7. Align fertilized eggs with chorions at the edge of the agarose ramp.
8. Inject approximately 1 nL of the mixture into the cytoplasm of the cell (**Fig. 4b**).
9. Incubate the injected embryos at 28°C.

3.5. Excision Assay: Detection of Excision of Tol2 from a Donor Plasmid

The transposase synthesized from mRNA catalyzes transposition in the injected zebrafish embryos. First, the transposon construct on the plasmid DNA is excised. After excision, the cleavage at the excision site is repaired and re-ligated. Such an empty site can be

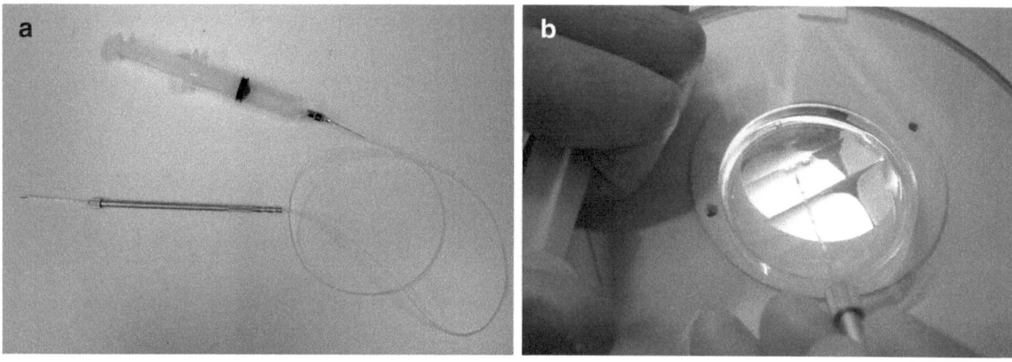

Fig. 4. Microinjection setup. (**a**) A microinjection capillary that is loaded with the injection mixture is attached to a micropipet holder. The holder is connected to a 20-mL syringe with a teflon tube. (**b**) Fertilized eggs are aligned on an agarose ramp made on 6-cm plastic dish. Manual injection is performed by giving air pressure to the injection mixture from a syringe.

detected by PCR (**Fig. 5**). To ensure that this reaction occurs in the injected cells, the excision assay should be performed.

1. Transfer 4–8 injected embryos at 10 hpf to wells in a PCR eight-strip tube. Put one embryo per well.
2. Remove as much water as possible and add 50 μL of embryo lysis buffer.
3. Incubate the sample at 50°C for 3 h to overnight. Inactivate the enzyme at 95°C for 5 min.
4. Mix 1 μL of the lysed sample, 2 μL of 10X PCR buffer, 2 μL of 25 mM $MgCl_2$, 2.5 μL of dNTP mix (2.5-mM each), 1 μL of 20 μM exR and exL primers, 0.25 μL of Expand HiFi Taq polymerase (Roche), and 11.3 μL of water. Perform PCR of

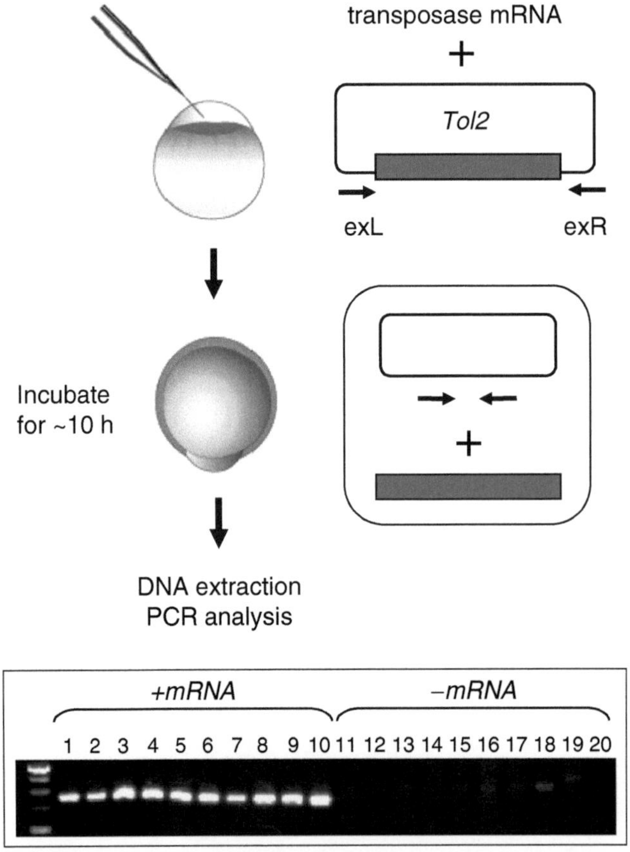

Fig. 5. A scheme for embryonic excision assay. The *Tol2* construct is excised from the injected donor plasmid during embryogenesis. After excision, the donor plasmid is repaired and religated. This structure can be detected by PCR using the exL and exR primers. The gel shows a result of an excision assay. In all of the embryos injected with a donor plasmid and the transposase mRNA, the excision products are detected (*lanes* 1–10). However, in the embryos injected only with a donor plasmid, no excision products are detected (*lanes* 11–20).

94°C for 2 min; 35 cycles of 94°C for 30 s, 58°C for 30 s, 72°C for 30 s.

5. Analyze 10 μL of the PCR samples by 1.5–2% agarose gel electrophoresis.

3.6. Transient and Stable Transgenesis

Figure 6 provides a schematic overview of using *Tol2* constructs for transient and stable transgenesis in zebrafish.

3.6.1. Transient Transgenesis

The activity of genes and genome can be rapidly analyzed in the injected embryos (**Fig. 6**). An injected *Tol2* construct is integrated in the genome of somatic cells during embryogenesis, causing a fairly uniform distribution of the construct throughout the body and sustained expression of a gene cloned in the construct. Fisher et al. demonstrated that enhancer activities of a genomic DNA were detectable by such a transient transgenesis approach *(15)*.

1. Inject the transposase mRNA and pT2H-fgf8DCR-hspGFP (or a plasmid of interest) into fertilized eggs.
2. Analyze the inject embryos under a fluorescent stereoscope at desired time points (e.g., 16, 24, 36, 48 hpf) (**Fig. 3c**).
3. At the best, approximately 20% of the injected embryos show highly representative non-mosaic reporter expression.
4. Anesthetize the embryo with 3-aminobenzoic acid ethyl ester. Mount the embryo on 2–3% methylcellulose (*see* **Note 2**) and take photos.

3.6.2. Stable Transgenesis

Stable transgenic fish show reliable and reproducible expression of a reporter gene. Also, the transgenes are transmitted to offspring in a Mendelian fashion. *Tol2*-mediated transgenesis is a powerful tool to create stable transgenic fish since germline transmission frequencies are extremely high; (i.e., 50–70% of injected fish can transmit the injected DNA to the F1 generation) *(9, 11)* (**Fig. 6**). Stable transgenic fish that show GFP expression in specific tissues have been created by *Tol2*-mediated transgenesis ((i.e., melanophore-specific GFP expression by using the Fugu tyrosinase-related protein1 gene promoter *[17]*, primordial germ cell-specific expression by using the *askopos* gene promoter and *nanos1* 3′-untranslated region element *[18])*.

1. Inject the transposase mRNA and pT2H-fgf8DCR-hspGFP (or a plasmid of interest) into fertilized eggs.
2. Raise the injected fish to sexual maturity. This step usually takes 2.5–3 months.
3. Cross the injected fish with wild-Type fish. Based on the high transgenic efficiency, it is expected that more than five out of ten injected fish give rise to between 5 and 50 GFP-positive embryos when one hundred F1 embryos are analyzed for each cross.

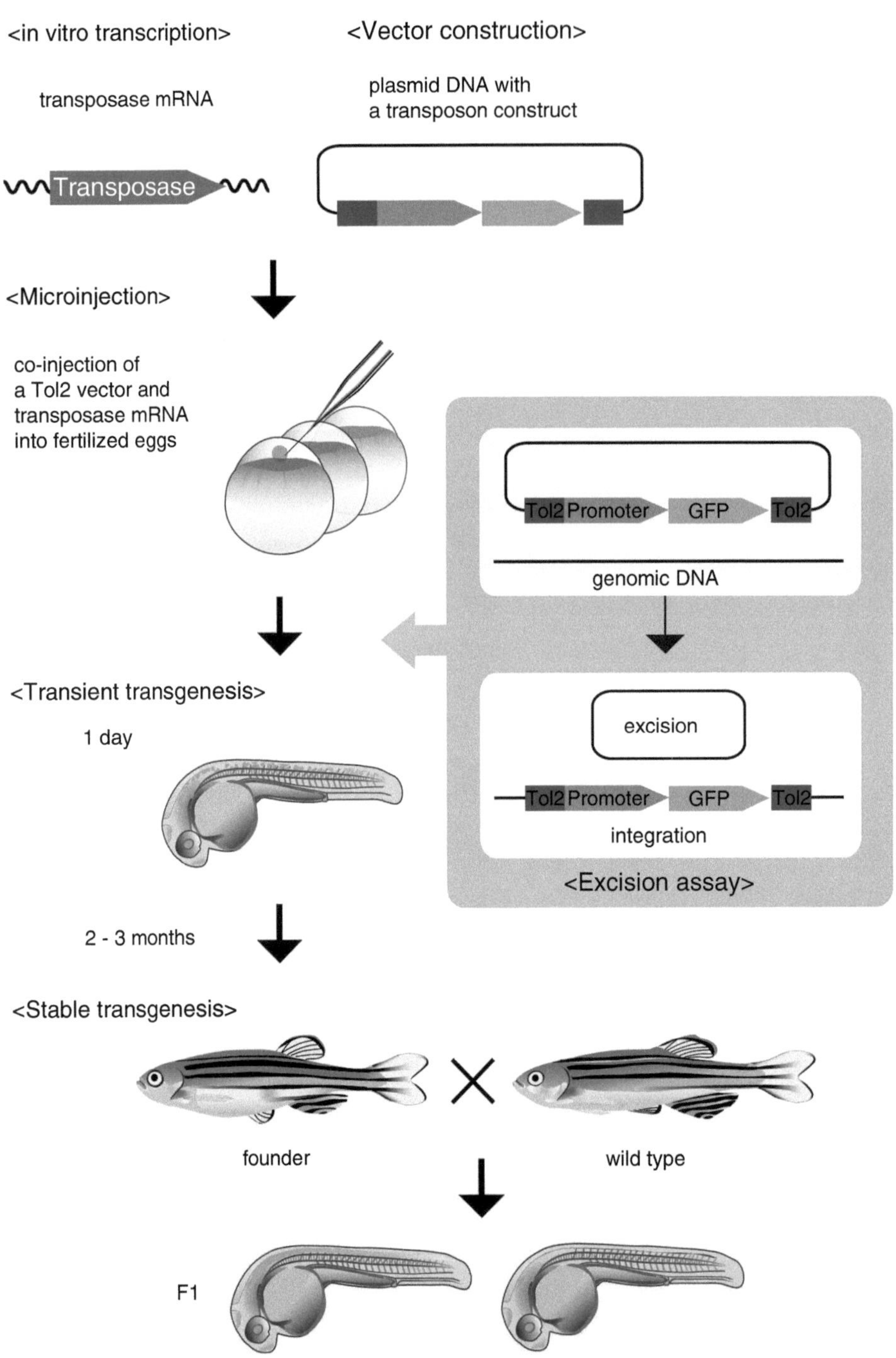

Fig. 6. A scheme for transient and stable transgenesis in zebrafish. The transposase mRNA synthesized in vitro and a plasmid DNA containing a *Tol2* construct are coinjected into zebrafish fertilized eggs. In zebrafish cells during embryogenesis, the *Tol2* construct is excised from the donor plasmid and integrated in the genome. In transient transgenesis, the *Tol2* construct is integrated in the genome of somatic cells. These cells express GFP under the control of the promoter activity cloned on the construct. Since integration does not occur in all somatic cells, mosaic expression of the reporter gene is observed. In stable transgenesis, the *Tol2* construct is integrated in the genome of germ cells. This occurs with high efficiency, and more than 50% of the injected fish transmit the *Tol2* insertions to a proportion of their F1 generation. The F1 embryos that carry *Tol2* insertions from their parent are heterozygous for such an insertion, and therefore display non-mosaic expression of the reporter gene (*see Color Plates*).

4. Analyze the embryo under a fluorescent stereoscope at desired time points (e.g., 16, 24, 36, 48 hpf, etc.) (**Fig. 3d**).
5. Pick GFP-positive F1 embryos up and raise them. The F1 fish often contain multiple insertions. We highly recommend analyzing these F1 fish by Southern blot hybridization (*see* Chapter on "Analysis of Genes and Genome by the *Tol2*-Mediated Gene and Enhancer Trap Methods"). This is important to determine the number of transposon insertions carried by the transgenic fish and to establish transgenic lines carrying single-copy transposon insertions (*see* **Note 3**). In the situation when all of the fish analyzed contain multiple insertions, pick the fish with the smallest number of insertions and outcross these fish with wild-Type fish to isolate fish with single insertions in the next generation.

4. Notes

1. To amplify a genomic DNA fragment by PCR, it is important to use primers of longer sizes, namely 25–30-mer.
2. In some injected embryos, specific reporter gene expression patterns may be observed against high background expression.
3. Expression of a reporter gene in stable transgenic fish may be affected by position effects. Therefore, it is important to isolate fish that harbor insertions in different loci and to compare reporter expression in these fish. Some should show reporter expression patterns that are consistent with the expression pattern of the enhancer/promoter.

Acknowledgments

We would like to thank Dr. Chi-Bin Chien for helpful comments on the manuscript; a postdoctoral fellowship from National Institute of Genetics (to HK); and the National BioResource Project and grants from the Ministry of Education, Culture, Sports, Science and Technology of Japan.

References

1. Stuart, G. W., McMurray, J. V., and Westerfield, M. (1988). Replication, integration and stable germ-line transmission of foreign sequences injected into early zebrafish embryos. *Development* **103**, 403–12.
2. Amsterdam, A., Lin, S., and Hopkins, N. (1995). The *Aequorea victoria* green fluorescent protein can be used as a reporter in live zebrafish embryos. *Dev Biol* **171**, 123–9.

3. Long, Q., Meng, A., Wang, H., Jessen, J. R., Farrell, M. J., and Lin, S. (1997). GATA-1 expression pattern can be recapitulated in living transgenic zebrafish using GFP reporter gene. *Development* **124**, 4105–11.
4. Higashijima, S., Okamoto, H., Ueno, N., Hotta, Y., and Eguchi, G. (1997). High-frequency generation of transgenic zebrafish which reliably express GFP in whole muscles or the whole body by using promoters of zebrafish origin. *Dev Biol* **192**, 289–99.
5. Raz, E., van Luenen, H. G., Schaerringer, B., Plasterk, R. H., and Driever, W. (1998). Transposition of the nematode *Caenorhabditis elegans Tc3* element in the zebrafish *Danio rerio. Curr Biol* **8**, 82–8.
6. Fadool, J. M., Hartl, D. L., and Dowling, J. E. (1998). Transposition of the *mariner* element from *Drosophila mauritiana* in zebrafish. *Proc Natl Acad Sci U S A* **95**, 5182–6.
7. Kawakami, K., Koga, A., Hori, H., and Shima, A. (1998). Excision of the *Tol2* transposable element of the medaka fish, *Oryzias latipes*, in zebrafish, *Danio rerio. Gene* **225**, 17–22.
8. Davidson, A. E., Balciunas, D., Mohn, D., Shaffer, J., Hermanson, S., Sivasubbu, S., Cliff, M. P., Hackett, P. B., and Ekker, S. C. (2003). Efficient gene delivery and gene expression in zebrafish using the *Sleeping Beauty* transposon. *Dev Biol* **263**, 191–202.
9. Kawakami, K., Takeda, H., Kawakami, N., Kobayashi, M., Matsuda, N., and Mishina, M. (2004). A transposon-mediated gene trap approach identifies developmentally regulated genes in zebrafish. *Dev Cell* **7**, 133–44.
10. Kawakami, K. (2005). Transposon tools and methods in zebrafish. *Dev Dyn* **234**, 244–54.
11. Urasaki, A., Morvan, G., and Kawakami, K. (2006). Functional dissection of the *Tol2* transposable element identified the minimal *cis*-sequence and a highly repetitive sequence in the subterminal region essential for transposition. *Genetics* **174**, 639–49.
12. Balciunas, D., Wangensteen, K. J., Wilber, A., Bell, J., Geurts, A., Sivasubbu, S., Wang, X., Hackett, P. B., Largaespada, D. A., McIvor, R. S., and Ekker, S. C. (2006). Harnessing a high cargo-capacity transposon for genetic applications in vertebrates. *PLoS Genet* **2**, e169.
13. Kwan, K. M., Fujimoto, E., Grabher, C., Mangum, B. D., Hardy, M. E., Campbell, D. S., Parant, J. M., Yost, H. J., Kanki, J. P., and Chien, C. B. (2007). The Tol2kit: a multisite gateway-based construction kit for *Tol2* transposon transgenesis constructs. *Dev Dyn* **236**, 3088–99.
14. Villefranc, J. A., Amigo, J., and Lawson, N. D. (2007). Gateway compatible vectors for analysis of gene function in the zebrafish. *Dev Dyn* **236**, 3077–87.
15. Fisher, S., Grice, E. A., Vinton, R. M., Bessling, S. L., and McCallion, A. S. (2006). Conservation of RET regulatory function from human to zebrafish without sequence similarity. *Science* **312**, 276–9.
16. Fisher, S., Grice, E. A., Vinton, R. M., Bessling, S. L., Urasaki, A., Kawakami, K., and McCallion, A. S. (2006). Evaluating the biological relevance of putative enhancers using *Tol2* transposon-mediated transgenesis in zebrafish. *Nat Protoc* **1**, 1297–305.
17. Zou, J., Beermann, F., Wang, J., Kawakami, K., and Wei, X. (2006). The Fugu tyrp1 promoter directs specific GFP expression in zebrafish: tools to study the RPE and the neural crest-derived melanophores. *Pigment Cell Res* **19**, 615–27.
18. Blaser, H., Reichman-Fried, M., Castanon, I., Dumstrei, K., Marlow, F. L., Kawakami, K., Solnica-Krezel, L., Heisenberg, C. P., and Raz, E. (2006). Migration of zebrafish primordial germ cells: a role for myosin contraction and cytoplasmic flow. *Dev Cell* **11**, 613–27.

Chapter 6

Analysis of Genes and Genome by the *Tol2*-Mediated Gene and Enhancer Trap Methods

Akihiro Urasaki and Koichi Kawakami

Summary

The *Tol2* transposon system can create insertions in the zebrafish genome efficiently. By using this system, the gene trap and enhancer trap methods have been developed. The gene trap and enhancer trap constructs contain the green fluorescent protein (GFP) reporter gene or the yeast Gal4 transcription activator gene. By creating random integrations of these constructs in the genome, transgenic fish expressing the GFP gene or the Gal4 gene in specific cells, tissues or organs are generated. These fish are valuable resources for developmental biology. Especially, the Gal4-expressing transgenic fish can be used to ectopically express any gene of interest placed downstream of the Gal4 recognition sequence, UAS, and thereby allow visualization, modification or ablation of the Gal4-expressing cells. In this chapter, we will describe how the gene trap and enhancer trap screens can be performed and how the transposon insertions created by these methods can be analyzed.

Key words: Transposon, *Tol2*, Gene trapping, Enhancer trapping, Gal4-UAS, GFP.

1. Introduction

Tol2-mediated transgenesis is highly efficient and is widely used as a method to create transgenic zebrafish *(1, 2)*. Because it is now possible for a small lab to generate hundreds or thousands of transposon insertions, gene trapping *(1)* and enhancer trapping *(3, 4)* have been performed in zebrafish by using the *Tol2* transposon system. A gene trap construct contains a splice acceptor and the green fluorescent protein (GFP) gene. When the construct is integrated within a gene and trapped an endogenous transcript by the splice acceptor, GFP is expressed. An enhancer trap construct contains a minimal promoter such as the zebrafish

Graham J. Lieschke et al. (eds.), *Zebrafish,* Methods in Molecular Biology, vol. 546
DOI: 10.1007/978-1-60327-977-2_6

hsp70 promoter, and the GFP gene. The *keratin8* or the *gata2* promoter have also been used as a minimal promoter for enhancer trapping in zebrafish *(3, 5)*. When an enhancer trap construct is integrated into the genome and the minimal promoter is activated by a chromosomal enhancer, GFP is expressed.

The gene trap and enhancer trap methods are important and powerful tools for studying development and genetics. First, transgenic fish expressing GFP in specific cells, tissues and organs created by these methods are useful for studying morphogenesis and organogenesis of those cells, tissues and organs. Second, by analyzing genomic DNA surrounding transposon insertions by inverse polymerase chain reaction (PCR) or adaptor-ligation PCR, genes expressed in specific patterns and enhancers/promoters controlling such specific expression patterns can be identified rapidly. Third, these insertions may disrupt important developmental genes. By crossing heterozygous parents that carry the same transposon insertions, homozygous embryos that show mutant phenotypes can be isolated.

Furthermore, gene trap and enhancer trap constructs containing the yeast Gal4 transcription activator gene were developed recently *(6–8)*. When transgenic fish expressing Gal4 in specific cells and tissues are crossed with transgenic fish that contain a reporter gene or an effecter gene downstream of the Gal4-recognition sequence, UAS, transcription of the reporter or the effecter gene is induced in the Gal4-expressing cells. Therefore the Gal4-expressing cells are visualized, modified or ablated. In this chapter, we will describe how the gene trap and enhancer trap screens can be performed and how the transposon insertions created by the gene trap and enhancer trap methods should be analyzed.

2. Materials

2.1. Gene Trap and Enhancer Trap Screens

2.1.1. GFP Gene Trap and Enhancer Trap Constructs

1. T2KSAG *(1)*.
2. T2KHG *(4)*.
3. ET *(3)*.

2.1.2. Construction and Identification of Gene Trap and Enhancer Trap Fish

1. Transposase mRNA and microinjection apparatus (*see* Chapter 5).
2. E3 solution: 5 mM NaCl, 0.17 mM KCl, 0.33 mM $CaCl_2$, 0.33 mM $MgSO_4$, 0.00001% Methylene blue (Waldeck, Germany).
3. 90-mm Plastic dish

4. Nine-well glass depression plate (Corning).
5. MULTIWELL six-well dish (Becton Dickinson Labware).
6. Fluorescent stereoscope LEICA MZ 16 FA (Leica Microsystems, Germany) or equivalent.

2.2. Gal4 Gene Trap and Enhancer Trap Screens

2.2.1. Gal4 Gene Trap and Enhancer Trap Constructs

1. SAGVG *(6)*.
2. hsp(1.5 kb):Gal4 *(7)*.
3. hsp(600 base pairs [bp]):Gal4 *(7)*.
4. T2KSAGFF *(8)*.
5. T2KhspGFF *(8)*.

2.2.2. Construction and Identification of Gal4 Gene Trap and Enhancer Trap Fish

1. UAS: GFP line *(8)*.
2. UAS: red fluorescent protein (RFP) line *(8)*.

2.3. Analysis of Transposon Insertions by Southern Blot Hybridization

1. DNA extraction buffer: 10 mM Tris–HCl (pH 8.2), 10 mM ethylenediamine tetraacetic acid (EDTA), 200 mM NaCl, 0.5% sodium dodecyl sulfate (SDS), 200 ng/μL proteinase K.
2. Phenol–chloroform solution.
3. 3 M sodium acetate.
4. 1X TE buffer: 10 mM Tris–HCl (pH 7.5), 1 mM EDTA.
5. 0.1N HCl.
6. 0.5N NaOH.
7. 10X SSC: 1.5 M NaCl, 0.15 M sodium citrate.
8. Hybridization buffer: 0.25 M Na_2HPO_4 (adjust pH to 7.4 with H_3PO_4), 1 mM EDTA, 10 g/L bovine serum albumin (BSA), 7% SDS.
9. Wash solution: 0.1X SSC, 0.1% SDS.
10. *Bgl*II (Takara, Japan).
11. Primers to amplify the Gal4FF probe: GAL4FF-f2 (5′-ATG AAG CTA CTG TCT TCT-3′); GAL4FF-r2 (5′-TCT AGA TTA GTT ACC CGG-3′).
12. Prime-it II random primer labeling kit (Stratagene).
13. Amersham Hybond-N^+ (GE Healthcare, England).
14. Vacuum blotter and vacuum pump (BE-600, BIO CRAFT, Japan).
15. Hybridization oven (Model 400 hybridization incubator, Robbins Scientific).
16. Plastic wrap (SaranWrap, Asahi Kasei, Japan).
17. Imaging plate and image analyzer for radioactivity (BAS2500, FUJIFILM, Japan).

2.4. Analysis of Integration Sites by PCR-Based Techniques

2.4.1. Cloning of Junction Fragments by Adaptor-Ligation PCR

1. Adaptor: AL (5′-CTA ATA CGA CTC ACT ATA GGG CTC GAG CGG CCG CGG GGG CAG GT-3′); GATC-AS (5′-GAT CAC CTG CCC CCG CTT-3′).
2. Primers for first PCR: Ap1 (5′-GGA TCC TAA TAC GAC TCA CTA TAG GG-3′); 175L-out (5′-TTT TTG ACT GTA AAT AAA ATT G-3′); 150R-out (5′-AAT ACT CAA GTA CAA TTT TA-3′).
3. Primers for second PCR and sequencing: Ap2 (5′-CAC TAT AGG GCT CGA GCG G-3′); 150L-out (5′-GAG TAA AAA GTA CTT TTT TTT CT-3′); 100R-out (5′-AGA TTC TAG CCA GAT ACT-3′); 100L-out (5′-AGT ATT GAT TTT TAA TTG TA-3′).
4. SAGFF specific primers: SAGFF-r1 (5′-AGC ACG TTG CCC AGG AGC TGT AGG A-3′); SAGFF-r2 (5′-GAA CAT GGT TAG CAG AGG GG-3′).
5. *Mbo*I (Takara Bio, Japan).
6. T4 DNA ligase (Takara Bio, Japan).
7. Expand High Fidelity PCR system (Roche Diagnostics, Germany).
8. QIAquick gel extraction kit (Qiagen, Germany).
9. BigDye terminator v3.1 cycle sequencing kit (Applied Biosystems).
10. 5X sequencing buffer (Applied Biosystems).
11. HiDi formamide (Applied Biosystems).
12. DNA sequencer (ABI PRISM 3130Xl DNA sequencer, Applied Biosystems).

2.4.2. Cloning of Junction Fragments by Inverse PCR

1. Primers for first PCR: SAGFF-f1 (5′-GGT TGT TGT GCT GTC TCA TCA TTT-3′); SAGFF-r1 (5′-AGC ACG TTG CCC AGG AGC TGT AGG A-3′).
2. Primers for second PCR: SAGFF-f2 (5′-TCA TTT TGG CAA AGA ATT CCT CG-3′); SAGFF-r2 (5′-GAA CAT GGT TAG CAG AGG GG-3′).

2.5. Analysis of Trapped Genes by 5′RACE or RT-PCR

1. A primer for reverse transcription: Gal4-r3 (5′-TTC AGA CAC TTG GCG CAC TTC GG-3′).
2. Primers for PCR: Gal4-r2 (5′-TTT AAG TCG GCA AAT ATC GCA TG-3′); Gal4-r1 (5′-TTC GAT AGA AGA CAG TAG CTT CA-3′); AAP (5′-GGC CAC GCG TCG ACT AGT ACG GGG GGG GGG-3′); AUAP (5′-GGC CAC GCG TCG ACT AGT AC-3′).
3. TRIzol regent (Invitrogen).

4. 5′ RACE system for rapid amplification of cDNA ends (Invitrogen).
5. SuperScript first-strand synthesis system for RT-PCR (Invitrogen).
6. Ex Taq (Takara Bio, Japan).
7. TOPO TA cloning dual promoter (Invitrogen).
8. Teflon tissue grinder (Wheaton Science Products).
9. QIAprep spin miniprep kit (QIAGEN).

3. Methods

3.1. Gene Trap and Enhancer Trap Screens

3.1.1. GFP Gene Trap and Enhancer Trap Constructs

The *Tol2* gene trap and enhancer trap constructs containing the GFP gene are shown in **Fig. 1a**. The T2KSAG gene trap construct contains a splice acceptor (SA) from the rabbit β-globin intron, the EGFP gene and the SV40 polyA signal *(1)*. The T2KHG enhancer trap construct contains the zebrafish *hsp70* promoter, the EGFP gene and the SV40 polyA signal *(4)*. The ET enhancer trap construct contains the EGFP gene downstream of 460-bp DNA of the zebrafish *keratin8* promoter *(3)*.

3.1.2. Construction and Identification of Gene Trap and Enhancer Trap Fish

A scheme for gene trap and enhancer trap screens are shown in **Fig. 2a**.

1. A transposon-donor plasmid DNA containing the *Tol2* construct and the transposase mRNA are injected into zebrafish fertilized eggs. A detailed protocol for microinjection is described in the chapter by Kikuta and Kawakami.
2. Raise the injected fish to adulthood. This usually takes 2.5–3 mo.
3. Incross the injected fish each other or outcross them with wild-type fish.
4. Collect embryos, submerged in E3 solution, and clean up to remove debris and unfertilized eggs.
5. Place embryos on nine-well glass plates, observe embryos at time points of interest (e.g., 16, 24, 36 hpf) under a fluorescent stereoscope, and pick GFP-positive embryos.
6. Transfer the GFP-positive embryos to six-well plastic dishes and incubate at 28°C.

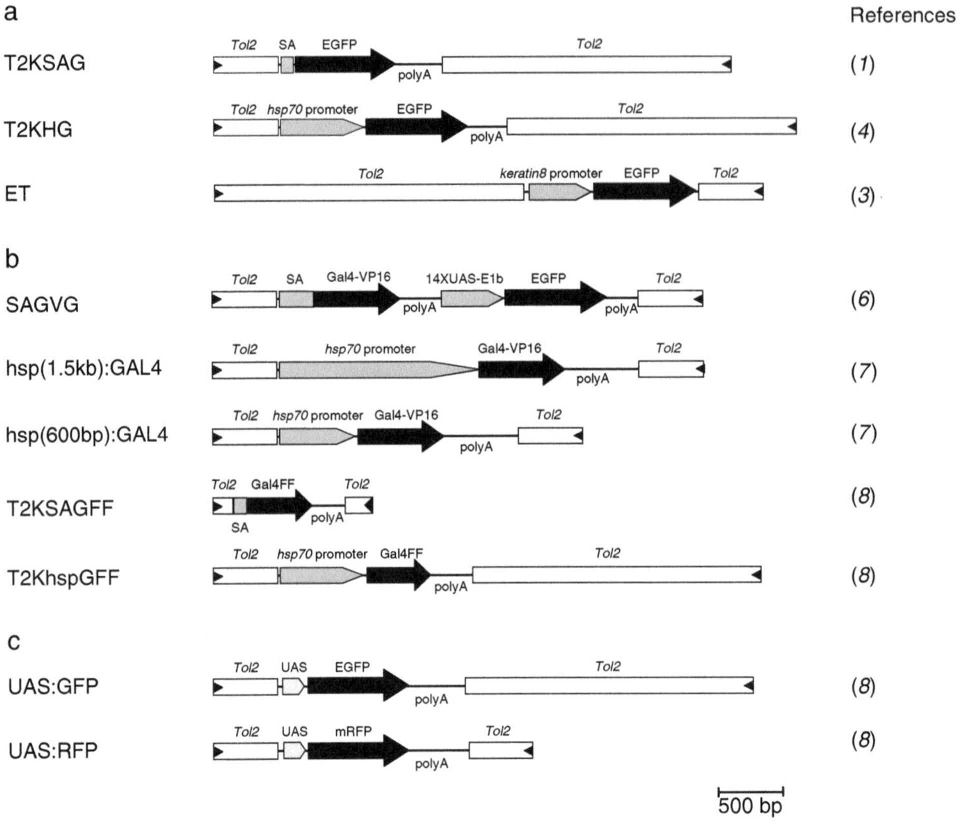

Fig. 1. The structures of the gene and enhancer trap constructs and the UAS reporter fish. (**a**) The structures of T2KSAG (gene trap), T2KHG (enhancer trap) and ET (enhancer trap) that contain the EGFP gene. (**b**) The structures of SAGVG (enhancer trap), hsp(1.5 kb):Gal4 (enhancer trap), hsp(600 bp):Gal4 (enhancer trap), T2KhspGFF (enhancer trap) and T2KSAGFF (gene trap) that contain Gal4-VP16 or Gal4FF. The *Tol2* sequences are shown as *open boxes. Black triangles* at the ends of *Tol2* show terminal inverted repeats. SA indicates the splice acceptor of an intron of the rabbit β-globin gene. (**c**) The structures of UAS:GFP and UAS:RFP reporter fish. UAS:GFP and UAS:RFP contain the EGFP gene and the mRFP gene downstream of five-times repeats of the Gal4 recognition sequence UAS, respectively.

Fig. 2. Gene trap and enhancer trap screens in zebrafish. (**a**) Gene trap and enhancer trap screens using GFP. A transposon-donor plasmid that contains either the gene trap or the enhancer trap construct using the EGFP gene is injected with transposase mRNA into fertilized eggs. The injected fish are raised to adulthood and crossed with wild-type fish. Embryos expressing GFP in specific regions are identified and collected in the F1 generation. (**b**) Gene trap and enhancer trap screens using Gal4FF. A transposon-donor plasmid that contains either the gene trap or the enhancer trap construct using Gal4FF is injected with transposase mRNA into fertilized eggs. The injected fish are raised to adulthood and crossed with the UAS:GFP line. Double transgenic embryos expressing GFP in specific regions are identified in the F1 generation. (**c**) Expression of UAS:GFP and UAS:RFP. The same Gal4FF transgenic fish was crossed with the UAS:GFP or the UAS:RFP reporter fish. Similar GFP and RFP expression patterns are observed in the double transgenic embryos. (**d**) Various GFP expression patterns in Gal4FF;UAS:GFP double transgenic embryos (*see Color Plates*).

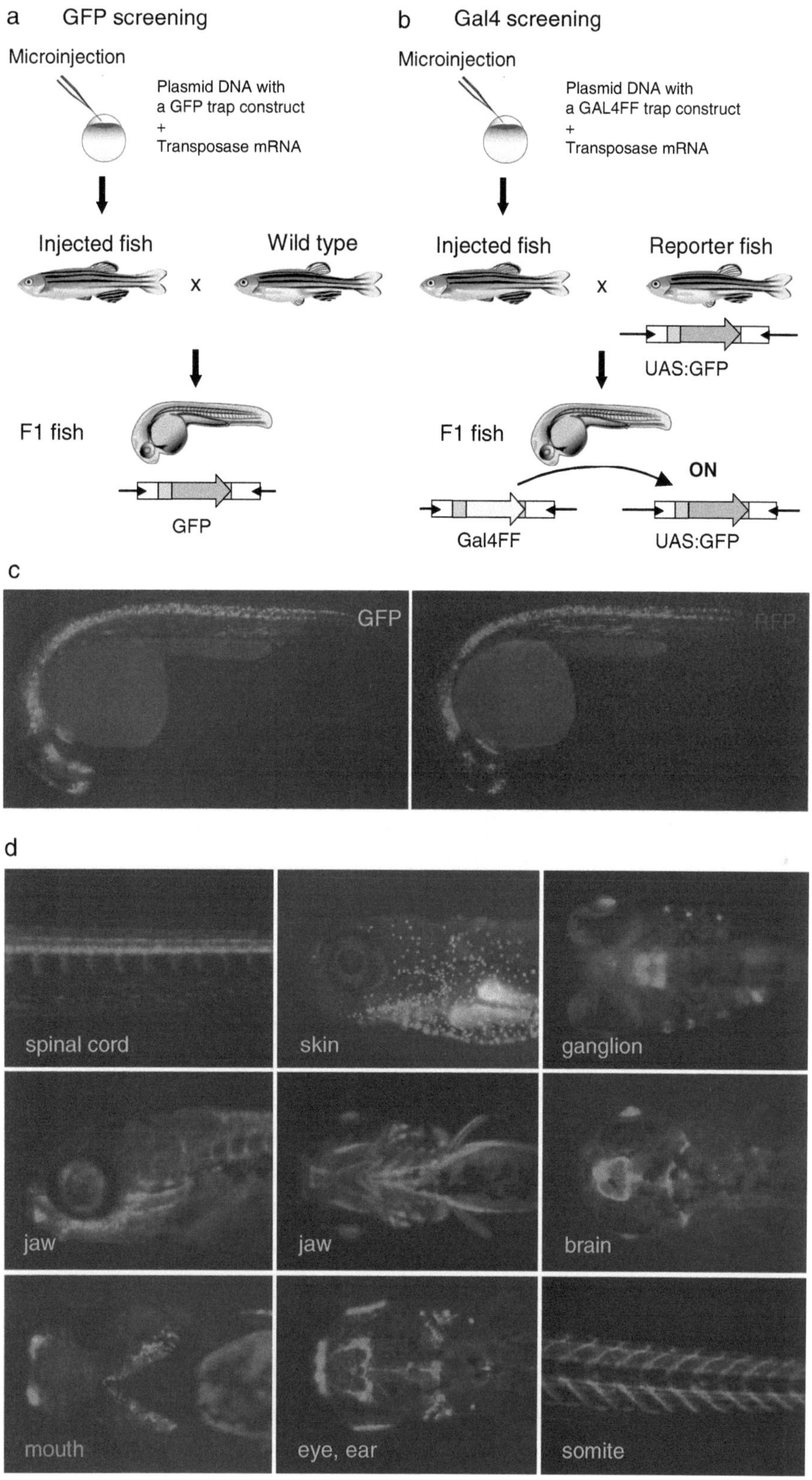
a
GFP screening
Microinjection
Plasmid DNA with
a GFP trap construct
+
Transposase mRNA
Injected fish
x
Wild type
F1 fish
GFP
b
Gal4 screening
Microinjection
Plasmid DNA with
a GAL4FF trap construct
+
Transposase mRNA
Injected fish
x
Reporter fish
UAS:GFP
F1 fish
ON
Gal4FF
UAS:GFP
c
GFP
RFP
d
spinal cord
skin
ganglion
jaw
jaw
brain
mouth
eye, ear
somite

3.2. Gal4 Gene Trap and Enhancer Trap Screens

3.2.1. Gal4 Gene Trap and Enhancer Trap Constructs

The *Tol2* gene trap and enhancer trap constructs containing the *Gal4* gene are shown in **Fig. 1b**. The SAGVG gene trap construct contains the transcriptional activator Gal4-VP16 gene placed downstream of the rabbit β-globin splice acceptor and the EGFP gene placed downstream of 14-times repeat of UAS (14X UAS) and adenovirus E1b minimal promoter element (E1b) *(6)*. The hsp(1.5 kb):Gal4 and hsp(600 bp):Gal4 enhancer trap constructs contain the Gal4-VP16 gene downstream of 1.5 kb and 600-bp DNA of the *hsp70* promoter, respectively *(7)*. The T2KSAGFF gene trap construct contains the Gal4FF gene downstream of the rabbit β-globin splice acceptor *(8)*. Gal4FF is a modified version of transcription activator Gal4, which has the DNA-binding domain of Gal4 and two short transcriptional activation motifs from VP16. Gal4FF shows a weaker transcriptional activity and less toxic than Gal4-VP16. The T2KhspGFF enhancer trap construct contains the Gal4FF gene downstream of the 650-bp *hsp70* promoter *(8)*.

3.2.2. Construction and Identification of Gal4 Gene Trap and Enhancer Trap Fish

A scheme for Gal4 gene trap and enhancer trap screens using T2KSAGFF and T2KhspGFF are shown in **Fig. 2b**. Gal4 activates a gene placed downstream of its recognition sequence UAS. The UAS:GFP fish contains the EGFP gene downstream of 5X UAS, and the UAS:RFP fish contains the mRFP gene downstream of 5X UAS *(8)*. Expression of Gal4 can be visualized by crossing T2KSAGFF and T2KhspGFF fish with the reporter fish harboring the UAS:GFP or UAS:RFP (**Fig. 2c**).

1. Inject a plasmid DNA harboring either Gal4FF gene trap or enhancer trap construct into fertilized eggs with transposase mRNA.
2. Raise the injected fish to the adulthood and cross them with the UAS:GFP reporter fish.
3. Place embryos on nine-well glass plates, observe embryos at time points of interest (e.g., 16, 24, 36 hpf) under a fluorescent stereoscope, and pick GFP-positive embryos (**Fig. 2d**).
4. Transfer the GFP-positive embryos to six-well plastic dishes and incubate at 28°C.
5. When Gal4FF transgenic fish are crossed either with UAS:GFP fish or with UAS:RFP fish, similar expression patterns are observed (**Fig. 2c**).

3.3. Analysis of Transposon Insertions by Southern Blot Hybridization

F1 fish created by *Tol2*-mediated transgenesis often carry multiple insertions. To determine the number of transposon insertions and to identify insertions that are responsible for the identified GFP or Gal4FF expression patterns, it is important to analyze the GFP-positive F1 fish by Southern blot hybridization.

1. Tail fins are cut from adult fish, soaked in 200 μL DNA extraction buffer, incubate at 50°C for 3 h to overnight, and dissolve completely.
2. 200 μL of phenol–chloroform are added and mixed well with the sample. After centrifugation at 15,000 rpm (20400 × g) for 5 min, transfer the upper aqueous phase to a clean 1.5-mL microcentrifuge tube.
3. Precipitate the DNA samples with 1 mL ethanol, rinse once with 70% ethanol and suspend in 50 μL TE. Measure DNA concentrations by a spectrometer.
4. Digest 5 μg of the genomic DNA sample with *Bgl*II in 30 μL reaction solution at 37°C for 1 h, and perform electrophoresis on 1% Tris-acetate EDTA (TAE)-agarose gel containing 0.25 ng/μL ethidium bromide.
5. After taking a photo of the gel under a ultraviolet (UV) trans-illuminator, soak the gel in 0.1N HCl for 15 min, rinse briefly with deionized water, soak in 0.5N NaOH for 30 min, rinse again briefly with deionized water and soak in 10X SSC.
6. Place the gel on a transfer apparatus with Hybond-N^+ presoaked with 10X SSC. Several vacuum transfer devices are available commercially. Transfer should be carried out according to the manufacturer's instructions.
7. After transfer, rinse the membrane briefly in 2X SSC, and dry at 55°C for 1 h to overnight (*see* **Note 1**).
8. Soak the membrane in 25 mL of preheated hybridization buffer on a tray, and together with the buffer transfer into a glass tube and incubate at 65°C for 1 h.
9. Amplify a DNA fragment containing the Gal4FF sequence by PCR using GAL4FF-f2 and GAL4FF-r2 and label it with ^{32}P-dCTP and Prime-it II random primer labeling kit according to the manufacturer's instructions.
10. Discard the hybridization buffer and replace with a new 10 mL hybridization buffer containing the ^{32}P-dCTP labeled DNA probe. Incubate at 65°C overnight.
11. After hybridization, rinse the membrane in the tube twice with approximately100 mL wash solution at 65°C for 30 min. Transfer the membrane to a tray, rinse with 0.1X SSC, and wrapped with a plastic wrap. Place the membrane on an imaging plate in a cassette for several hours to overnight and analyze the image by an image analyzer (**Fig. 3**).

3.4. Analysis of Integration Sites by PCR-Based Techniques

The integration sites of the *Tol2* constructs can be determined mainly by two approaches. One is adaptor-ligation PCR and the other is inverse PCR (**Figs. 4** and **5**). Both of these are PCR-based methods to clone unknown genomic DNA by using known

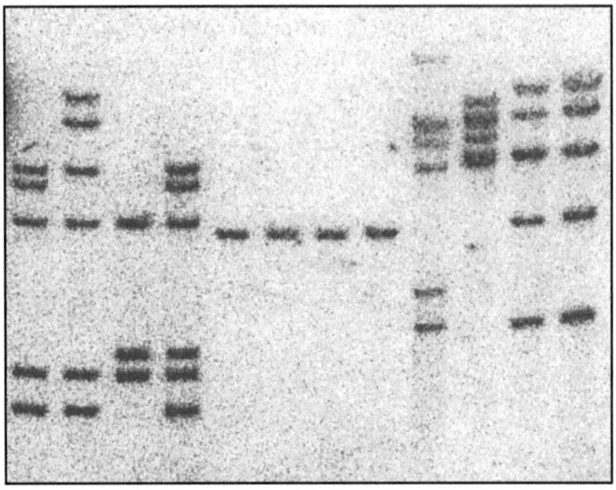

Fig. 3. Southern blot hybridization analysis of the gene trap and enhancer trap fish. An example of Southern blot hybridization analysis of F1 fish harboring the Gal4FF transgene. Genomic DNA prepared from F1 fish were digested with *BglII*, separated on 1% agarose gel, transferred to Hybond-N+, and hybridized with the ^{32}P-labeled Gal4FF DNA probe. Lanes 1–4, 5–8 and 9–12 are F1 fish from the same injected fish that show the same GFP expression patterns. Fish harboring multiple insertions and single insertions are detected. Note that F1 fish from the same injected fish contain at least one insertion in common.

DNA sequences, in this case the sequence of the *Tol2* constructs, as a tag (for the relative advantages of the methods, *see* **Note 2**).

3.4.1. Cloning of Junction Fragments by Adaptor-Ligation PCR

A scheme for adaptor-ligation PCR is shown in **Fig. 4**. First, genomic DNA is digested with an appropriate restriction enzyme and ligated with an adaptor (*see* **Note 3**). Then the transposon-genome junction DNA is amplified by using a primer in the *Tol2* sequence and a primer in the sequence in the adaptor (*see* **Note 4**).

Adaptor-Ligation PCR for T2KSAGFF

1. Digest 500–1,000 ng of genomic DNA prepared from tails fins of GFP-positive fish with *MboI* in 10 μL reaction solution at 37°C for 1 h, heat at 70°C for 15 min, and ligate with the GATC adaptor in a ligation mixture at 16°C for 3 h to overnight.
2. Inactivate the ligase at 70°C for 15 min. Dilute the sample to 1/10 by adding 180 μL of H_2O. Then by using 2 μL of the sample, perform the first PCR (30 cycles of 94°C for 30 s, 57°C for 30 s, 72°C for 1 min) in 50 μL 1X PCR buffer containing 1 μM Ap1 primer and 1 μM 175L-out (for amplification of the left junction) or 150R-out (for amplification of the right junction) primer.

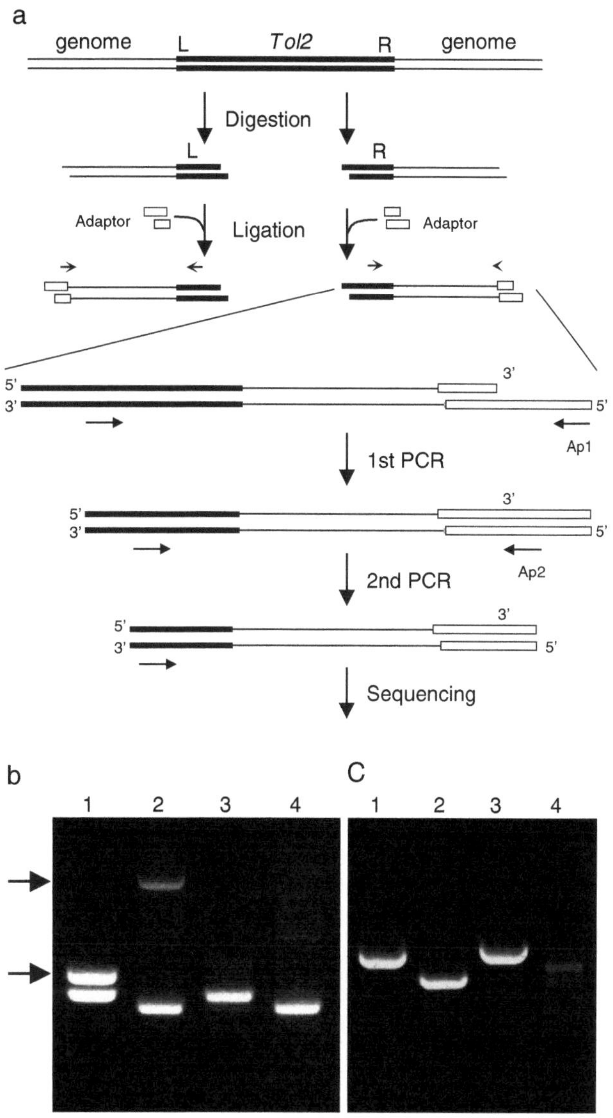

Fig. 4. Analysis of *Tol2* integration sites by adaptor-ligation PCR. (**a**) A scheme for adapter-ligation PCR. Genomic DNA that contains the *Tol2* sequence (*thick lines*) and flanking genomic sequence (*thin line*) is digested with a restriction enzyme and ligated with adaptors (*open boxes*). Then, the junction fragments are amplified by two-rounds of PCR using primers in the *Tol2* sequence toward the outside and in the adaptor sequence. Arrows show positions and directions of primers. (**b**) An example of agarose gel electrophoresis of adaptor-ligation PCR products. *Lane 1*: the left junction fragments amplified from the Gal4;UAS:GFP double transgenic fish. *Lane 2*: the right junction fragments amplified from the Gal4;UAS:GFP double transgenic fish. *Lane 3*: the left junction fragment amplified from the UAS:GFP fish. Lane 4: the right junction fragment amplified from the UAS:GFP fish. *Lanes 1* and *2* show two bands (i.e., one derived from the Gal4 transgene (*arrows*) and the other derived from the UAS:GFP transgene). (**c**) Adaptor-ligation PCR for the right end of T2KSAGFF by using the T2KSAGFF specific primers. *Lanes 1–3*: PCR products from three gene trap lines. The same gene trap lines are used in **Fig. 5b**. *Lane 4*: negative control: only a faint artifact band is detected from the UAS:GFP fish.

3. Dilute the first PCR sample to 1/10 and, by using 1 μL of the diluted sample, perform the second PCR (30 cycles of 94°C for 30 s, 57°C for 30 s, 72°C for 1 min) in 50 μL 1X PCR buffer containing 1 μM Ap2 primer and 1 μM 150L-out (for amplification of the left junction) or 100R-out (for amplification of the right junction) primer.
4. Analyze 10 μL of the sample by 1.5–2% agarose-TAE electrophoresis (**Fig. 4b**). GFP positive fish created by the Gal4FF gene trap and enhancer trap method are double transgenic for the Gal4FF construct and UAS:GFP transgene. Because the UAS:GFP transgene was also constructed by *Tol2*-mediated transgenesis, two bands are detected.
5. Cut DNA bands that are unique to the Gal4 transgenic fish from the gel and purify with QIAquick gel extraction kit according to the manufacturer's instructions. Final volume of the DNA sample will be approximately 40 μL.
6. 4 μL of the DNA sample is subjected to sequencing in 10 μL sequencing reaction mix containing 0.2 μM 100R-out primer. After performing PCR (25 cycles of 96°C for 10 s, 50°C for 5 s, 60°C for 4 min), mix the sample with 1/10 vol of 3 M sodium acetate and 3 vol of ethanol, incubate at room temperature for 5 min and centrifuge at 15,000 rpm (20400 × g) for 20 min at 4°C. Rinse the precipitated DNA once with 30 μL 70% ethanol, dry and suspend in 20 μL of Hi-Di formamide. Analyze DNA sequence using a DNA sequencer.

Adaptor-Ligation PCR for the Right End of T2KSAGFF by Using T2KSAGFF Specific Primers

Based on the structural difference between T2KSAGFF and UAS:GFP, PCR primers specific to the T2KSAGFF insertions can be designed.

1. Digest 500–1,000 ng of genomic DNA sample with *Mbo*I in 10 μL reaction solution at 37°C for 1 h, heat at 70°C for 15 min, and ligate with the GATC adaptor in a ligation mixture at 16°C for 3 h to overnight. Inactivate the ligase at 70°C for 15 min. Dilute the sample to 1/10 by adding 180 μL of H_2O.
2. By using 2 μL of the ligated sample, perform the first PCR (30 cycles of 94°C for 30 s, 57°C for 30 s, 72°C for 1 min) in 50 μL 1X PCR buffer containing 1 μM SAGFF r1 and Ap1 primers.
3. By using 1 μL of the first PCR product, perform the second PCR (30 cycles of 94°C for 30 s, 57°C for 30 s, 72°C for 1 min) in 50 μL 1X PCR buffer containing 1 μM SAGFF r2 and Ap2 primers.
4. Analyze 10 μL of the sample by 1.5–2% agarose-TAE electrophoresis (**Fig. 4c**). Unlike PCR using the 150R-out and 100R-out primers, single bands can be detected.
5. Extract DNA bands from the gel with QIAquick gel extraction kit and perform sequencing as described above.

3.4.2. Cloning of Junction Fragments by Inverse PCR

A scheme for inverse PCR is shown in **Fig. 5a**. First, genomic DNA is digested with an appropriate restriction enzyme and self-ligated to form a circular DNA. Then PCR using primers toward outsides of the *Tol2* sequence is performed. It is

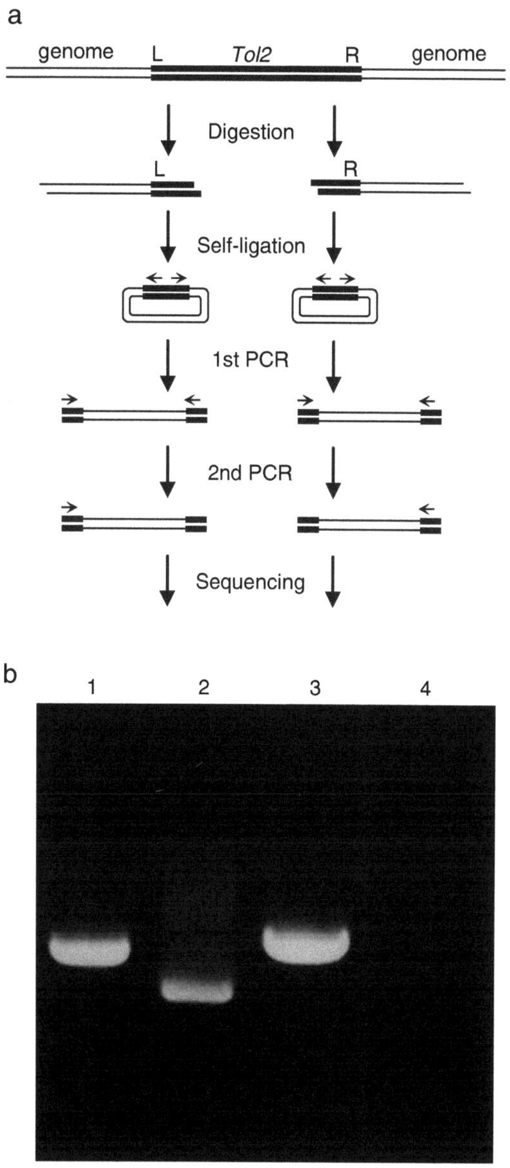

Fig. 5. Analysis of *Tol2* integration sites by inverse PCR. (**a**) Genomic DNA that contains the *Tol2* sequence (*thick lines*) and flanking genomic sequence (*thin line*) is digested with a restriction enzyme and self-ligated. Then, DNA fragments that contain junctions of *Tol2* and genomic DNA are amplified by PCR using primers toward the outsides of the *Tol2* sequence. *Arrows* show positions and directions of the primers. (**b**) An example of agarose gel electrophoresis of inverse PCR products that contain junctions of the right end of T2KSAGFF and genomic DNA. *Lanes 1–3*: PCR products from three gene trap lines. The same gene trap lines are used in **Fig. 4c**. *Lane 4*: negative control: no PCR product is detected from the UAS:GFP fish.

necessary to use different restriction enzymes and different sets of primers when different types of constructs are integrated in the genome. We describe a protocol to clone the right end of T2KSAGFF insertions.

1. Digest 500 ng of genomic DNA with *MboI* in 20 μL reaction solution at 37°C for 1 h, and heat at 70°C for 15 min. Add 430 μL water to the sample, incubate at 70°C for 2–3 min, and cool to 16°C.
2. Add 50 μL 10X T4 DNA ligase buffer and 2 μL T4 DNA ligase and incubate at 16°C for 3 h to overnight.
3. Add 50 μL 3 M sodium acetate and 1 mL ethanol to the sample. Chill the sample at –20°C for 30 min, centrifuge at 15,000 rpm for 20 min at 4°C, rinse once with 70% ethanol and suspend in 20 μL H_2O.
4. By using 10 μL of the ligated sample, perform the first PCR (30 cycles of 94°C for 30 s, 57°C for 30 s, 72°C for 1 min) in 50 μL 1X PCR buffer containing 1 μM SAGFF f1 and SAGFF r1 primers.
5. By using 2 μL of the first PCR product, perform the second PCR (30 cycles of 94°C for 30 s, 57°C for 30 s, 72°C for 1 min) in 50 μL 1X PCR buffer containing 1 μM SAGFF f2 and SAGFF r2 primers.
6. Analyze 10 μL of the PCR products by 1.5–2% agarose-TAE electrophoresis (**Fig. 5b**). Extract DNA bands from the gel with QIAquick gel extraction kit and sequence by using 100R-out primer.

3.5. Analysis of Trapped Genes by 5′ RACE or RT-PCR

Endogenous transcripts trapped by the gene trap construct can be identified by the 5′ RACE analysis (**Fig. 6a**). Alternatively, genomic sequence adjacent to the transposon insertions can be mapped on the genome by searching the Ensemble database, and a candidate trapped gene, either a known or a predicted gene, may be found. In such a case, the transcript can be analyzed by reverse transcriptase (RT)-PCR (**Fig. 6b**) (*see* **Note 5**).

3.5.1. Preparation of RNA from Embryos and Adult Fish

1. Transfer approximately 50 GFP-positive embryos into a teflon tissue grinder, add 1 mL Trizol reagent, homogenize well, and incubate at room temperature for 5 min. Add 0.2 mL chloroform to the sample, mix vigorously by hand for 15 s and incubate at room temperature for 2–3 min.
2. Centrifuge the sample at 12,000 rpm (13,000 × *g*) for 15 min at 4°C. Transfer the upper aqueous phase to a new tube, mix with 0.5 mL isopropyl alcohol, incubate at room temperature for 10 min and centrifuge at 10,000 rpm (7,500 × *g*) for 10 min at 4°C.

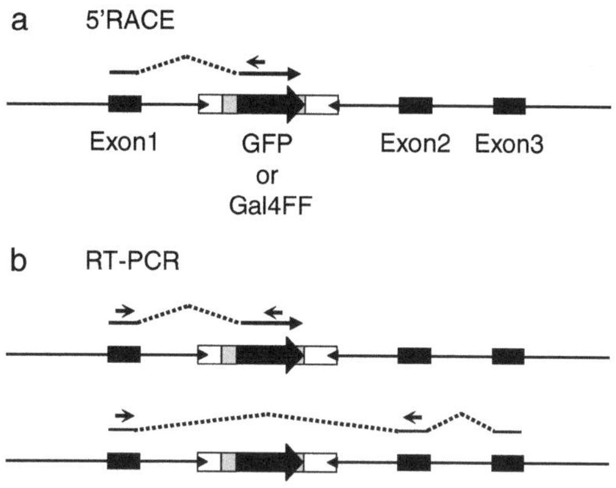

Fig. 6. The 5′ RACE and RT-PCR analyses of transcript surrounding *Tol2* insertions. (**a**) A scheme for identification of a gene trapped by an insertion of a gene trap construct by 5′ RACE. (**b**) A scheme for identification of a gene trapped by an insertion of a gene trap construct by RT-PCR. When a candidate gene is predicted from a database search, its putative transcript can be analyzed by RT-PCR using primers in the candidate gene (exon) and in the GFP or Gal4FF gene. In some cases, such a transcript may skip the insertion. In such a case, a candidate transcript can be analyzed by RT-PCR using primers in the predicted upstream and downstream exons.

3. Add 1 mL of 75% ethanol, mix by vortexing and centrifuge at 10,000 rpm (7,500 × *g*) for 5 min at 4°C.
4. Dissolve the precipitated RNA in ~50 μL RNase-free water by pipeting, and incubate at 55–60°C for 10 min.
5. Submerge an adult fish into liquid nitrogen in a grinder, and freeze completely. Grind the fish sample into powder and transfer into a 50-mL plastic tube.
6. After evaporating the liquid nitrogen, add 5 mL Trizol reagent and mix well. Incubate the sample at room temperature for 5 min, add 1 mL chloroform, mix vigorously by hand for 15 s, and incubate at room temperature for 2–3 min.
7. Centrifuge the sample at 3,500 rpm at 4°C for 15 min. Transfer the upper aqueous phase to five to six 1.5-mL microtubes. Add equal volumes of isopropyl alcohol to the samples and purify RNA as described above.

3.5.2. 5′ RACE

1. By using ~5 μg of total RNA, synthesize cDNA with 5′ RACE system for rapid amplification of cDNA ends and 1 μM of Gal4-r3 according to the manufacturer's instructions.
2. By using 5 μL of dC-tailed cDNA, perform the first PCR (30 cycles of 94°C for 30 s, 55°C for 30 s, 72°C for 2 min) in 50 μL 1X PCR buffer containing 0.4 μM Gal4-r2 and Abridged Anchored primers.

3. Analyze 5–20 μL of the 5′ RACE product by 1.5–2% agarose-TAE electrophoresis. Should a band not be detected, 5 μL of the first PCR product is diluted 1/100 by adding 495 μL of H_2O and used for the second PCR. By using 1–2 μL of the diluted first PCR sample, perform the second PCR (30 cycles of 94°C for 30 s, 55°C for 30 s, 72°C for 2 min) in 50 μL 1X PCR buffer containing 0.4 μM Gal4-r1 primer and UAP primers.
4. Analyze the 5′ RACE product by 1.5–2% agarose-TAE electrophoresis. Purify DNA bands from the gel, and clone with TA cloning kit.
5. Prepare plasmid DNA from transformed bacteria with QIAprep spin miniprep kit. Use 100–1,000 ng of the plasmid DNA for DNA sequencing.

3.5.3. RT-PCR

1. By using 5 μg of total RNA, synthesize cDNA with SuperScript first-strand synthesis system for RT-PCR and Oligo(dT)$_{12-18}$ as a primer (*see* **Note 6**).
2. By using 5 μL of the cDNA sample, perform the first PCR (30 cycles of 94°C for 30 s, 55°C for 30 s, 72°C for 2 min) in 50 μL 1X PCR buffer containing 0.4 μM Gal4-r2 and exon-specific primers.
3. Analyze 10 μL of the RT-PCR product by 1.5–2% agarose-TAE electrophoresis. Should a band not be detected, 5 μL of the first PCR product is diluted to 1/100 by adding 495 μL of H_2O and used for the second PCR. By using 1–2 μL of the diluted first PCR sample, perform the second PCR (30 cycles of 94°C for 30 s, 55°C for 30 s, 72°C for 2 min) in 50 μL 1X PCR buffer containing 0.4 μM Gal4-r1 and a nested gene-specific primers.
4. Analyze the RT-PCR products by 1.5–2% agarose-TAE electrophoresis, purify from the gel, clone with TA cloning kit and use for sequencing.

4. Notes

1. Hybond-N$^+$ is positively charged and no treatment after blotting is required. However from our experience, the result is better when the membrane was dried completely.
2. Adaptor-ligation PCR may be suitable for analyzing many samples at a time and inverse PCR may be suitable for analyzing a small number of samples carefully and cloning transposon-genomic DNA junction fragments of longer lengths. Inverse

PCR and adaptor-ligation PCR may not be easy when the transposon has integrated within repetitive sequences.

3. It is necessary to use a different adaptor when a different restriction enzyme is used. *Mbo*I, *Bgl*II and *Bam*HI, that create 5′-GATC-3′ cohesive ends, can be used with the GATC adaptor, as in the example given. *Spe*I, *Xba*I or *Nhe*I, that create 5′-CTAG-3′ cohesive ends, should be used in combination with a CTAG adapter.
4. It is necessary to design different sets of primers when insertions of different constructs are analyzed. Also, as described above, it should be useful to design primer sets that are specific to the construct when junction DNA is cloned from DNA of Gal4FF; UAS:GFP double transgenic fish as the UAS:EGFP transgene also contains *Tol2* sequence.
5. 5′ RACE does not always work especially when the amount of transcript is small. When 5′ RACE or RT-PCR do not work, purify and use polyA +RNA for cDNA synthesis instead of total RNA.
6. When the amount of a transcript is small and the Oligo$(dT)_{12-18}$ primer does not work for the RT-PCR amplication, design a reverse primer in a downstream exon and try to use it for the first strand cDNA synthesis.

Acknowledgments

We thank the National BioResource Project and grants from the Ministry of Education, Culture, Sports, Science and Technology of Japan.

References

1 Kawakami, K., Takeda, H., Kawakami, N., Kobayashi, M., Matsuda, N., and Mishina, M. (2004). A transposon-mediated gene trap approach identifies developmentally regulated genes in zebrafish. *Dev Cell* 7, 133–44.
2. Kawakami, K. (2005). Transposon tools and methods in zebrafish. *Dev Dyn* **234,** 244–54.
3. Parinov, S., Kondrichin, I., Korzh, V., and Emelyanov, A. (2004). *Tol2* transposon-mediated enhancer trap to identify developmentally regulated zebrafish genes in vivo. *Dev Dyn* **231**, 449–59.
4. Nagayoshi, S., Hayashi, E., Abe, G., Osato, N., Asakawa, K., Urasaki, A., Horikawa, K., Ikeo, K., Takeda, H., and Kawakami, K. (2008). Insertional mutagenesis by the *Tol2* transposon-mediated enhancer trap approach generated mutations in two developmental genes: *tcf7* and *synembryn-like. Development* **135**, 159–69.
5. Ellingsen, S., Laplante, M. A., Konig, M., Kikuta, H., Furmanek, T., Hoivik, E. A., and Becker, T. S. (2005). Large-scale enhancer detection in the zebrafish genome. *Development* **132**, 3799–811.
6. Davison, J. M., Akitake, C. M., Goll, M. G., Rhee, J. M., Gosse, N., Baier, H., Halpern, M. E., Leach, S. D., and Parsons, M. J. (2007). Transactivation from Gal4-VP16 transgenic

insertions for tissue-specific cell labeling and ablation in zebrafish. *Dev Biol* **304**, 811–24.

7. Scott, E. K., Mason, L., Arrenberg, A. B., Ziv, L., Gosse, N. J., Xiao, T., Chi, N. C., Asakawa, K., Kawakami, K., and Baier, H. (2007). Targeting neural circuitry in zebrafish using GAL4 enhancer trapping. *Nat Methods* **4**, 323–6.
8. Asakawa, K., Suster, M. L., Mizusawa, K., Nagayoshi, S., Kotani, T., Urasaki, A., Kishimoto, Y., Hibi, M., and Kawakami, K. (2008). Genetic dissection of neural circuits by *Tol2* transposon-mediated Gal4 gene and enhancer trapping in zebrafish. *Proc Natl Acad Sci U S A* **105**, 1255–60.

Chapter 7

Bacterial Artificial Chromosome Transgenesis for Zebrafish

Zhongan Yang, Hong Jiang, and Shuo Lin

Summary

Transgenesis using bacterial artificial chromosomes (BAC) allows greater fidelity in directing desirable expression of transgenes. Application of this technology in the optically transparent zebrafish with fluorescent protein reporters enables unparalleled visual analysis of gene expression in a living organism. We have developed a streamlined procedure of directly selecting multiple BAC clones based on public sequence databases followed by rapid modification with green fluorescent protein or red fluorescent protein for transgenic analysis in zebrafish. In this chapter, experimental procedures for BAC DNA preparation and generation of transgenic zebrafish lines by microinjection are described.

Key words: Transgenic zebrafish, Bacterial artificial chromosomes, BAC modification, Fluorescent reporter genes, GFP/RFP.

1. Introduction

Zebrafish is a superb model organism allowing a unique opportunity to integrate the tools of forward genetics, experimental embryology, functional genomics, and molecular biology toward understanding vertebrate development. In addition, the zebrafish genome-sequencing project will be completed soon. The challenge facing us is to understand the expression, regulation, and function of the more than 30,000 genes that are typically contained in a vertebrate genome. At present, methods involving whole mount RNA *in situ* hybridization, or transgenesis using plasmid-based reporter genes, and forward genetics are popular tools to address these fundamental issues in zebrafish and other organisms. Regarding expression analysis, whole mount RNA

Graham J. Lieschke et al. (eds.), *Zebrafish,* Methods in Molecular Biology, vol. 546
DOI: 10.1007/978-1-60327-977-2_7

in situ hybridization requires fixed embryos, whereas current transgenic technologies often lack fidelity in their recapitulation of endogenous gene expression patterns resulting from use of insufficient regulatory genomic sequences *(1–5)*.

Numerous examples obtained in various organisms demonstrate that bacterial artificial chromosome (BAC) transgenesis is a better approach than plasmid vector transgene expression. BACs are cloning vectors that can accommodate large genomic DNA inserts of up to 300 kilobases (kb). In most cases, BAC transgenic constructs can direct transgene expression more accurate than smaller insert vectors because they can include distal regulatory sequences that spread across tens or hundreds of kbsses *(6, 7)*. Over the past few years, several methods have been established for modifying BACs based on homologous recombination that allows insertion of transgenes into large DNA fragments at precise locations *(8–10)*. Previously an efficient method has been employed for modifying BACs to perform high throughput transgenic studies of neuronal gene expression in mouse *(11)*. In parallel, we used this approach in zebrafish and modified more than 100 BAC clones for different zebrafish genes. This chapter describes detailed procedures for BAC transgenesis, including selection of BAC clones based on genomic sequence, insertion of green fluorescent protein (GFP), red fluorescent protein (RFP), and other reporter genes into BACs by modification in *Escherichia coli*, and generation of transgenic zebrafish by microinjection of modified BAC DNA in fertilized eggs.

2. Materials

2.1. BAC Clones and Primers

1. BAC clones are obtained from zebrafish genomic BAC library CHORI-211.
2. Three primers are designed for each modification: Primer A3R, the 3′ reverse primer, is located at the 5′ end adjacent to the translational start codon (ATG) of gene of interest. Primer A5Asc, the 5′ forward primer, is located at 300–500 base pairs (bp) upstream of the reverse primer and contains digestion site of restriction enzyme *Asc*I (*see* **Note 1**). This polymerase chain reaction (PCR) fragment is named “A box” for the modification procedure whereas the corresponding genomic sequence on BAC is called “A′ box”. Another 5′ forward primer (primer A5F) located about 100 bp upstream of the A′ box is designed for detecting successfully modified BAC clones after homologous recombination.

3. Reverse primer GFPr (5′-GCTGCTTCATGTGGTCG-GGGTA-3′) is located in the GFP reporter gene and used for screening of shuttle vector carrying "A box" and successfully modified BAC clones.

2.2. Molecular Biology Kits and Reagents

1. Nucleotide triphosphates (NTPs) mix (5 mM each) for polymerase chain reaction (PCR) grade.
2. Fast Start PCR kit (Roche).
3. MiniElute PCR purification kit and QuickSpin PCR purification kit (Qiagen).
4. Restriction enzymes and their respective working buffers (New England Biolabs).
5. The Quick Ligation Kit and T4 DNA ligase and its working buffer (New England Biolabs).
6. Agarose for electrophoresis grade.
7. Competent cells: DH5α (Invitrogen).
8. Competent cells: One Shot *pir2* (Invitrogen).
9. Luria Bertani (LB) medium and agar.
10. *SOC medium*. 0.5% yeast extract, 2.0% tryptone, 10 mM NaCl, 2.5 mM KCl, 10 mM $MgCl_2$, 10 mM $MgSO_4$, 20 mM glucose.
11. QuickSpin Miniprep Kit and Hispeed Maxiprep kit (Qiagen, USA).

2.3. Solutions

1. *Tris-acetate ethylenediamine tetraacetic acid (TE)TE buffer*. 10 mM Tris–HCl (pH 8.0), 1 mM EDTA.
2. Ethidium bromide (EB): 10 mg/mL (20,000X stock).
3. 70% Ethanol.
4. 100% Ethanol.
5. 5 mg/mL tetracycline in 100% Ethanol (1000X stock).
6. 25 mg/mL chloramphenicol (1000X stock).
7. 50 mg/mL ampicillin (1000X stock).
8. 10% glycerol in sterile water.
9. 50 mM $CaCl_2$ solution in sterile water.
10. 50 mM $CaCl_2$ in 10% glycerol solution.

2.4. Devices

1. Spectrophotometer
2. PCR system (GeneAmp® PCR system 9700, Applied Biosystems)
3. Thin wall PCR tubes (Applied Biosystems)
4. 50-mL Centrifuge tube

5. Agarose gel electrophoresis apparatus
6. Pulsed Field Agarose gel electrophoresis apparatus: FIGE Mapper Electrophoresis System (Bio-Rad Inc, USA) or equivalent.
7. Electroporator: Electroporator 2510 (Brinkmann Instruments) or equivalent

3. Method

3.1. Identification of BAC Clones Containing Genes of Interest

The zebrafish genomic BAC library CHORI-211 is one of the BAC libraries used in the zebrafish genome sequencing project at the Sanger Institute (http://www.sanger.ac.uk/Projects/D_rerio/library_details.shtml). We focus on how to select desirable BAC clones from this library because many laboratories have it in house, or alternatively, individual clones are available for purchase from the BACPAC Resource Center (http://bacpac.chori.org/). Sequences from cDNAs or partial genomic fragments of genes of interest are used to "Run a BLAST search" using the most current assembly of the zebrafish genome. The current assembly is the Zv7 database for zebrafish (http://www.ensembl.org/Danio_rerio/). ContigView is first selected to show the location of the gene on a contig. Either clone names or the BAC ID number of corresponding genes are then identified by refreshing the webpage after selecting "BAC Ends" in the pull-down menu of "Decorations" and expanding the sequence region to cover the entire sequence or approximately 400 kb of a single contig at "Jump to region". If no BAC clones directly covering the gene are found, several nearby BAC clones showing ends located within 200 kb of the targeting genes can be selected for analysis by PCR. Once confirmed, BAC clones containing a gene of interest can be used for further modification.

If clone names are available, they can be used as acces numbers to search the Clone Status database (http://www.sanger.ac.uk/cgi-bin/humpub/clone_status?species = Zebrafish). This allows the user to identify BAC ID numbers from the CHORI-211 library. If BAC clones identified through this search are not from CHORI-211 library, a further online search can be performed using the zebrafish BAC fpc database (http://www.sanger.ac.uk/Projects/D_rerio/WebFPC/zebrafish/large.shtml or http://www.sanger.ac.uk/Projects/D_rerio/WebFPC/zebrafish/small.shtml) in order to identify additional CHORI-211 BAC clones corresponding to the gene of interest. If a CHORI-211 BAC clone cannot be identified, then BAC clones from CHORI-73, DKEYP or DKEY BAC libraries can also be used for direct modification.

In the BlastView window, select approximately 1 kb of genomic sequence upstream of the ATG start codon of each gene and select "export sequence" in order to design PCR primers. PCRs are performed with these gene specific primers to confirm the identification of desirable BACs. To date we have been able to obtain CHORI-211 BAC clones for approximately 80% of genes that have been modified (**Fig. 1**).

3.2. Construction of the Shuttle Vector Carrying GFP Reporter Gene

We have successfully modified BAC clones with a number of transgenes including GFP, RFP, and oncogenes. Here we use GFP as an example to describe the procedure.

In this example, the shuttle vector pLD53.SC2 is used as the GFP carrier *(10)*. The shuttle vector is amplified in One Shot *pir2* cells (Invitrogen) that is compatible with the R6Kγ origin of replication, and the DNA is then prepared using the QuickSpin Miniprep Kit (Qiagen). A PCR amplified 300- to 500-bp DNA fragment based on the BAC sequence will be inserted into the shuttle vector, this provides an homologous arm for the subsequent homologous recombination (**Fig. 2**).

1. Each corresponding BAC clone needs to be further confirmed by PCR using primers A3R and A5Asc which are designed

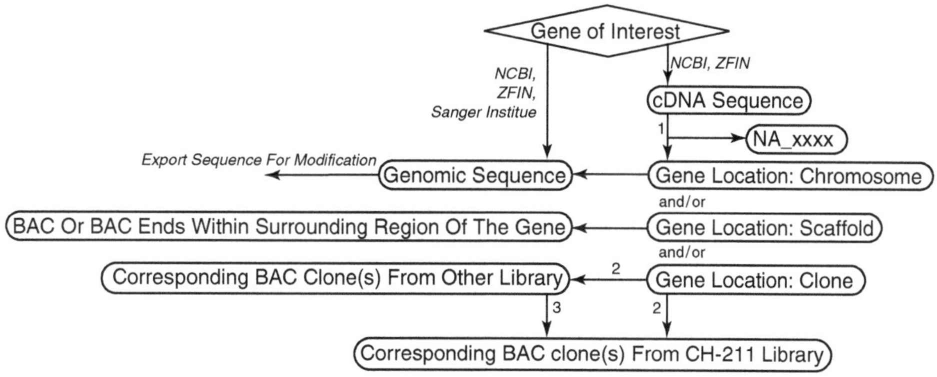

Fig. 1. Flow chart for identifying corresponding BAC clone(s) from CHORI-211 library. cDNA sequence of the gene of interest is normally available or can be obtained by searching ZFIN and NCBI databases. Genomic sequences of these genes are then obtained from the websites of Sanger Institute, UCSC, ZFIN or NCBI. In some cases, genomic sequence of the gene of interest could be obtained from the websites directly by searching with the name of the gene. The location of the gene on the appropriate chromosome, scaffold and chunk is obtained from the Sanger Institute website (1, http://www.ensembl.org/Multi/blastview?species = Danio_rerio) by performing a Blast/SSAHA search with the cDNA sequence. As described in the text, corresponding BAC clones or BAC ends located in the surrounding region of the gene of interest are obtained by selecting "ContigView", then checking "BAC Ends" in the pull-down menu of "Decorations". The corresponding BAC clones from the CHORI-211 library are then selected. Using the clone name as the Accession number, BAC clones are obtained by searching the Clone Status database (2, http://www.sanger.ac.uk/cgi-bin/humpub/clone_status?species = Zebrafish). In a case where a BAC clone from a library other than the CHORI-211 library is found, the corresponding CHORI-211 BACs can be obtained by searching against the Zebrafish fpc (Zebrafish Genome Fingerprinting project database) at http://www.sanger.ac.uk/Projects/D_rerio/WebFPC/zebrafish/small.shtml, or http://www.sanger.ac.uk/Projects/D_rerio/WebFPC/zebrafish/large.shtml).

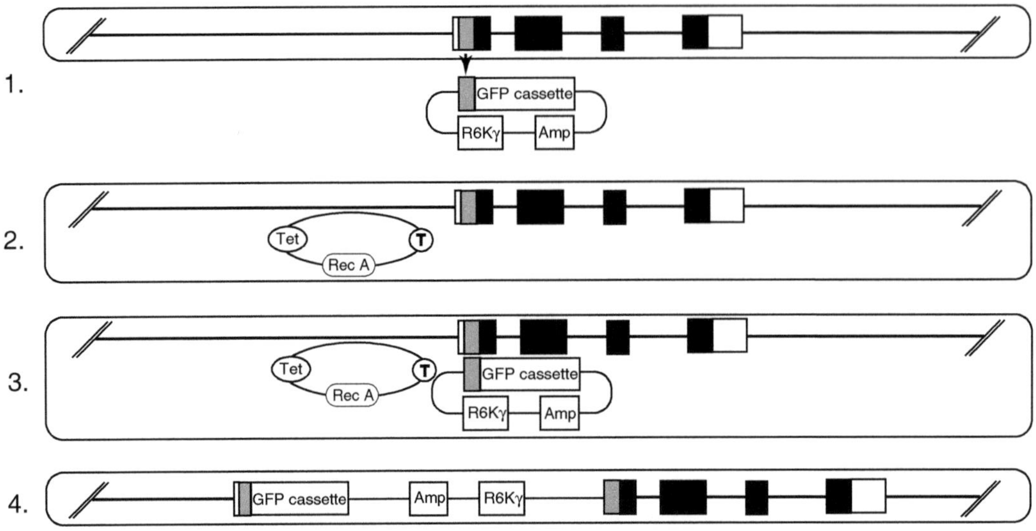

Fig. 2. Modification of zebrafish BAC clones. (1) A BAC contains a fragment of genomic sequence of interest (*Gray box*, called A′ box) adjacent to the 5′ translation start codon of the targeted gene, which is is PCR amplified and ligated into the targeting vector. In the targeting vector, this PCR fragment (*Gray box*, called the A′ box) is located upstream of the GFP reporter gene. The BAC carries a chloramphenicol resistance gene. (2) A plasmid carrying the *RecA* gene is transformed into the BAC host cells. The plasmid contains a tetracycline resistance gene (Tet) and a temperature sensitive origin of replication (*T*). (3) The targeting vector carrying GFP and A box is electroporated into the *E. coli.* cells containing both the BAC clone and *RecA* plasmid. With transient expression of Rec A, homologous recombination occurs between the A and A′ boxes. (4) By homologous recombination, the targeting vector is integrated into the BAC clones. Modified BACs are first screened by double selection with the antibiotics chloramphenicol and ampicillin. *Black boxes*: coding regions of the targeted gene; Open box: UTR; *Gray box*: A box in targeting vector or A′ box in BAC insert; (*T*): Temperature sensitive origin of replication of *RecA* plasmid.

according to the genomic sequence of each specific gene as described above. For PCR reactions, the original BAC clone should be streaked on a LB/agar plate containing 50 μg/mL chloramphenicol. Single *E. coli* colonies can be resuspended in sterile water and used as the PCR template (*see* **Note 2**). Set up a PCR mixture as follows:

(a) Templates: 5 μL *E. coli* suspension

(b) 10X Fast Start PCR buffer with $MgCl_2$: 5 μL

(c) dNTPs (5 mM each): 2 μL

(d) Fast Start Taq DNA polymerase: 0.4 μL

(e) Forward primer (A5Asc): 0.5 μM

(f) Reverse primer (A3R): 0.5 μM

(g) Add sterile water to 50 μL

2. Perform PCR thermal cycling: preheat at 94°C for 5 min; 25 cycles of 94°C for 30 s, 55°C (*see* **Note 3**) for 30 s and 72°C for 45 s; additional extension at 72°C for 7 min.

3. Analyze 5–10 µL of each PCR product by electrophoresis on 1% agarose gel containing 0.5 µg/mL ethidium bromide.
4. PCR product (A box) is purified using the QuickSpin PCR Purification Kit (Qiagen).
5. Digest the purified PCR product (A box) with restriction enzyme *Asc*I at 37°C for overnight.
6. Digest the plasmid DNA (~2 µg) of shuttle vector pLD53. SC2 with restriction enzymes *Asc*I and *Sma*I at 37°C for overnight.
7. Analyze the digestion reaction by electrophoresis on 1% agarose gel containing 0.5 µg/mL ethidium bromide (EB). Cut the gel slice containing the desired digested DNA fragment (~3.1 kb).
8. Use the MiniElute Gel Purification Kit (Qiagen) to purify the digested shuttle vector DNA.
9. Use the MiniElute PCR Purification Kit (Qiagen) to purify the digested PCR product (A box) and reduce the volume of DNA solution to the minimum.
10. Set up DNA ligation reaction with digested shuttle vector DNA (100–200 ng) and digested PCR product (A box, 50–100 ng), following the Quick Ligation Kit manufacturer's instructions at room temperature for 10 min, or using T4 ligase (NEB) at 16°C for overnight.
11. Perform a transformation:
    (a) Transfer 3 µL of the ligation reaction to a tube containing 50 µL One Shot *pir2* competent cells (Invitrogen) and incubate on ice for 30 min.
    (b) Heat shock at 42°C for 45 s and chill on ice; add 1 mL SOC medium (*see* **Note 4**) and incubate at 37°C for 45 min with 225 rpm shaking.
    (c) Collect *E. coli* cells by centrifugation with a bench top centrifuge (e.g. Eppendorf) at 5,500 rpm for 5 min; resuspend *E. coli* in 100 µL LB medium and spread on a LB/Agar plate containing 50 µg/mL ampicillin.
    (d) Incubate the LB/Agar plate at 37°C overnight.
12. Screen for desired clone by PCR using the A5Asc and GFPr primers. Set up the PCR reactions as described above (**Step 1** of this section).
13. Inoculate the correct colony into 200 mL of LB medium containing 50 µg/mL ampicillin. The shuttle vector DNA is isolated following the protocol provided in Hi-Speed Maxi Prep Kit (Qiagen) with some modifications (*see* **Note 5**).

3.3. Transformation of RecA Plasmid into E. coli Cells Containing the BAC Clone

A plasmid containing the bacterial *RecA* gene is transformed into the original BAC host cells to transiently express *RecA* enzyme to facilitate homologous recombination. The plasmid carries a tetracycline resistance gene and a temperature sensitive origin of replication (**Fig. 2**).

1. Transform approximately 20 ng *RecA* plasmid DNA into DH5α competent cells and incubate the competent cells on ice for 30 min.
2. Heat shock at 42°C for 90 s and chill on ice.
3. Add 250 μL SOC medium and culture at 37°C for 1 h with shaking of 225 rpm (*see* **Note 6**).
4. Spread the *E. coli* onto an LB/Agar plate containing tetracycline (5 μg/mL) and incubate at 30°C overnight.
5. Inoculate a single colony into LB medium containing tetracycline (5 μg/mL) and culture at 30°C with 250–300 rpm shaking.
6. Perform DNA miniprep with the QuickSpin Miniprep Kit (Qiagen).
7. Inoculate a single *E. coli* colony carrying target BAC clone into 3–4 mL LB with chloramphenicol (25 μg/mL), incubate at 37°C with 300 rpm shaking.
8. When the optical density (OD) reaches 0.3–0.5, chill the culture on ice for at least 10 min and collect the culture in two Eppendorf tubes (1.5 mL) and centrifuge at 4°C, 5,500 rpm for 10 min.
9. Resuspend the cells in each Eppendorf tube with 1.5 mL ice-cold 50 mM $CaCl_2$ solution and incubate on ice for 5 min.
10. Collect the *E. coli* cells at the same condition as above. Resuspend the *E. coli* in 150 μL ice-cold 50 mM $CaCl_2$ solution containing 10% glycerol and aliquot into 3 pre-chilled Eppendorf tubes.
11. Add 2–3 μL of the *RecA* plasmid DNA from **step 6** to one of the aliquots of the chemically competent cells containing the BAC clone from the last step, then incubate on ice for 30 min.
12. Heat shock at 42°C for 90 s and chilled on ice.
13. Add 800 μL of SOC medium to the tube and incubate at 37°C with shaking at 225 rpm for 1 h (*see* **Note 6**).
14. Collect *E. coli* cells at the same condition as above and resuspend the cells in 100 μL LB medium.
15. *E. coli* cells are spread onto LB/agar plates containing chloramphenicol (25 μg/mL) and tetracycline (5 μg/mL) and incubated at 30°C for overnight. The extra aliquots of competent cells can be stored at −80°C.

3.4. Homologous Recombination and Confirmation of the Modified BAC Clones

E. coli cells containing both the targeting BAC clone and *RecA* plasmid are electroporated with the shuttle vector carrying an A box followed by GFP reporter gene. With the transiently expressed *RecA* protein, homologous recombination may occur between the A box of the targeting vector and A′ box in the BAC clone. As a result, the targeting vector is integrated into the BAC clone at a precise location and the GFP/RFP reporter gene is under the control of regulatory elements contained in the modified BAC. The *RecA* plasmid is subsequently eliminated by incubation at high temperature (43°C) because it has a temperature-sensitive origin of replication (Ts-*ori*) (**Fig. 2**).

To screen for correctly modified BAC clones, DNA from PCR-confirmed modified BAC colonies, as well as the original unmodified BAC clone, are prepared by standard miniprep procedure and then double digested with *Not*I/*Sal*I. The digestion patterns of modified BAC and original unmodified BAC are analyzed using pulsed field electrophoresis (**Fig. 3**).

With a *Not*I/*Sal*I digest, one of the bands from the digestion of the original unmodified BAC DNA is expected to shift up about 3.5 kb compared to that of modified BAC DNA whereas all other bands should remain the same (**Fig. 3**). The precise integration can be confirmed by sequencing using the A5F primer.

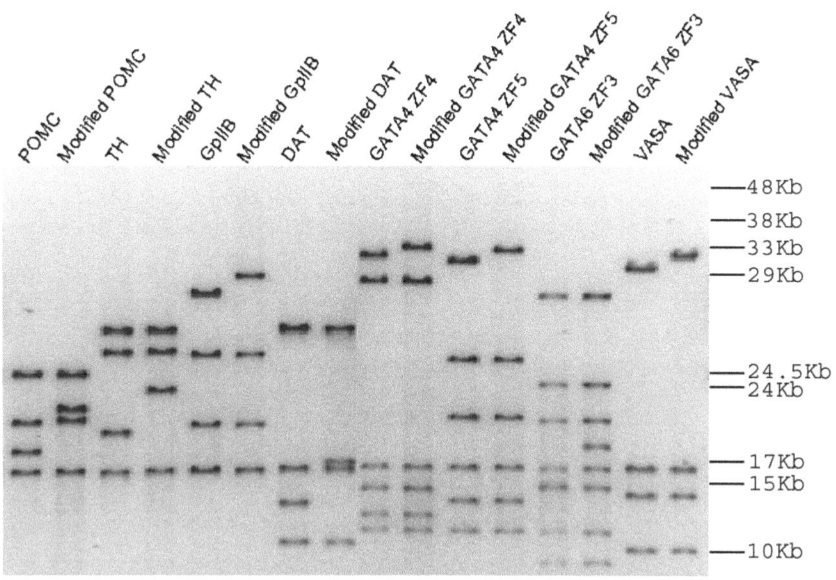

Fig. 3. Confirmation of modified BAC clones by restriction enzyme analysis and pulsed-field electroporesis. Both original unmodified BACs and modified BACs are double digested with *NotI* and *SalI* and separated by pulsed-field gel eletrophoresis. After correct modification, one of the digestion bands will shift up approximately 3.5 kb in the modified BAC DNA as compared with the original BAC DNA. This corresponds to the size of integrated targeting vector. Modification of eight different BAC clones is shown; clones are named as follows: POMC, TH, GpIIB, DAT, GATA4 ZF4, GATA4 ZF5, GATA6 ZF3, VASA.

In our studies, the efficiency of homologous recombination is between 15 and 100%.

1. Inoculate a single colony from the last step into 2 mL of LB medium with chloramphenicol (25 μg/mL) and tetracycline (5 μg/mL) and incubate at 30°C overnight with shaking.
2. Transfer 1 mL of the overnight LB culture into 100 mL LB with the same antibiotics and culture at 30°C with shaking at 300 rpm.
3. When the 100-mL culture reaches the OD of 0.5–0.7, chill it on ice for at least 15 min.
4. Collect *E. coli* cells by centrifugation at 5,500 rpm for 10 min at 4°C and resuspend in 100 mL ice-cold 10% glycerol.
5. Repeat **step 4**, then collecting the cells by centrifugation with the same conditions. The final pellet is resuspended in 200 μL 10% glycerol.
6. Transfer a 40 μL aliquot into a pre-chilled 1.5-mL Eppendorf tube and mix with 2 μL pre-made targeting vector (~1 μg of DNA).
7. Transfer the mixture to a 1-mm cuvet (Brinkmann Instruments), and then electroporate using Electroporator 2510 (Brinkmann Instruments) at 1800 kV.
8. Mix *E. coli* cells with 1 mL SOC medium and transfer into a 17 mm ´ 100 mm bacteria culture tube.
9. Incubate the *E. coli* cells at 37°C with 225 rpm shaking, then collect the cells by centrifugation at 5500 rpm for 5 min.
10. Remove the supernatant. Transfer the *E. coli* cells into 5 mL LB medium containing chloramphenicol (25 μg/mL), ampicillin (50 μg/mL), and tetracycline (5 μg/mL), followed by incubation at 30°C with 300 rpm shaking overnight.
11. Spread 10, 50 and 100 μL aliquots of the overnight culture onto LB/agar plates containing chloramphenicol (25 μg/mL) and ampicillin (50 μg/mL) and incubate at 43°C overnight.
12. Perform PCR with the A5F and GFPr primers. Set up the PCR reactions as described above using single colonies as a template (*see* **Note 2**).
13. Inoculate the PCR-positive colonies into 2 mL LB medium containing ampicillin (50 μg/mL) and chloramphenicol (25 μg/mL). An original unmodified BAC colony is also inoculated in 2 mL LB medium containing chloramphenicol (25 μg/mL) only. Incubate all *E. coli* cells at 37°C overnight with 250 rpm shaking.
14. Collect the overnight-cultured *E. coli* cells in 1.5-mL Eppendorf tubes by centrifugation at 5500 rpm for 10 min. DNA is

	Forward	Reverse
Voltage	180 V	120 V
Initial switch time	0.1 s	0.1 s
Final switch time	0.4 s	0.4 s
Ramp	50% linear	50% linear

isolated using buffers P1, P2, and P3 from QuickSpin Miniprep Kit (Qiagen), without using the column (*see* **Note 7**).

15. Double digest ~300 ng of each DNA sample with restriction enzymes *Not*I and *Sal*I. Set up the digestion reaction as follows:
    (a) 300 ng DNA sample
    (b) 2 μL Restriction enzyme buffer 3 (NEB)
    (c) 0.2 μL Restriction enzyme *Not*I
    (d) 0.2 μL Restriction enzyme *Sal*I
    (e) Add sterile water to 20 μL and incubate at 37°C for 4 h.
16. 10 μL of each digestion reactions is analyzed by pulsed field electrophoresis. In our laboratory, we use FIGE Mapper Electrophoresis System (Bio-Rad) and run 1% Agarose Gel in 0.25X TAE for approximately 16 h with the following program.

4. Generation of Transgenic Zebrafish with Modified BACs

1. Inoculate a single *E. coli* colony containing the modified BAC into 5 mL of LB medium with ampicillin (50 μg/mL) and chloramphenicol (25 μg/mL) and culture at 37°C with shaking overnight.
2. Transfer the 5 mL of overnight LB culture into 250 mL of LB medium with the same antibiotics and incubate at 37°C with shaking at 300 rpm for another 6 h.
3. BAC DNA is prepared essentially the same way as the protocol of the Hi-Speed Maxi Prep Kit from Qiagen with some modifications (*see* **Note 5**).
4. Digest the modified BAC DNA with restriction enzyme *Not*I at 37°C overnight. Set up the digestion reaction as follows:

(a) 5 μg DNA

(b) 5 μL Buffer 3 (NEB)

(c) 1 μL Restriction enzyme *Not*I

(d) Add sterile water 50 μL

5. Float a piece of 0.025 μm membrane filter (Millipore Mixed Cellulose Ester Membrane Filter) on the surface of the 2 L 0.5X TE and keep the glassy side up.
6. Load the digestion reaction on the floating membrane filter and dialyze at 4°C overnight.
7. Recover the dialyzed DNA with a wide-opening pipet tip and adjust the DNA concentration to 100–150 ng/μL in 0.1 M KCl solution.
8. Inject the linearized BAC DNA into fertilized wild-type zebrafish embryos at the one cell stage based on the laboratory setting *(6, 12)*. The dose delivered to each embryo should be 0.1–0.2 ng.
9. Grow the injected eggs to adult fish and out cross with wild-type fish.
10. Observe the offspring embryos at the stages when the target gene is expressed under an ultraviolet (UV) microscope (*see* **Note 8**).

5. Notes

1. To design the A5Asc primer, add a digestion site for restriction enzyme *Asc*I. Usually, we add 2 additional nucleotides at the 5′ of the *Asc*I site and resulting in: 5′-TTGGCGCGCCN$_{18}$–3′. The *Asc*I site is underlined and N$_{18}$ stands for at least 18 bp primer sequence based on the genomic sequence.
2. To make *E. coli* cell suspension for PCR test, pick 5–10 medium-sized single colonies and resuspend in 20 μL sterile water separately, pipet up and down to ensure complete suspension. 3–5 μL of this suspension will be used as template for PCR depending on the concentration.
3. The annealing temperature has to be adjusted according to the T$_m$ of each primer.
4. SOC medium is provided by Invitrogen in the *pir2* competent cell package. Lab-made SOC medium can also be used.

5. For preparation of shuttle vector and modified BAC DNA samples with Hi-Speed Maxiprep Kit (Qiagen), we modified the protocol provided by the company as follows:
   (a) Double the volumes of resuspending (P1), lysis (P2) and neutralization (P3) buffers
   (b) Elute DNA from the cartridge with 65°C pre-heated elution buffer
   (c) After eluting from the Qiaprecipitator with TE, BAC DNA is re-precipitated with 0.1X volume of 3 M sodium acetate (pH 5.2) and 0.7X volume of isopropanol.
6. Alternatively, culture of the transformed and/or electroporated competent cells carrying *RecA* plasmid can be performed at 30°C for 1.5 h.
7. To prepare BAC DNA samples for small scale digestion, the overnight cultured *E. coli* cells are collected in 1.5-mL Eppendorf tubes. *E. coli* cells are then resuspended, lysed, and neutralized with 200 μL of buffer P1, P2, and P3. The DNA is precipitated by adding 0.7X volume of isopropanol and washed with 70% ethanol. The DNA is air dried and resuspended in 10–20 μL sterile water.
8. In a transient transgenic assay, mosaic and non-specific GFP expression are often present. The non-specific GFP expression is significantly reduced by 48 hpf.

References

1. Bradshaw, M. S., Shashikant, C. S., Belting, H. G., Bollekens, J. A., and Ruddle, F. H. (1996). A long-range regulatory element of Hoxc8 identified by using the pClasper vector. *Proc Natl Acad Sci U S A* **93**, 2426–30.
2. Amacher, S. L. (1999). Transcriptional regulation during zebrafish embryogenesis. *Curr Opin Genet Dev* **9**, 548–52.
3. Heintz, N. (2001). BAC to the future: the use of bac transgenic mice for neuroscience research. *Nat Rev Neurosci* **2**, 861–70.
4. Gong, S., Yang, X. W., Li, C., and Heintz, N. (2002). Highly efficient modification of bacterial artificial chromosomes (BACs) using novel shuttle vectors containing the R6Kgamma origin of replication. *Genome Res* **12**, 1992–8.
5. Jamsai, D., Orford, M., Nefedov, M., Fucharoen, S., Williamson, R., and Ioannou, P. A. (2003). Targeted modification of a human beta-globin locus BAC clone using GET Recombination and an I-Scei counterselection cassette. *Genomics* **82**, 68–77.
6. Jessen, J. R., Meng, A., McFarlane, R. J., Paw, B. H., Zon, L. I., Smith, G. R., and Lin, S. (1998). Modification of bacterial artificial chromosomes through chi-stimulated homologous recombination and its application in zebrafish transgenesis. *Proc Natl Acad Sci U S A* **95**, 5121–6.
7. Jessen, J. R., Willett, C. E., and Lin, S. (1999). Artificial chromosome transgenesis reveals long-distance negative regulation of rag1 in zebrafish. *Nat Genet* **23**, 15–6.
8. Pfeffer, P. L., Payer, B., Reim, G., di Magliano, M. P., and Busslinger, M. (2002). The activation and maintenance of *Pax2* expression at the mid-hindbrain boundary is controlled by separate enhancers. *Development* **129**, 307–18.
9. Warming, S., Costantino, N., Court, D. L., Jenkins, N. A., and Copeland, N. G. (2005).

Simple and highly efficient BAC recombineering using galK selection. *Nucleic Acids Res* **33**, e36.

10. Yang, X. W., Model, P., and Heintz, N. (1997). Homologous recombination based modification in Escherichia coli and germline transmission in transgenic mice of a bacterial artificial chromosome. *Nat Biotechnol* **15**, 859–65.
11. Gong, S., Zheng, C., Doughty, M. L., Losos, K., Didkovsky, N., Schambra, U. B., Nowak, N. J., Joyner, A., Leblanc, G., Hatten, M. E., and Heintz, N. (2003). A gene expression atlas of the central nervous system based on bacterial artificial chromosomes. *Nature* **425**, 917–25.
12. Meng, A., Tang, H., Ong, B. A., Farrell, M. J., and Lin, S. (1997). Promoter analysis in living zebrafish embryos identifies a cis-acting motif required for neuronal expression of GATA-2. *Proc Natl Acad Sci U S A* **94**, 6267–72.

Chapter 8

Simple and Efficient Transgenesis with Meganuclease Constructs in Zebrafish

Daniele Soroldoni, Benjamin M. Hogan, and Andrew C. Oates

Summary

In the past, microinjection of plasmid DNA into early embryos represented the state of the art to generate transgenic zebrafish. However, this approach suffers significant drawbacks (mosaic distribution of the injected transgene, late transgene integration at high copy numbers, low transgenesis frequency), making the generation of transgenic lines a laborious task. Coinjection of *I-SceI* meganuclease with a reporter construct flanked by *I-SceI* sites overcomes these problems by earlier transgene integration into the host genome. Here, we provide an optimized protocol for *I-SceI* meganuclease-mediated transgenesis in zebrafish. This simple protocol provides a reliable method to transiently test tissue-specific reporter expression of meganuclease constructs in injected embryos (F0). Furthermore, it substantially facilitates the generation of multiple stable transgenic lines increasing transgenesis frequencies up to 45%, compared with 5% without *I-SceI*. The reliable reporter activity in F0 and the improved transgenesis frequency make this protocol a powerful tool for use in gain- and loss-of-function, cell tracing, and cell labeling experiments.

Key words: *I-SceI*, Meganuclease, Transgenesis, Transient transgenesis, Zebrafish, Transgenesis frequency, Germline transmission.

1. Introduction

Historically, transgenic zebrafish strains have been produced by the direct microinjection of circular or linearized plasmid DNA into early embryos *(1–3)*. Commonly, this traditional approach results in the formation of long DNA concatamers, which are replicated during initial cleavages, maintained in an episomal state, and eventually degraded during gastrulation *(4, 5)*. Because of its extrachromosomal location, the injected DNA is distributed

Graham J. Lieschke et al. (eds.), *Zebrafish,* Methods in Molecular Biology, vol. 546
DOI: 10.1007/978-1-60327-977-2_8

unevenly, leading to highly mosaic and transient reporter expression in injected embryos (F0) *(4–7)*. Stable transgene integration into the host genome occurs during late cleavages. Therefore, injected embryos are highly mosaic and the chance of germline integration (transgenesis frequency) is relatively low (5% or less) *(4)*. Nevertheless, due to its ease, DNA microinjection has been used extensively to assess reporter activity in F0 animals (transient transgenesis) and to establish transgenic lines (stable transgenesis). Both the mosaic reporter expression and the low germline integration in F0 made transgenesis a laborious task in the past. Hundreds of embryos had to be injected and screened in order to transiently test reporter constructs or to identify multiple independent founder fish, which is desirable during stable transgenesis experiments in order to exclude positional effects *(4, 8, 9)*.

The recent development of transgenesis approaches mediated by coinjection of transposase or *I-SceI* meganuclease substantially improves transient and stable transgenesis in zebrafish *(10–13)*. These approaches promote earlier integration of transgenes into the host genome improving both the extent and specificity of transient expression in F0 and increasing the transgenesis frequency *(14, 15)*. Transposase allows a large number of single-copy integrations in host genomes, while *I-SceI* meganuclease mediates a single integration event at low copy number *(12, 15–18)*. Single insertions of transgenes have advantages such as better long-term maintenance (no loss of insertions by outcrosses), easy determination of the insertion site, and selection of specific lines with no positional effects.

The mechanism by which *I-SceI* meganuclease mediates enhanced transgenesis is not yet fully understood. *I-SceI* is a homing endonuclease isolated from yeast *S. cerevisiae* *(19)* with no known restriction sites (18 bp in length) in any sequenced vertebrate genome. When coinjected with a plasmid containing the transgene cassette flanked by *I-SceI* sites, the meganuclease counteracts concatamerization, probably by remaining attached to its recognition sites and inhibiting endogenous ligases present in the early embryo *(2, 10, 15, 20)*. This activity of *I-SceI* might be responsible for decreasing mosaicism of transient expression and enhancing transgenesis frequency *(15)*.

Here, we provide an optimized protocol for *I-SceI* meganuclease transgenesis in zebrafish. We introduce a reliable method to determine bolus size in order to calibrate the volume injected. We used this approach to compare different parameters such as DNA concentration, composition of the injection mix, and time point of injection side by side. We suggest keeping the bolus size (100 μm or less) and concentration of the buffer (1×), $MgCl_2$ (5 mM), and *I-SceI* (20 U) constant and optimizing the DNA concentration for individual constructs. This simple protocol greatly reduces mosaicism and enhances tissue-specificity of reporter expression in injected embryos. These features enable researchers to quickly and reliably test various meganuclease constructs for reporter activity in the F0 generation. The improved transgenesis frequency (up to 45%) for our *I-SceI* meganuclease protocol means that this protocol reliably delivers multiple transgenic insertions from one injection session for any given construct, reducing time- and space-consuming screening for founder fish. Additionally, we report a quick in vitro assay that allows the testing of *I-SceI* activity in individual experimental injection mixes. Neither the *in vitro* results nor the reporter activity *in vivo* support earlier observations that preincubation of the injection mix enhances transgenesis mediated by *I-SceI* meganuclease *(10, 12)*.

2. Materials

2.1. Zebrafish

1. Zebrafish AB strain (*see* **Note 1**).

2.2. Instrumentation

1. Needle puller (e.g., P-97 Flaming/Brown Micropipette Puller, Sutter Instruments).
2. Microinjection apparatus (e.g., Pneumatic Picopump PV820, World Precision Instruments).
3. Micromanipulator (e.g., MN153, Narishige).
4. Stereomicroscope (e.g., Olympus SZ40).
5. Fluorescence stereomicroscope (e.g., Olympus SZX12) with suitable filter sets (e.g., GFP, RFP2).
6. Agarose gel electrophoresis equipment.

2.3. Small Equipment

1. Injection tray mold (*see* **Note 2**).
2. Borosilicate filamented injection needles: 1.0-mm OD, 0.78-mm ID, 75-mm length (Harvard Apparatus).
3. Standard metric hemocytometer or scaled microscope slide: Hemocytometer; depth 0.1 mm, 1/400 mm^2 (Hawksley, Lancing, UK). Scaled microscope slide; 0.01-mm scaled microscope slide with reservoir **(Fig. 1)**.
4. Fine dissecting forceps: NeoLab-Dumont (Number 5).
5. Sequencing gel pipette tips (Eppendorf microloader tips; 20 μL).
6. Petri dish or plastic box with Plasticine/modeling clay.
7. Micropipette (P20, Gilson).
8. Plastic Pasteur pipettes.

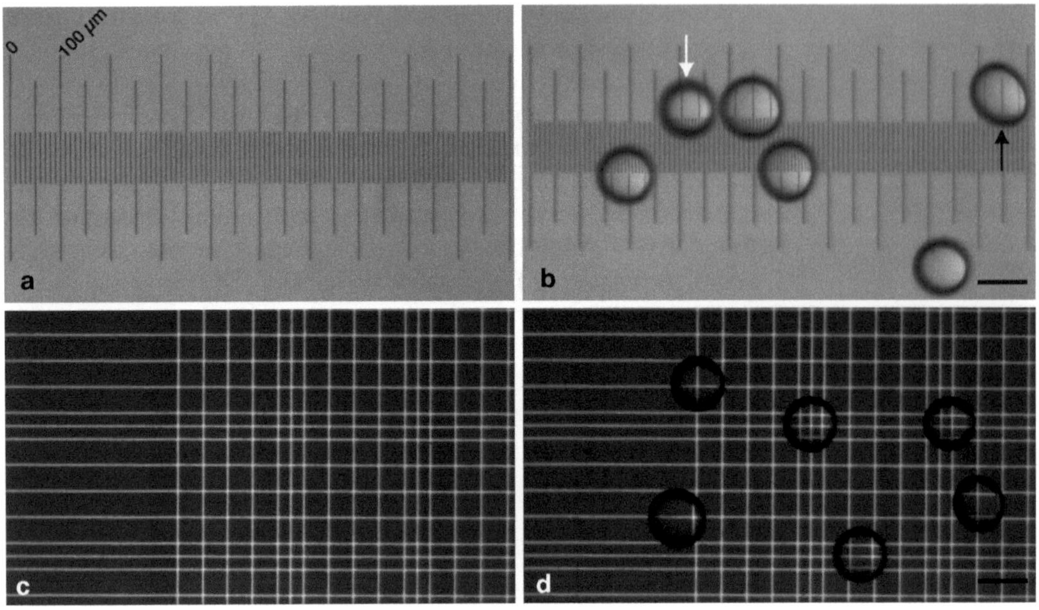

Fig. 1. Bolus size calibration: In order to guarantee reproducibility and minimize toxicity the bolus size should be calibrated prior to injection. In our hands, a bolus of approximately 100 μm diameter gives best results. The optimal DNA concentration for each construct is determined by injection of various dilutions at the given bolus size. (**a**) We use a scaled microscope slide for accurate calibration. (**b**) The scale bar is covered with a droplet of oil (*see* **Subheading 2.4**). Once the needle is broken and the back pressure is adjusted (*see* **Note 11**) several boli of constant size are injected into the oil. Calibration of floating (*white arrow*) or flattened (*black arrow*) boluses can influence the calibration procedure and enhance toxicity. Therefore, these should not be considered during the calibration of the needle. (**c, d**) Alternatively, a standard metric hemocytometer can be used for calibration. Scale bar = 100 μm.

2.4. Reagents, Media, and Solutions

1. Injection mix:
   (a) Reporter plasmid: Transgene cassette in *I-SceI*-pBSII-SK + vector backbone (Genbank accession number DQ836146) (15), e.g., pαact-GFPI2 (7.9 kb): αact promoter driving muscle-specific GFP expression; αact-GFPI2 cassette is flanked by two *I-SceI* sites.
   (b) Plasmid DNA stock solution: DNA of high quality (midiprep or maxiprep) stored at a concentration of 100 ng/μL (*see* **Note 3**).
   (c) Injection solution premix on ice (without *I-SceI*):

DNA	10–50 ng/μL
Buffer without $MgCl_2$ (10× Roche)	2 μL (*see* **Note 4**)
$MgCl_2$ (50 mM)	2 μL
Phenol red 0.5% (Sigma)	1 μL
H_2O	Up to 18 μL

(d) *I-SceI* meganuclease (Roche): *I-SceI* enzyme should be stored at −80°C in individual aliquots and added to the premix (2 μL to earlier mix) immediately prior to injection (*see* **Note 5**).

2. Paraffin oil (relative density 0.865, viscosity 171 mPa s).
3. Egg water: Stock Sea Salts (Aquasonic), final concentration 60 μg/mL.
4. E3 embryo medium (without methylene blue): 5 mM NaCl; 0.17 mM KCl; 0.33 mM $MgSO_4$; 0.33 mM $CaCl_2$; pH 7.5.
5. Proteinase K 10 mg/mL.

3. Methods

3.1. Preparation of Needles for Injection

1. Load filamented capillaries into a commercially supplied micropipette puller.
2. Pull capillaries using settings that result in needles that are long and thin with a tapered tip (*see* **Note 6**).
3. Multiple needles can be produced using the desired settings and stored in rows on Plasticine or modeling clay in a Petri dish.

3.2. Preparation of Injection Tray

1. Prepare a 1–2% agarose solution with embryo medium; dissolve it by heating and pour into a Petri dish.
2. Carefully place a prewet standard plastic injection mold (*see* **Note 7**) face down onto the cooling agarose so that the mold floats.
3. Allow the agarose to set and then remove the plastic injection mold.
4. Store injection tray at 4°C submersed in E3 medium or water.
5. Prewarm the injection tray to 28.5°C prior to use.

3.3. Preparation of Injection Mix and Mounting of Embryos for Injection

1. Prepare injection premix containing no *I-SceI* meganuclease and store it on ice.
2. Collect embryos from pairwise matings (*see* **Note 8**) in E3 medium immediately upon spawning (*see* **Note 9**). Inspect clutch quality and ensure that the embryos are of the desired stage (**Fig. 2**).
3. Remove *I-SceI* meganuclease aliquot from storage and add directly to the premix on ice immediately prior to injection (*see* **Note 5**). Mix without vortexing.

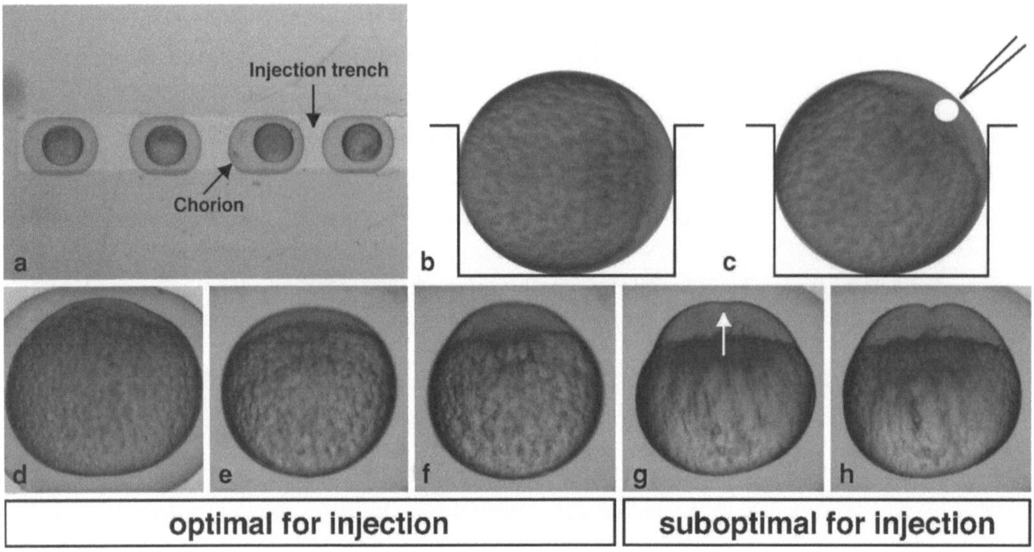

Fig. 2. Staging of one-cell embryos. The stage of injected embryos has a strong influence on the mosaicism of the germline. For the generation of transgenic lines it is optimal to harvest and inject embryos as early as possible to ensure early transgene integration at the one- or two-cell stage. (**a**) Immediately after spawning, clutches are harvested and embryos are aligned individually in the injection tray. The depicted injection trench is 0.95-mm wide and 0.8-mm high. (**b**) Prior to injection single embryos are oriented so that the animal pole faces the injection needle. (**c**) Subsequently embryos are tilted approximately 45° upward and the DNA is delivered at the indicated position (white point). Never inject more than 50–80 embryos or for a period of more than 15 min using the same needle (*see* **Note 15**). (**d–f**) For the generation of transgenic lines only early one-cell embryos should be considered. (**g**) As soon as a cleavage furrow is visible (*white arrow*) the embryos should be discarded. Injection at late one-cell stage (**g**) or early two-cell (**h**) stages results in late transgene integration and in a more mosaic animal and germline.

4. Array selected early one-cell embryos side by side in each trench in an injection tray with the animal pole (single cell) facing the needle for injection (*see* **Fig. 2** and **Note 10**).
5. Take a few microliters of injection mix and fill premade injection needles with microloaders.
6. Proceed immediately to calibration of bolus size.

3.4. Calibration of Injection Bolus and Volume

1. Attach loaded injection needle to the injection arm of the microinjection apparatus.
2. Break the tip of the needle using forceps so that the needle is sharp but not overly flexible (*see* **Note 11**).
3. Apply paraffin oil to the surface of a scaled slide or hemocytometer and inject several boli (3–5 is sufficient) immediately above the gridded surface (**Fig. 1**).
4. Adjust bolus size to deliver the desired volume (*see* **Note 12**) based on the following calculations used to estimate amount of DNA and injection mix delivered:

- Estimate bolus diameter based on the number of hemocytometers or scaled slide grids spanned (*see* **Fig. 1** and **Note 13**).
- Calculate the volume of the injection bolus, assuming that the bolus is a sphere, using the equation for the volume of a sphere (volume in $\mu m^3 = 1/6\pi d^3$, where d = bolus diameter).
- Calculate the amount in nanograms to be delivered based on the volume of the bolus and the concentration of the DNA injection mix.

5. For rapid use, the bolus size can be precalibrated on a hemocytometer/gridded slide and then referenced against an eyepiece with reticule. The required bolus can thus be rapidly set at a given volume (e.g., 0.5–1 nL) based on the number of scaled eyepiece bars spanned at a predetermined magnification.

3.5. Injection of Embryos

1. Inject DNA directly into the cytoplasm of carefully staged embryos (*see* **Fig. 2** and **Note 14**).
2. Inject approximately 50–80 embryos over a period of 15 min using the first injection needle.
3. Change needles, refreshing with injection mix that has been stored on ice for every consecutive batch of embryos to be injected (*see* **Note 15**).
4. Discard and do not inject any embryos that are at late one-cell or early two-cell stages throughout the injection process (**Fig. 2**).
5. Incubate embryos in an embryo medium at 28.5°C until the desired stage.
6. Retain uninjected embryos from every individual clutch as an injection control.

3.6. Assay Reporter Activity in Injected Embryos (F0)

1. Screen injected embryos for reporter activity at the desired stage using a fluorescence stereomicroscope.
2. Record strength, extent, and tissue specificity of obtained signals (*see* **Note 16**).
3. Record the number of phenotypes and dead embryos and compare directly to uninjected control embryos (*see* **Note 12**).

3.7. Assays to Control for Meganuclease Efficiency

In order to control for the activity of the *I-SceI* enzyme we have provided the following methods. The first is a simple gel electrophoresis approach to assay *I-SceI* activity in vitro, which can be performed at the bench rapidly after injection. The second is a longer term but simple genetic approach, which can be used to determine from the transmission rate both the integration efficiency and the insertion frequency achieved in injection experiments for scenarios where such readouts are desirable.

3.7.1. A Simple Assay for Meganuclease Activity

1. After injecting, aliquot the remaining injection solution into three prechilled and labeled Eppendorf tubes.
2. Incubate one aliquot at room temperature for 30 min.
3. Add 1 μL of Proteinase K (10 mg/mL) to a second aliquot and leave it together with the third and final untreated aliquot on ice.
4. Prepare an agarose gel suitable for separation of the expected fragments.

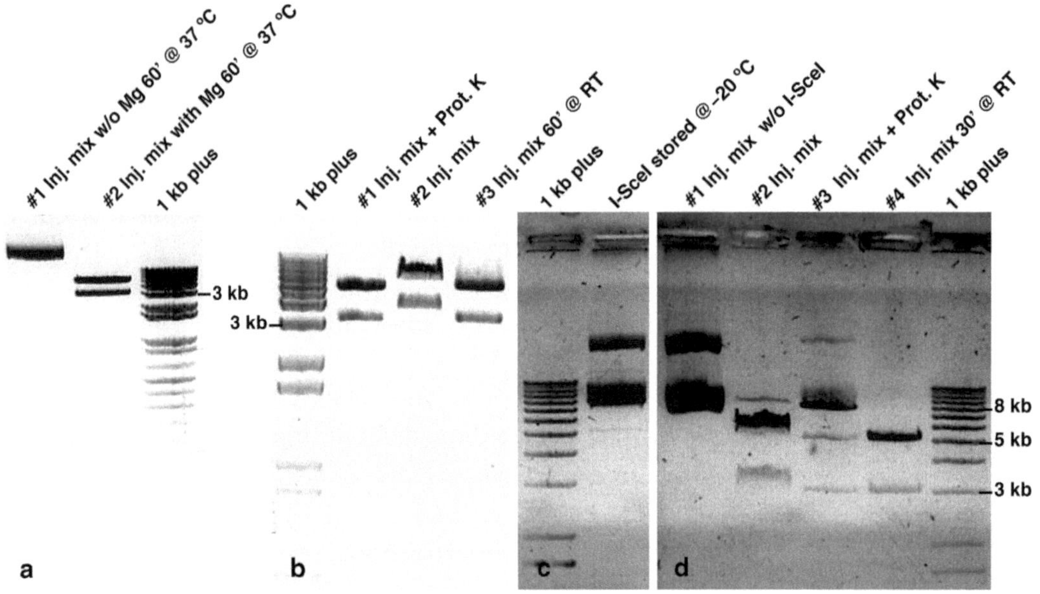

Fig. 3. In vitro assay for *I-SceI* activity. (**a**) Side-by-side comparison of injection mixes containing pαact-GFPI2 (7.9 kb; *see* **Subheading 2**) with and without (w/o) $MgCl_2$ shows that *I-SceI* meganuclease activity is Mg^{2+}-dependent. Therefore, rather than rely on endogenous Mg^{2+} levels, we recommend adding $MgCl_2$ to the injection mix. (**b**) In our hands, preincubation of the injection mix at room temperature (RT, 23°C) does not produce optimal reporter expression in F0 embryos. In order to compare *I-SceI* meganuclease activity under incubated and unincubated conditions, we loaded different injection mixes on an agarose gel. Plasmid DNA was digested in all samples. However, fragments from the mix incubated on ice (lane #2), rather than at RT (lane #3) showed a clear band shift, running at a higher molecular weight. Importantly, the addition of Proteinase K to the injection mix inhibited this band shift, demonstrating that this shift is *I-SceI*-dependent (lane #1). Therefore, *I-SceI* meganuclease probably retains a higher activity when incubated on ice because it stays attached to its recognition sites (leading to DNA running at a higher molecular weight), which is believed to enhance transgene integration. (**c**) *I-SceI* meganuclease stored incorrectly dramatically loses activity and does not digest plasmid DNA when incubated on ice, while enzyme stored at −80°C consistently digests the plasmid DNA even when stored on ice (lane #2 in **b**, **d**). (**d**) *I-SceI* meganuclease activity can be tested in vitro. Side-by-side comparison of identical injection mixes incubated under different conditions confirms that the *I-SceI*-dependent band shift is temperature-dependent. On ice, *I-SceI* retains its ability to both digest and attach to the DNA (lane #2). If Proteinase K is added both digestion and band shifting are inhibited (lane #3). Preincubation at RT results in complete digestion of the plasmid DNA; however, the *I-SceI* enzyme loses its ability to bind to the DNA and no longer causes band shift (lane #4). Considering these data, we suggest that cutting of the plasmid DNA alone is not a suitable readout for *I-SceI* meganuclease activity, but that cutting and band shifting is. All samples in (**c**) and (**d**) were incubated for 30 min. Samples in (**d**) originated from the same master mix and were incubated on ice (lane #2, #3) or RT (lane #4).

5. Load sample after incubation and run high-resolution gel (*see* **Note 17**).
6. Document and interpret fragment pattern (**Fig.** 3).

3.7.2. Genetic Determination of Germline Mosaicism and Integration Number

1. After the analysis of transient expression in injected embryos (F0), maintain injected embryos and grow to adulthood (approximately 2–3 months).
2. Outcross F0 animals to nontransgenic wild-type fish in pairs.
3. Collect clutches of embryos (F1) from outcrosses and score the transgenesis frequency from the number of positive clutches. Within these clutches, score the percentage that carries the transgene (e.g., fluorescent embryos) to estimate the mosaicism of the F0 germline (*see* **Table 1** and **Note 18**).
4. Retain fluorescent outcrossed F1 embryos and grow to adulthood (approximately 2–3 months).
5. Outcross F1 animals to nontransgenic wild-type fish in pairs.
6. Collect clutches of embryos (F2) from outcrosses and score the percentage that carries the transgene (e.g., fluorescent embryos) to determine the number of genomic insertions achieved per positive founder (*see* **Table 1** and **Note 19**).

Table 1
Transmission rates determined from F1 and F2 offspring (*see* Notes 18 and 19)

Transmission rates determined from F1 offspring			
Founder fish (F0)	*n*	T rates (%)	Integration at
#10	89	20	2-cell stage
#11	130	10	4-cell stage
#15	55	45	1-cell stage
Transmission rates determined from F2 offspring			
Clutches F1 #10	*n*	T rates (%)	
1	84	33	
2	171	46	
3	372	51	
4	248	44	
5	125	47	
Clutches F1 #11	*n*	T rates (%)	
1	114	57	
2	228	48	
3	75	47	
4	310	49	
5	142	48	

(continued)

Table 1 (continued)

Clutches F1 #15	*n*	T rates (%)
15	157	55
16	154	56
17	225	50
18	271	43
19	171	50

Injected construct: pherl-UbCherry-3′herl, *n* Total number, *T rates (%)* transmission rates in percent

4. Notes

1. Although the specific strain used appears not to be critical and a number of strains are available (www.zfin.org), the assays used to optimize methods described here were performed primarily using the AB strain.
2. The use of standard plastic injection tray molds is a routine approach in zebrafish research. One common example of these molds with full specifications can be found in *The Zebrafish Book* (http://www.zfin.org/zf_info/zfbook/zfbk.html). Recently, Rembold et al. provided another useful mold *(12)*.
3. Dilution of high-concentration midiprep DNA directly into the final injection mix will increase concentration variation that can occur due to small differences when pipetting. The storage and reuse in multiple experiments of a more diluted DNA stock solution is suggested here to reduce variations in DNA concentration introduced when pipetting and to improve the reproducibility observed between different experiments.
4. It is important to be aware of whether or not the buffer used contains $MgCl_2$. Injections performed in the absence of $MgCl_2$ rely on the presence of embryonic Mg^{2+} ions for the enzyme to function in vivo. This is evidenced by the inability of *I-SceI* (Roche) to digest DNA upon in vitro incubation in the absence of $MgCl_2$ (*see* **Fig. 3**).
5. *I-SceI* is commercially available from a number of companies. We have experienced surprising variation in the activity of enzymes supplied from different sources. This variation appears to be, at least in part, due to the method of shipment used. *I-SceI* enzymes should be shipped on dry ice. An observable reduction in the activity of *I-SceI* meganuclease can be seen when the enzyme is stored at –20°C when compared with the enzyme stored at –80°C (**Fig. 3**).

Hence, we suggest the storage of small individual aliquots (2 μL) of enzyme for single use to reduce freeze-thawing damage. The methodological optimizations described here are based on the use of *I-SceI* enzyme sourced from Roche.

6. Needles that are long and thin are ideal for injection into early one-cell stage embryos. Settings on a needle puller will vary dependent on the age of the filament. As a guide, we commonly use the following or similar settings: $P = 150$; heat = 635– (check by ramp test); pull = 100; vel = 170; time = 120. The heat can be determined first by a ramp test (see manufacturer's instructions for the needle puller). As a rule of thumb the indicated value from the ramp test should not be exceeded by more than 20. Subsequently, the other parameters can be adapted.
7. In order to avoid bubbles forming between the injection mold and agarose as the agarose is setting, we suggest wetting the injection tray mold with water before lowering it slowly one side first onto the hot agarose.
8. Although mass matings can be (and are commonly) used for microinjection, we recommend pairwise mating. This approach has an advantage when it comes to timing and staging early one-cell embryos. Once an injection premix has been prepared on ice, mating can be induced by combining male and female fish. *I-SceI* can then be added, allowing for bolus size calibration immediately upon the induction of spawning. Using this approach, embryo injections can be performed with fresh *I-SceI* meganuclease into the earliest possible one-cell stage embryos.
9. Although egg water and E3 medium are both used interchangeably by many laboratories for incubation of developing embryos, E3 medium is a more carefully buffered medium and provides more controlled laboratory conditions. Methylene blue is best omitted as it can complicate the detection of weak fluorescence.
10. The staging of early versus late one-cell stage embryos for injection (*see* **Fig. 2**) influences the distribution of transgene expression in injected embryos. It is intuitive that the earlier the genomic integration, the more ubiquitous the transgene expression will be in the target tissue in transiently expressing F0 embryos. Injection into early one-cell stage embryos gives less mosaic expression and the injection of late one-cell stage embryos gives a more mosaic pattern of expression in injected embryos. It should be noted that when a more mosaic expression pattern is desired (e.g., for cell tracing or fate mapping experiments), injection into single cells in late 1-, 2-, or 4-cell stage embryos can also be performed.
11. Needles with a very small opening at the tip will not cause damage to the injected embryo but will bend easily when

attempting to pierce the chorion and will become blocked regularly. Needles with a large opening will cause more damage but will bend less and pierce the chorion more readily. A balance can be found when breaking the tip of the needle. When using needles produced as described earlier (*see* **Note 6**) a correctly broken needle should produce boli of approximately 100 μm diameter at the following microinjector settings: 40 psi, 200–500 ms, with a back (hold) pressure that avoids backflow and free flow.

12. Toxicity and mortality of injected embryos depends on bolus size and DNA concentration. If either significantly exceeds more than 30% in an injection clutch, confirm that the bolus size is 100 μm or less and reduce DNA concentration in the injection mix. As a guide, we recommend the use of a standard bolus size of approximately 0.5–1 nL (1nL = 120 μm) to minimize developmental abnormalities.

13. In order to guarantee accurate estimations of bolus size between injections one should estimate bolus diameter at a consistent depth in the paraffin oil. We suggest focusing the microscope on the scale bar and making sure that the bolus is in the same focal plane when estimating diameter (**Fig. 1**).

14. We use a very steep angle (~45%) for injection into the cell. This helps to avoid injection into the yolk or the injection tray, which represent the most common errors for beginners. Furthermore, the angle ensures that the embryo does not move when injected right into the middle of the cell (**Fig. 2**).

15. *I-SceI* has been proposed to enhance transient expression levels and stable integration efficiency by remaining attached to the digested recognition sites on the injected transgene and thereby inhibiting the formation of concatamers *(10, 15, 20)*. We have found that once taken off ice during incubation, *I-SceI* digests and detaches from the injected DNA (**Fig. 3**). Hence, while the needle stands at room temperature during injection, the *I-SceI* will be progressively detaching from the injection DNA. To circumvent this, we recommend using the injection needle for only a limited time and changing to a fresh needle containing unincubated injection mix at regular intervals. We observe significant improvement in transient expression levels and the extent of transient expression using this approach.

16. We suggest scoring positive embryos according to extent and specificity of the signal. Maintaining a record of these data from multiple experiments provides a quality readout for individual injections. If expression intensity is consistently low for a given construct the GAL4.UAS system can be used to optimize signal intensity from weak promoters *(21)*.

17. To detect band shifts as shown in **Fig. 3**, run standard 1% agarose gels at a low voltage.

18. The transmission rate scores the percentage of transgenic F1 offspring that arise from an outcross between a founder fish (F0) and a wild-type fish. Transgene insertion at the one-cell stage ensures that the transgene is distributed equally during subsequent cleavages and results in a transmission rate of 50%. Integration events that occur later lower this percentage (Two-cell = 25%; Four-cell = 12.5%). Therefore, the transmission rate reflects the mosaicism of the germline of a given founder fish.
19. The transmission rate determined from transgenic F2 offspring reflects the transgene insertion number for a given transgenic line. Outcrossing of F1 transgenics should result in 50% of transgenic F2 offspring. If this percentage reaches 75%, the F1 carries more than one insertion.

Acknowledgments

We would like to thank F. Loosli and J. Wittbrodt for sharing reagents and knowledge concerning the *I-SceI* meganuclease. This work was supported by the Max Planck Society.

References

1. Houdebine, L. M., and Chourrout, D. (1991) Transgenesis in fish, *Experientia 47*, 891–897.
2. Iyengar, A., Muller, F., and Maclean, N. (1996) Regulation and expression of transgenes in fish – a review, *Transgenic Res 5*, 147–166.
3. Maclean, N. (1998) Regulation and exploitation of transgenes in fish, *Mutat Res 399*, 255–266.
4. Stuart, G. W., McMurray, J. V., and Westerfield, M. (1988) Replication, integration and stable germ-line transmission of foreign sequences injected into early zebrafish embryos, *Development 103*, 403–412.
5. Winkler, C., Vielkind, J. R., and Schartl, M. (1991) Transient expression of foreign DNA during embryonic and larval development of the medaka fish (Oryzias latipes), *Mol Gen Genet 226*, 129–140.
6. Culp, P., Nusslein-Volhard, C., and Hopkins, N. (1991) High-frequency germ-line transmission of plasmid DNA sequences injected into fertilized zebrafish eggs, *Proc Natl Acad Sci U S A 88*, 7953–7957.
7. Westerfield, M., Wegner, J., Jegalian, B. G., DeRobertis, E. M., and Puschel, A. W. (1992) Specific activation of mammalian Hox promoters in mosaic transgenic zebrafish, *Genes Dev 6*, 591–598.
8. Muller, F., Williams, D. W., Kobolak, J., Gauvry, L., Goldspink, G., Orban, L., and Maclean, N. (1997) Activator effect of coinjected enhancers on the muscle-specific expression of promoters in zebrafish embryos, *Mol Reprod Dev 47*, 404–412.
9. Muller, F., Chang, B., Albert, S., Fischer, N., Tora, L., and Strahle, U. (1999) Intronic enhancers control expression of zebrafish sonic hedgehog in floor plate and notochord, *Development 126*, 2103–2116.
10. Grabher, C., Joly, J. S., and Wittbrodt, J. (2004) Highly efficient zebrafish transgenesis mediated by the meganuclease I-SceI, *Methods Cell Biol 77*, 381–401.
11. Gupta, S., Zhu, H., Zon, L. I., and Evans, T. (2006) BMP signaling restricts hematovascular development from lateral mesoderm during somitogenesis, *Development 133*, 2177–2187.

12. Rembold, M., Lahiri, K., Foulkes, N. S., and Wittbrodt, J. (2006) Transgenesis in fish: efficient selection of transgenic fish by co-injection with a fluorescent reporter construct, *Nat Protoc 1*, 1133–1139.
13. Nyholm, M. K., Wu, S. F., Dorsky, R. I., and Grinblat, Y. (2007) The zebrafish zic2a-zic5 gene pair acts downstream of canonical Wnt signaling to control cell proliferation in the developing tectum, *Development 134*, 735–746.
14. Kawakami, K. (2007) Tol2: a versatile gene transfer vector in vertebrates, *Genome Biol 8* Suppl 1, S7.
15. Thermes, V., Grabher, C., Ristoratore, F., Bourrat, F., Choulika, A., Wittbrodt, J., and Joly, J. S. (2002) I-SceI meganuclease mediates highly efficient transgenesis in fish, *Mech Dev 118*, 91–98.
16. Kawakami, K. (2004) Transgenesis and gene trap methods in zebrafish by using the Tol2 transposable element, *Methods Cell Biol 77*, 201–222.
17. Ogino, H., McConnell, W. B., and Grainger, R. M. (2006) Highly efficient transgenesis in Xenopus tropicalis using I-SceI meganuclease, *Mech Dev 123*, 103–113.
18. Pan, F. C., Chen, Y., Loeber, J., Henningfeld, K., and Pieler, T. (2006) I-SceI meganuclease-mediated transgenesis in Xenopus, *Dev Dyn 235*, 247–252.
19. Jacquier, A., and Dujon, B. (1985) An intron-encoded protein is active in a gene conversion process that spreads an intron into a mitochondrial gene, *Cell 41*, 383–394.
20. Grabher, C., and Wittbrodt, J. (2007) Meganuclease and transposon mediated transgenesis in medaka, *Genome Biol 8* Suppl 1, S10.
21. Koster, R. W., and Fraser, S. E. (2001) Tracing transgene expression in living zebrafish embryos, *Dev Biol 233*, 329–346.

Part III

Tissue-Specific Manipulations

Color Plates

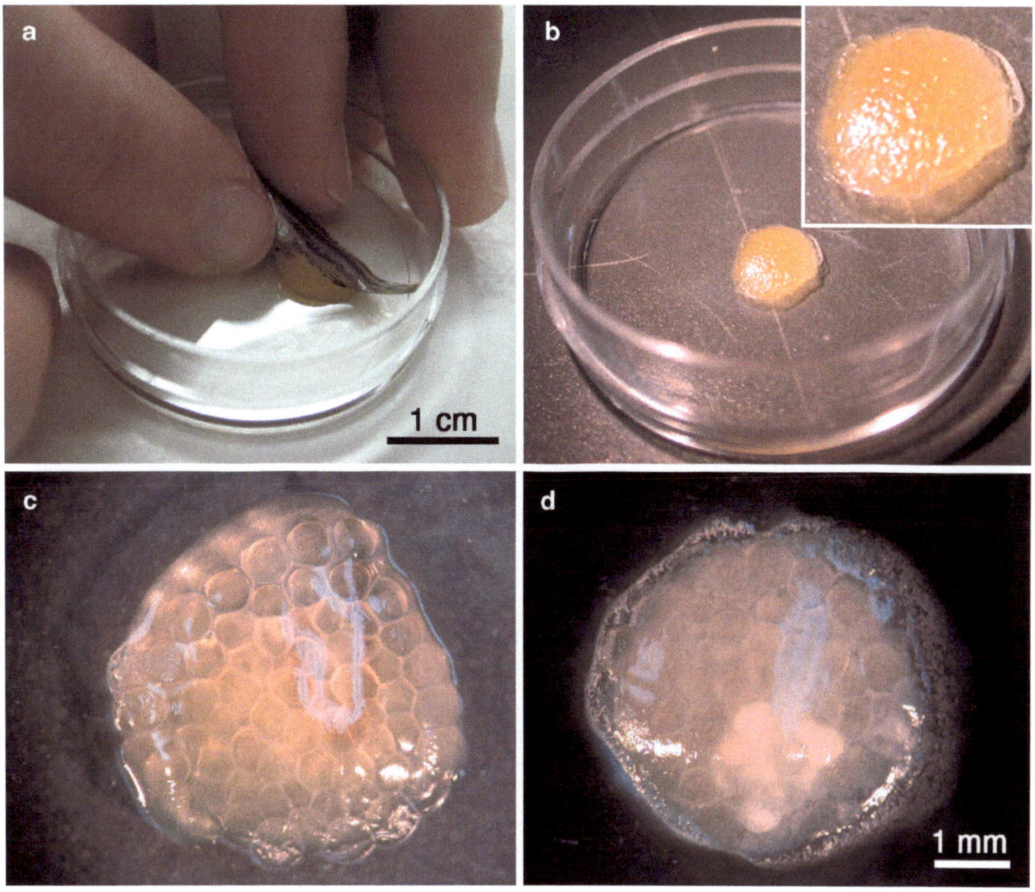

Chapter 4, Fig. 3. Visual assessment of egg quality. (**a**) Anesthetized female is squeezed gently to release eggs into 35 mm Petri dish. (**b**, **c**) "Good" quality batches of eggs. Eggs are yellowish in incident light and appear translucent with transmitted light. (**d**) Lower quality batch of eggs. Note the opaque, white eggs intermingled with normal looking ones. Scale bars: (**a**) 1 cm; (**b–d**) 1 mm.

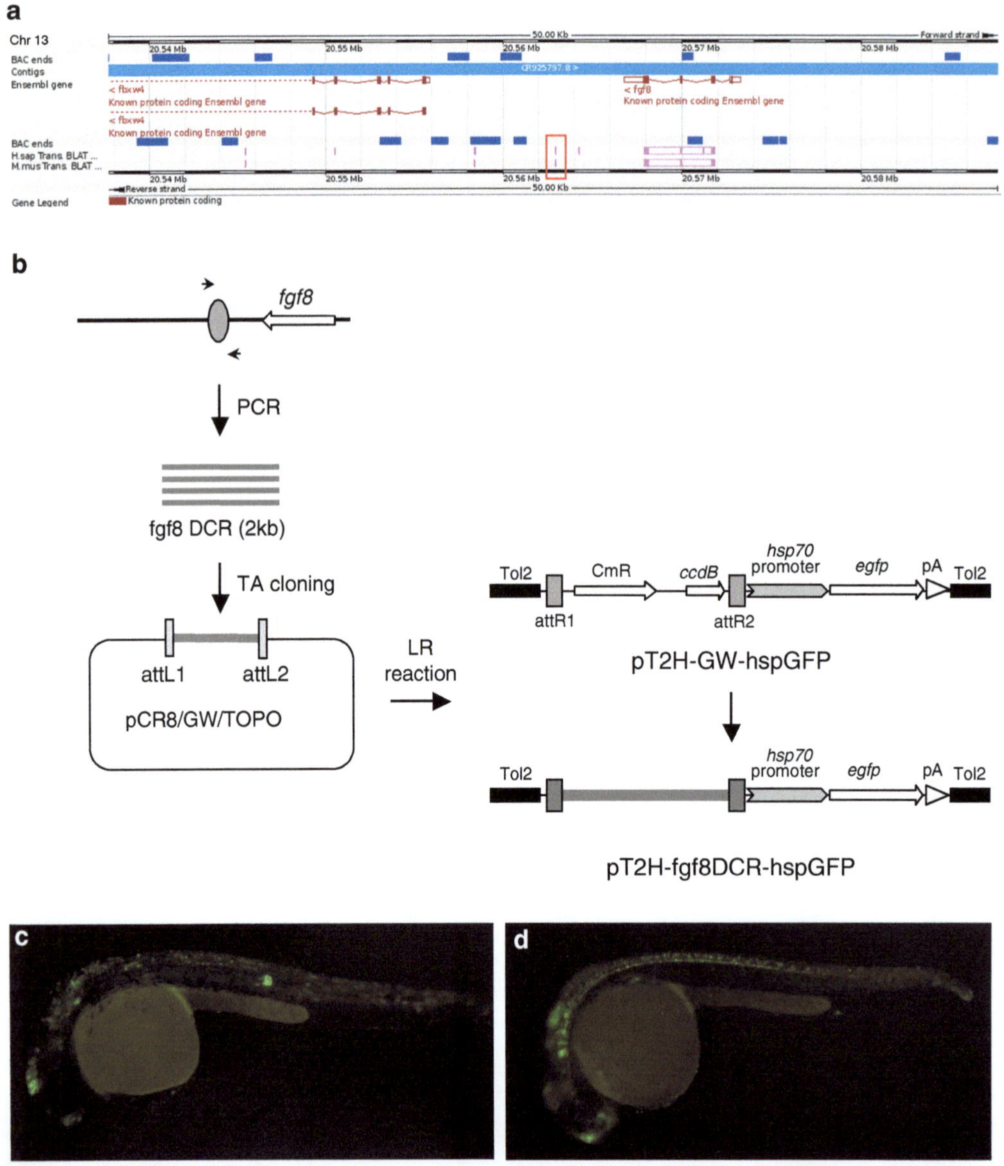

Chapter 5, Fig. 3. Analysis of a genomic enhancer activity with a *Tol2*-Gateway enhancer assay vector: an example of analysis of a genomic enhancer of the *fgf8* locus. (**a**) Comparison of the zebrafish *fgf8* locus on the chromosome 13 with human and mouse translated BLAT. A red box indicates DNA sequence conserved among these species downstream of the *fgf8* gene. (**b**) The fgf8DCR is amplified by PCR from genomic DNA and cloned into the pCR8/GW/TOPO plasmid. Then, the fgf8DCR fragment is transferred to the enhancer assay vector pT2H-GW-hspGFP through the LR reaction, resulting in pT2H-fgf8DCR-hspGFP. (**c**) Transient transgenesis using pT2H-fgf8DCR-hspGFP. A side view of a 24 hpf-embryo injected with the transposase mRNA and pT2H-fgf8DCR-hspGFP. (**d**) Stable transgenesis using pT2H-fgf8DCR-hspGFP. Fish injected with the transposase mRNA and pT2H-fgf8DCR-hspGFP at the one cell stage were raised to adulthood and crossed with wild-type fish. A side view of the GFP-positive F1 embryo at 24 hpf. Both transient and stable transgenic embryos show essentially similar GFP expression patterns including expression in forebrain, midbrain-hindbrain boundary, and spinal cord.

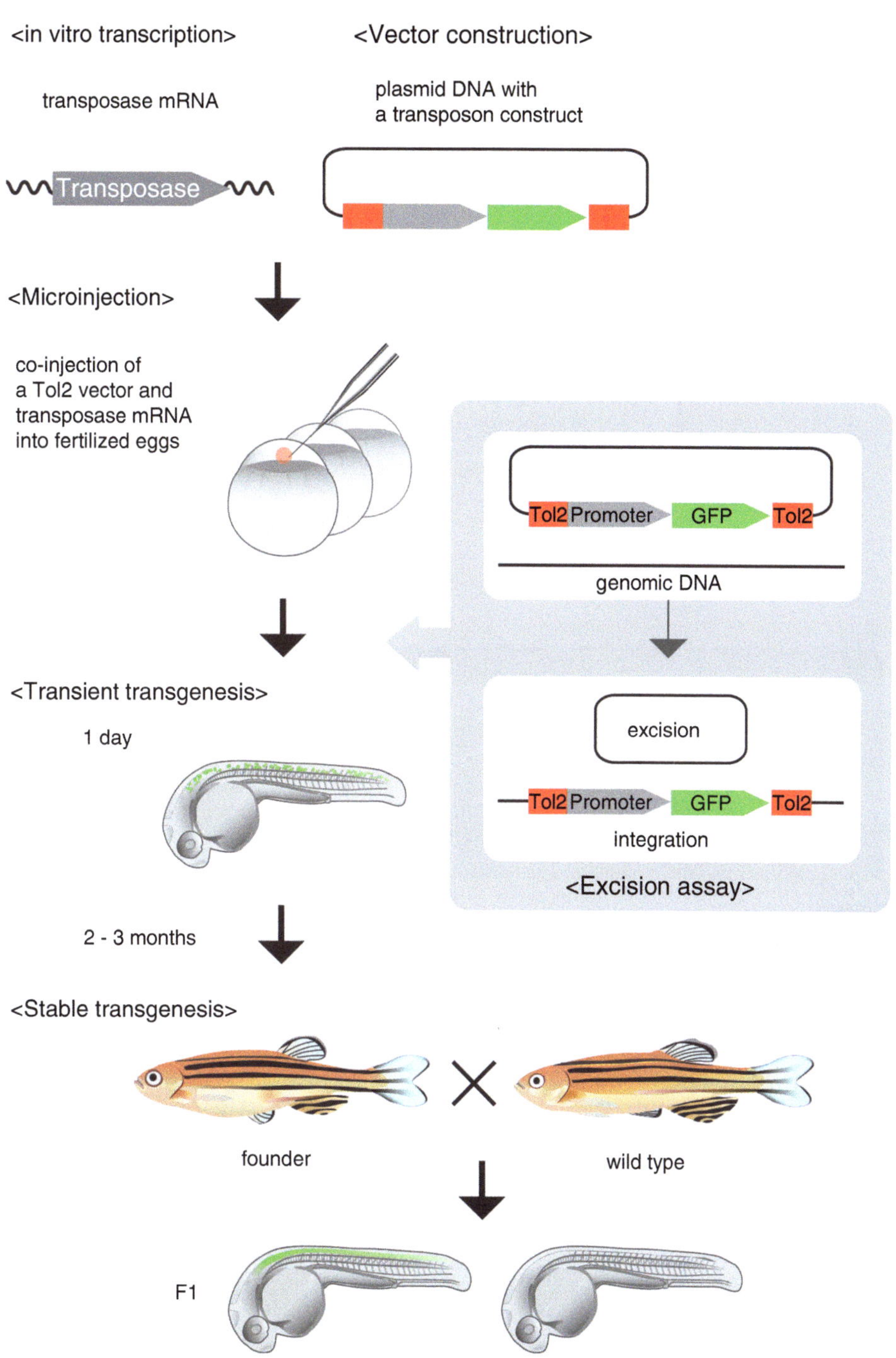

Chapter 5, Fig. 6. A scheme for transient and stable transgenesis in zebrafish. The transposase mRNA synthesized in vitro and a plasmid DNA containing a *Tol2* construct are coinjected into zebrafish fertilized eggs. In zebrafish cells during embryogenesis, the *Tol2* construct is excised from the donor plasmid and integrated in the genome. In transient transgenesis, the *Tol2* construct is integrated in the genome of somatic cells. These cells express GFP under the control of the promoter activity cloned on the construct. Since integration does not occur in all somatic cells, mosaic expression of the reporter gene is observed. In stable transgenesis, the *Tol2* construct is integrated in the genome of germ cells. This occurs with high efficiency, and more than 50% of the injected fish transmit the *Tol2* insertions to a proportion of their F1 generation. The F1 embryos that carry *Tol2* insertions from their parent are heterozygous for such an insertion, and therefore display non-mosaic expression of the reporter gene.

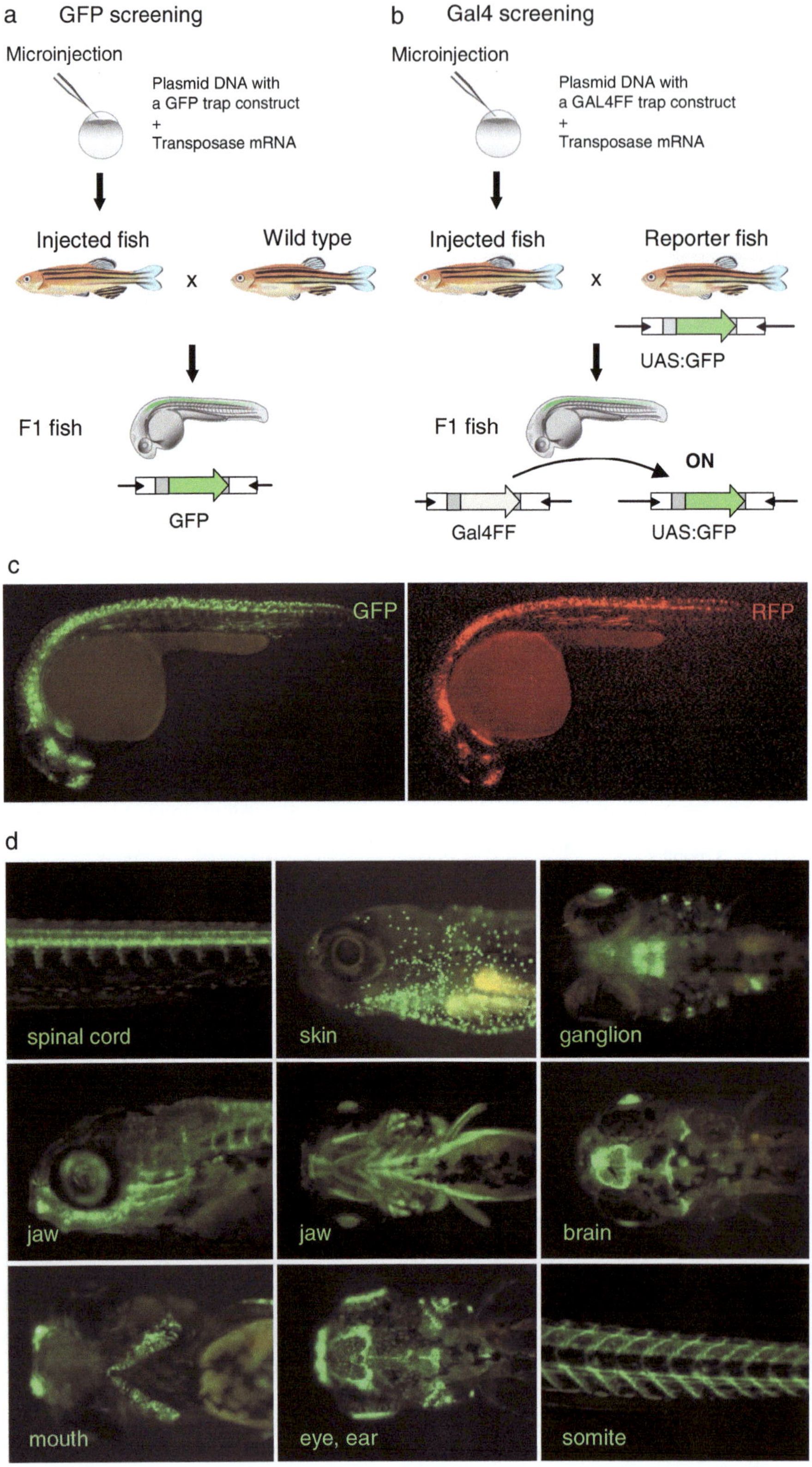

Chapter 6, Fig. 2. Gene trap and enhancer trap screens in zebrafish. (*For Complete Caption refer Page 90*)

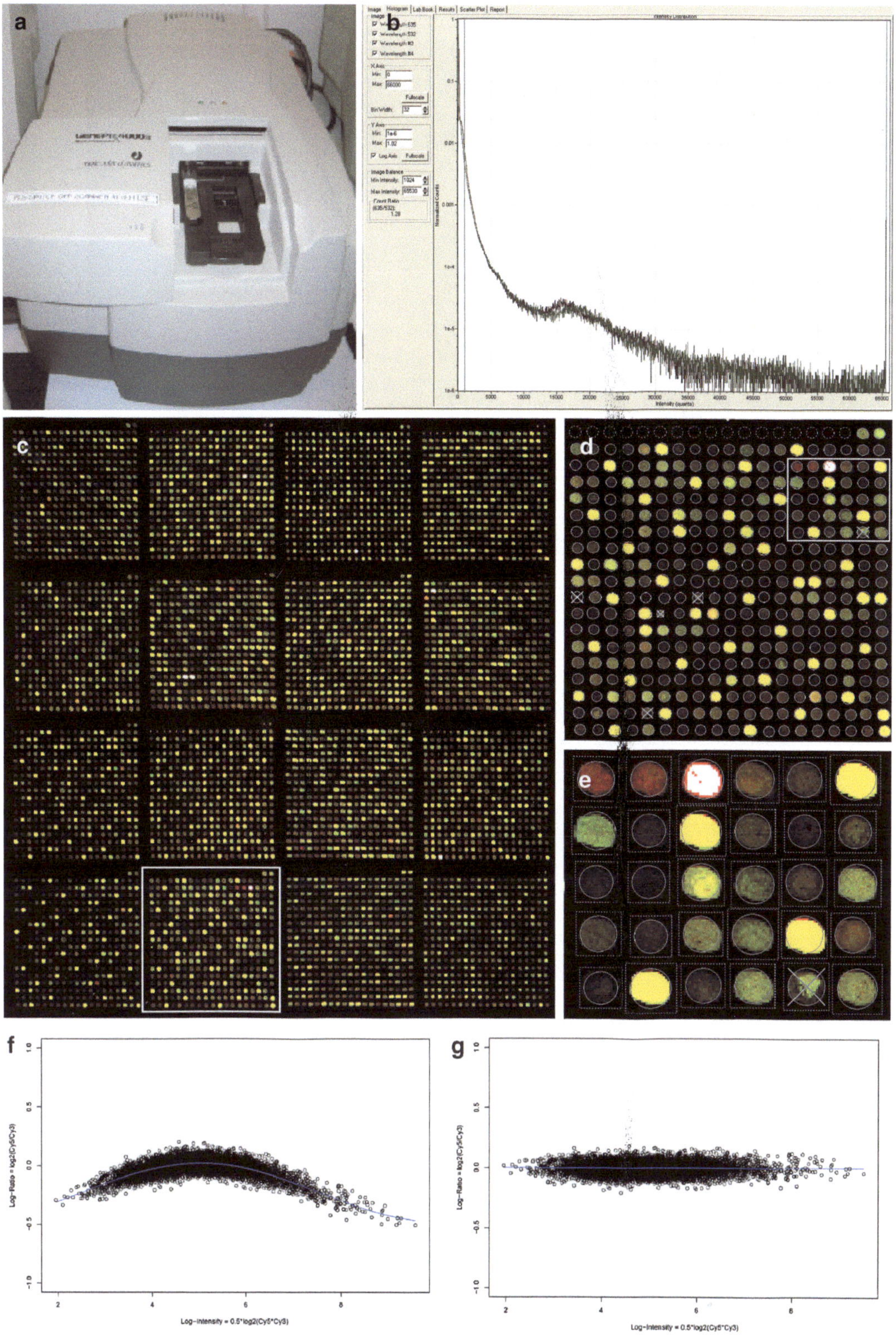

Chapter 13, Fig. 2. Microarray image-data acquisition, analysis, and normalization. (*For Complete Caption refer Page 203*)

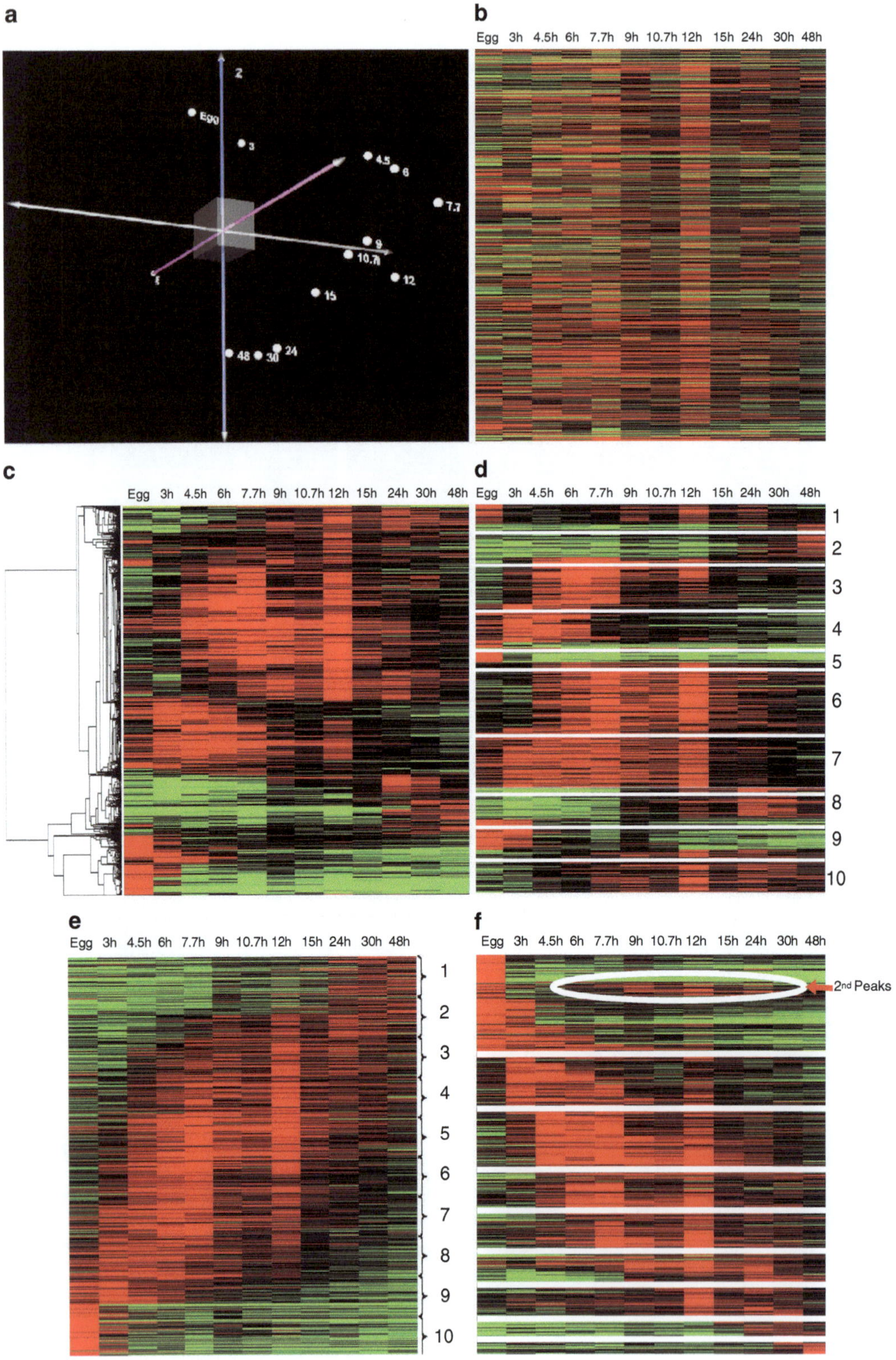

Chapter 13, Fig. 3. Gene expression profile of normal developing zebrafish at progressive developmental time-points viewed using different pattern visualization and discovery algorithms. (*For Complete Caption refer Page 212*)

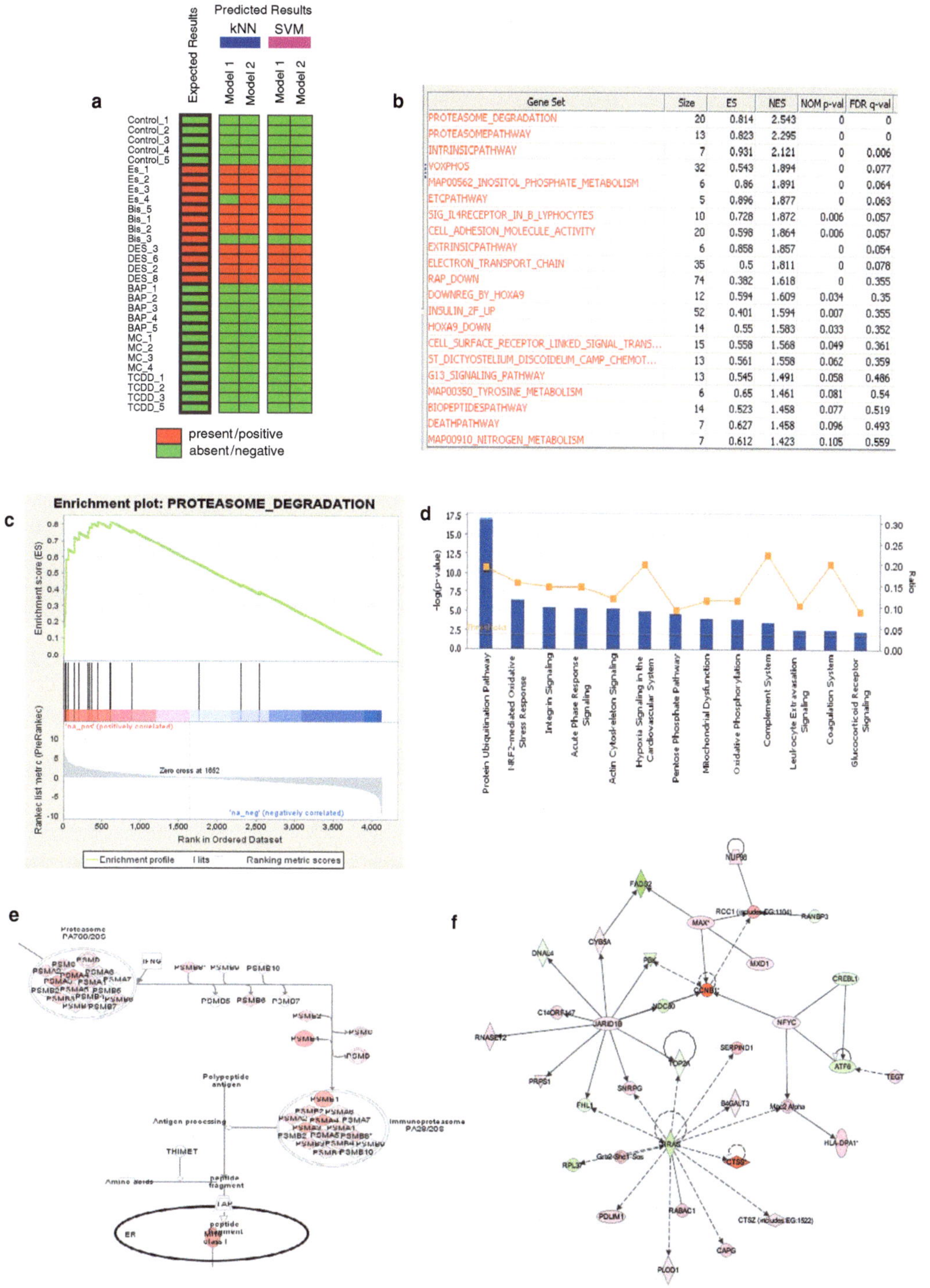

Chapter 13, Fig. 4. Data analysis outputs of toxicant-treated zebrafish for class prediction, gene set enrichment analysis, biologic function classification, inference of pathways, and networks (unpublished data). (*For Complete Caption refer Page 216*)

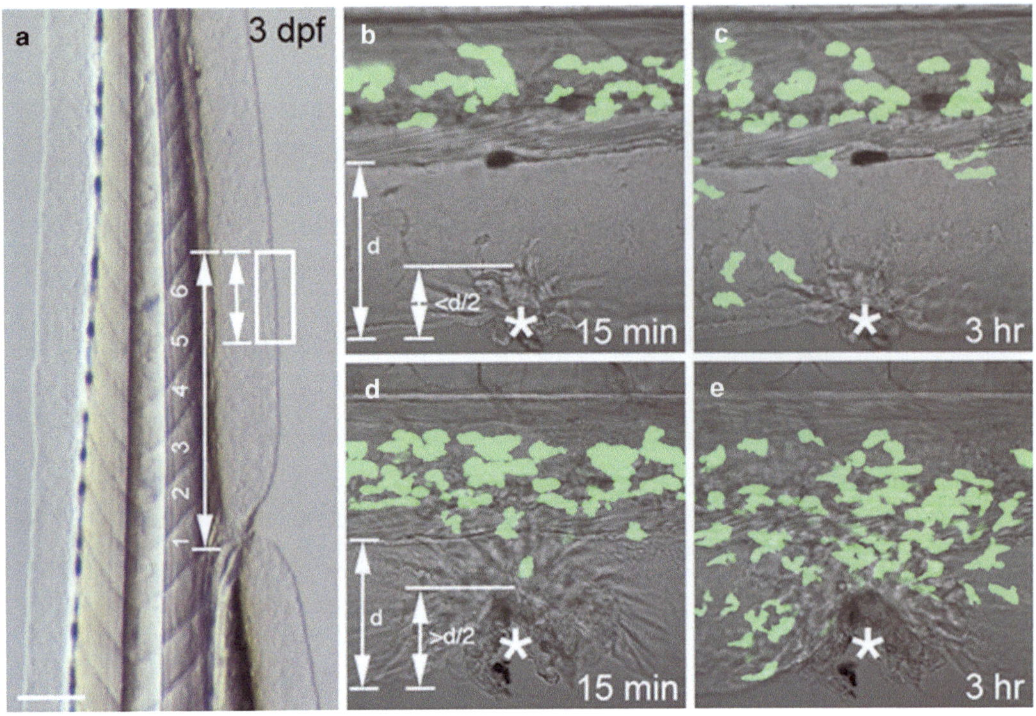

Chapter 16, Fig. 1. Magnitude of inflammatory response correlates with size of induced wound. (*For Complete Caption refer Page 260*)

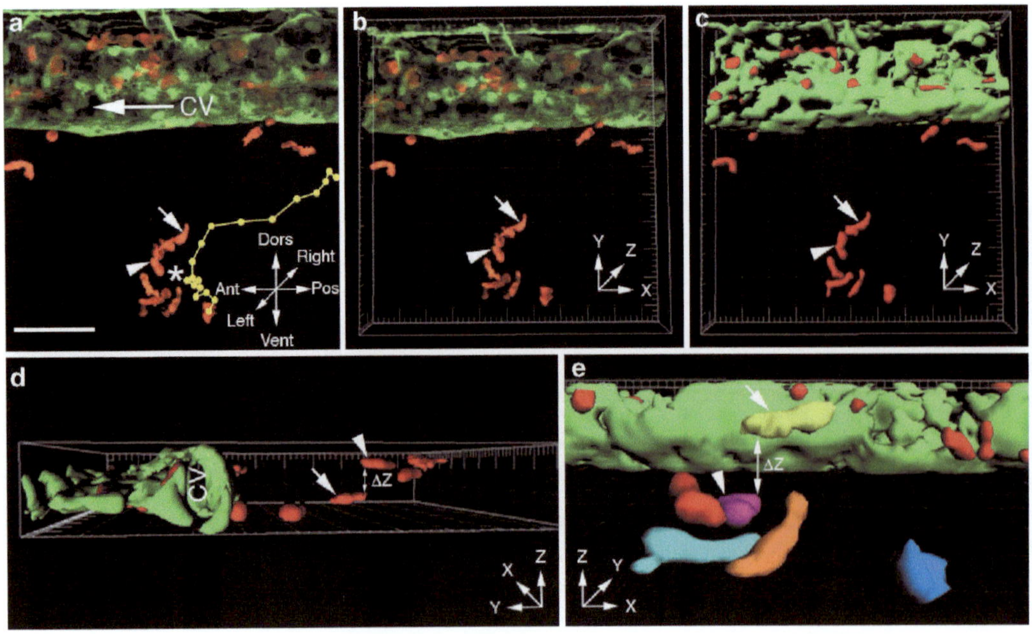

Chapter 16, Fig. 2. Multi-dimensional imaging of leukocytes during inflammation. (*For Complete Caption refer Page 262*)

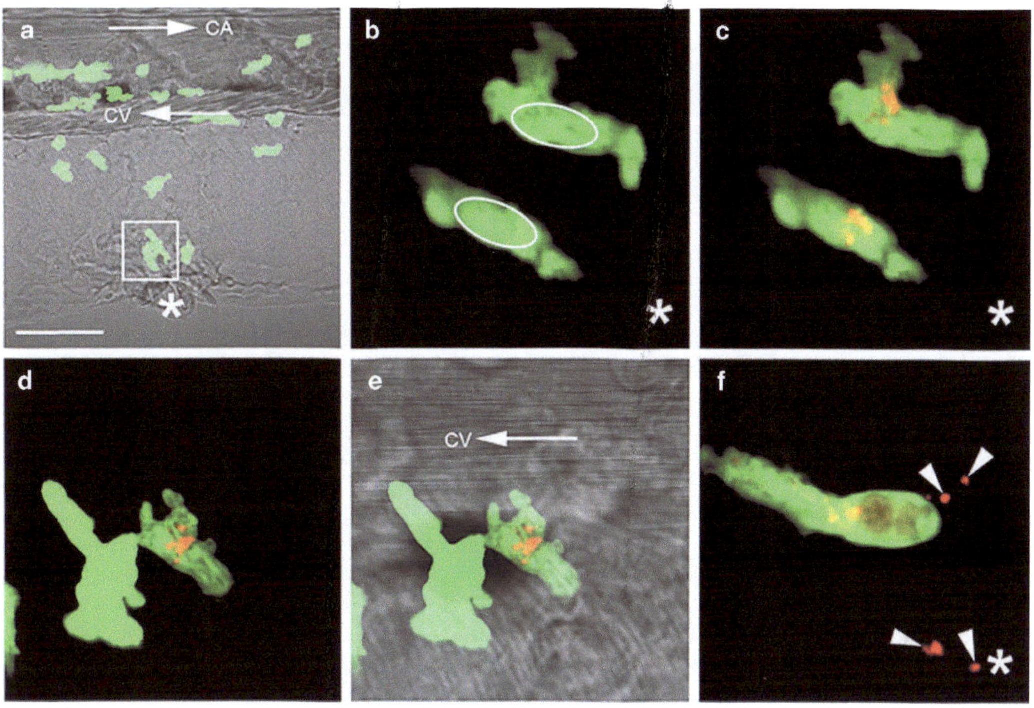

Chapter 16, Fig. 3. Tracking/lineage tracing uncaged leukocytes in Tg(lyz:EGFP) larvae. (**a–f**) Confocal images (merged with bright field in **a** and **e**) before (**a** and **b**) and after (**c–f**) uncaging of CMNCBZ-caged carboxy-Q-rhodamine within three dpf Tg(lyz:EGFP) larvae following wounding (all lateral views, anterior to left). * marks the wound region. *Arrows* highlight location and direction of circulation within the caudal artery (CA) and caudal vein (CV). (**b** and **c**) Higher magnification of boxed region in A before (**b**) and immediately following (**c**) uncaging. White boundaries in B mark the region of interest to be targeted with the 405 nm (uncaging) laser. (**d** and **e**) An uncaged leukocyte that has migrated back towards the CV demonstrating photobleaching of the EGFP signal relative to a leukocyte that has not been exposed to the 405 nm laser. (**f**) Rhodamine dye (*arrowheads*) uncaged outside of the targeted EGFP-expressing leukocyte. Scale bar: 50 μm in **a**.

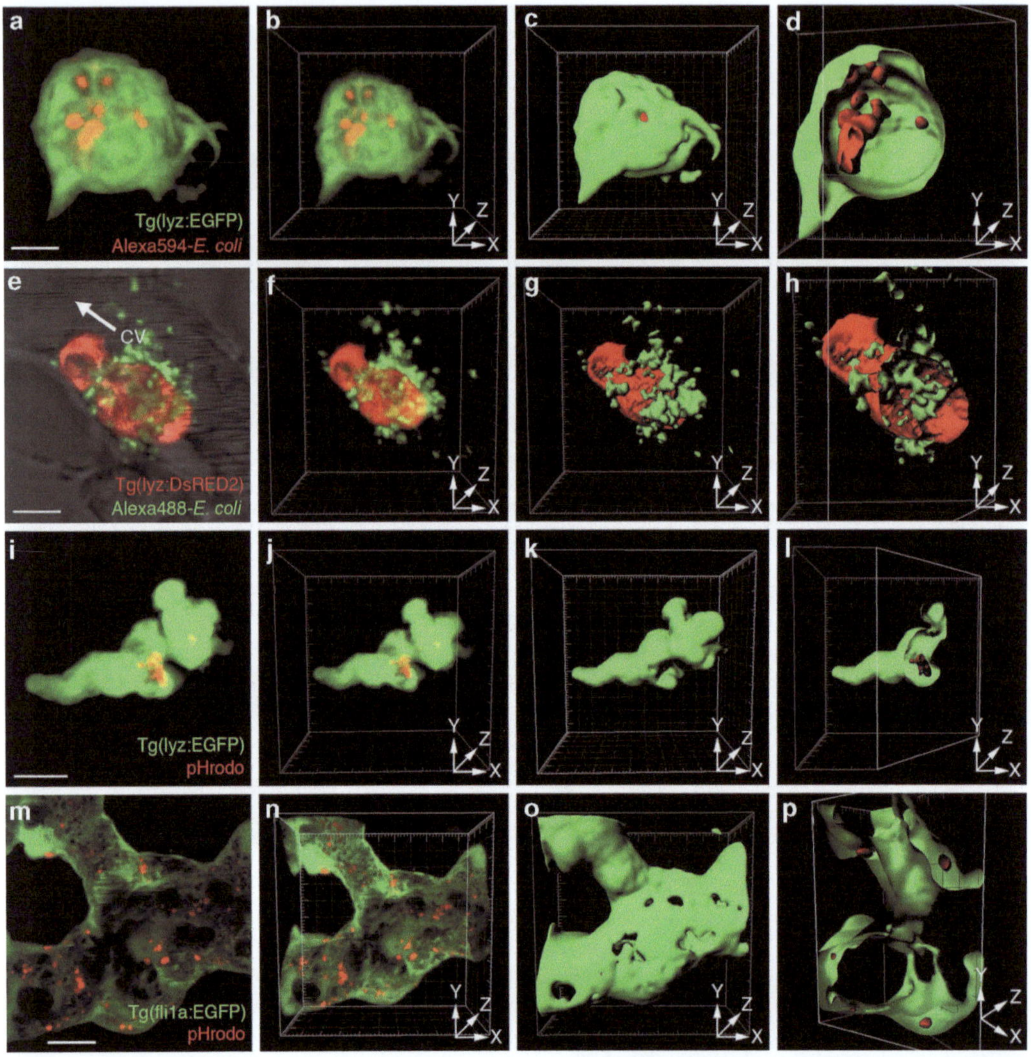

Chapter 16, Fig. 4. Live multi-dimensional imaging of phagocytosis by leukocytes. *z* projection (**a**, **e** (merged with bright field), **i** and **m**), volume rendered (**b**, **f**, **j** and **n**) and Isosurface (**c**/**d**, **g**/**h**, **k**/**l** and **o**/**p**) views of an EGFP-expressing leukocyte on the yolk containing Alexa Fluor 594 *E. coli* (**a–d**), a DsRED2-expressing leukocyte in the PBI surrounded by, and containing, Alexa Fluor 488 *E. coli* (**e–h**), an EGFP-expressing leukocyte within the PBI containing red-fluorescent pHrodo (**i–l**) and EGFP-expressing endothelial cells of the caudal vascular plexus containing red fluorescent pHrodo (**m–p**). *Arrow* in **e** marks the direction of circulation in the caudal vein (CV). Views in **d**, **h**, **l** and **p** represent cutting planes through the marked cells revealing the phagocytosed fluorescent contents. Scale bars: 3 μm in **a**; 5 μm in **e** and **i**; 10 μm in **m**.

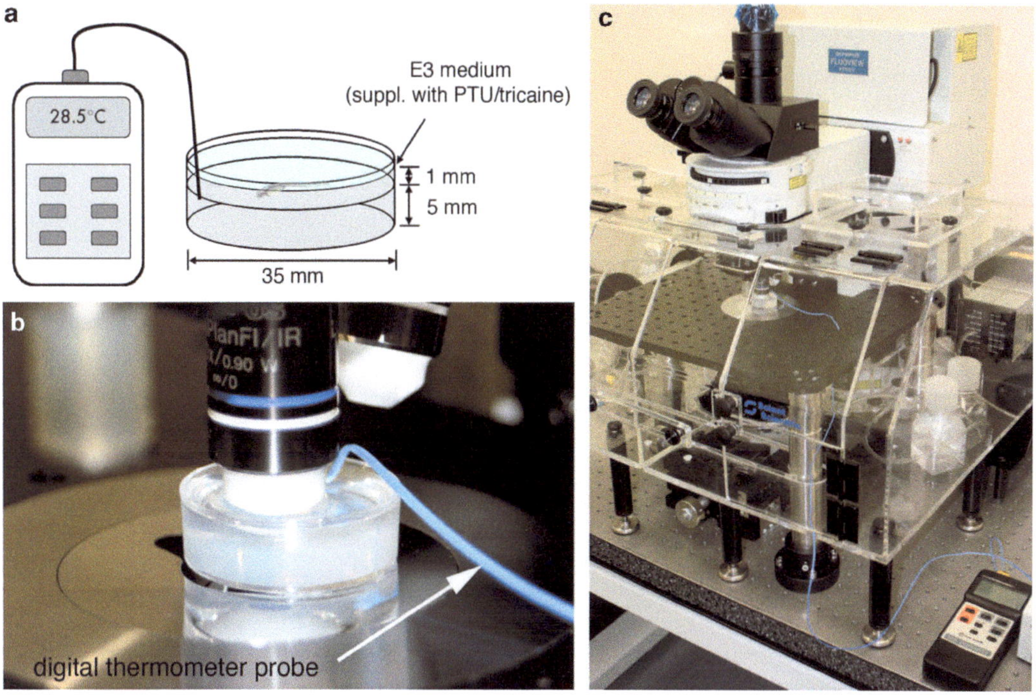

Chapter 16, Fig. 5. Live confocal imaging setup. (**a**) Schematic diagram of mounted embryo. (**b** and **c**) Views of mounted embryo within incubation chamber of an Olympus FluoView FV1000 confocal microscope.

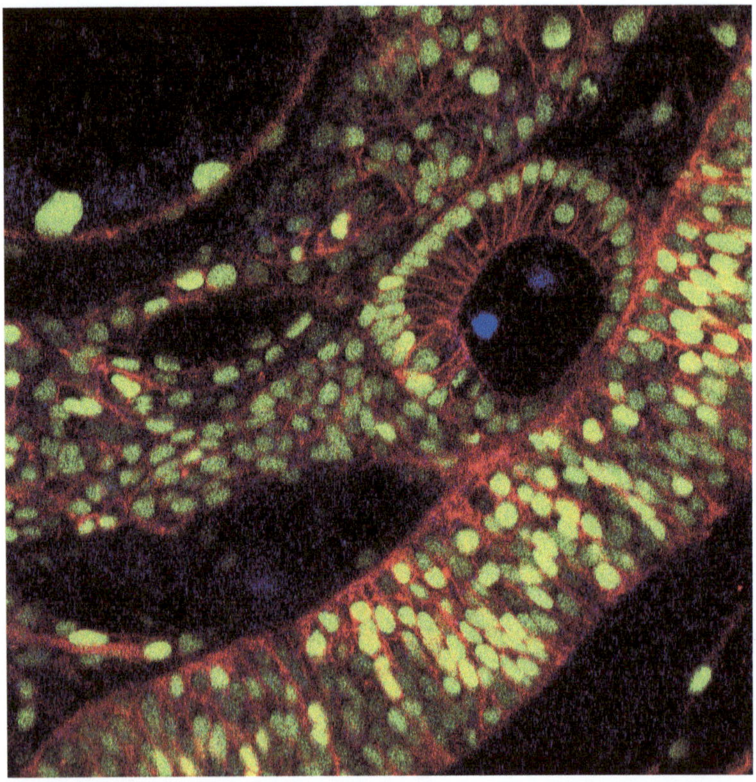

Chapter 19, Fig. 4. A single *xy* image extracted from a very large in toto image to show a typical image. This image is a dorsal-lateral view of the otic vesicle and hindbrain. Nuclei are marked in green with H2B-EGFP and cell membranes are marked in red with a membrane localized mCherry.

Chapter 9

Nitroreductase-Mediated Cell Ablation in Transgenic Zebrafish Embryos

Harshan Pisharath and Michael J. Parsons

Summary

Prodrug dependent cell ablation is a method that allows inducible and spatially restricted cell destruction. We describe transgenic methods to express the *Escherichia coli nfsB* in a tissue restricted manner in the zebrafish. This bacterial gene encodes a nitroreductase (NTR) enzyme that can render prodrugs such as metronidazole (Met) cytotoxic. Using the expression of NTR fused to a fluorescent protein, one can simultaneously make cells susceptible to prodrug treatment and visualize cell ablation as it occurs.

Key words: Ablation, Nitroreductase, Metronidazole, Prodrug, Transactivation, Gal4, UAS.

1. Introduction

Conditional cell ablation studies are instrumental in studying both regulative development and regeneration. Additions of cell-specific toxins allow the precise timing of cell ablation. Unfortunately, the list of known cell-specific toxins is fairly limited. One way around this restriction is to perform chemical screens on zebrafish larvae to isolate useful pharmaceutical agents *(1)*. Another option is to utilize transgenic methods to direct inducible cell death *(2–5)*. The majority of these methods involve using tissue-specific promoters to control expression of exogenous genes that encode prodrug converting enzymes. At the desired time an embryo/larvae can be treated with a nontoxic prodrug and any cell containing the exogenous enzyme will metabolize the prodrug into a lethal cytotoxin. The zebrafish is not only

Graham J. Lieschke et al. (eds.), *Zebrafish,* Methods in Molecular Biology, vol. 546
DOI: 10.1007/978-1-60327-977-2_9

a powerful genetic system, but also displays an impressive capability to regenerate many tissue types *(6)*. Furthermore, prodrugs can be directly added to the fish's water, facilitating uptake. For these reasons, the zebrafish is an ideal model to utilize prodrug dependent cell ablation techniques in tissue regeneration studies.

Nitroreductase is a bacterial homodimeric flavoenzyme encoded by the *nfsB* gene. The flavin mononucleotide associated with the enzyme is first reduced by NADH or NADPH *(7)*. Following this reduction, the enzyme transfers these reducing equivalents to various substrates such as the prodrug CB1954, creating cytotoxic metabolites *(4, 8, 9)*. CB1954-mediated ablation has been used successfully in rodents, to target luminal cells of the mammary gland *(3, 10)*, astrocytes *(11)*, neurons *(12)*, apidocytes *(13)*, and progenitors within both the adult brain *(14)* and developing prostate *(15)*.

Work in our lab and the Stainier lab has recently demonstrated the utility of using the NTR system in embryonic/larval zebrafish to allow cell specific ablation *(16, 17)*. Instead of the prodrug CB1954, another compound Metronidazole (Met) was used. We chose to use Met, as it had higher efficacy over CB1954 in our experiments. The prodrug Met only mediated cell ablation in cells expressing NTR; leaving neighboring cells unharmed. Other work has utilized the Gal4/UAS system to control spatial/temporal restricted NTR expression in zebrafish embryos/larvae *(18)*. Gal4-VP16 is a strong transactivator that directs expression of an upstream activator sequence (UAS) *(19)*. We created a fish line with an open reading frame (ORF) encoding NTR-mCherry cloned under the control of an UAS and minimal promoter. When these *Tg(UAS:nfsB-mCherry)* fish are crossed to a *Gal4*-VP16 driver line, any cell expressing this transactivator will simultaneously be labeled by fluorescence and be made susceptible to prodrug-mediated cell ablation *(18)*. We describe here methods used to perform such cell specific cell ablation.

2. Materials

2.1. NTR Expressing Transgenic Fish

A transgenic line expressing NTR in the desired spatial/temporal pattern is required to generate embryos/larva for ablation experiments. Alternatively, a *Gal4* driver line can be used in conjunction with a responder line, such as our *UAS:nfsB-mCherry*, to obtain tissue specific expression of NTR. The *UAS:nfsB-mCherry* has been submitted to the Zebrafish International Resource Center (http://zebrafish.org/zirc/home/guide.php) for distribution.

2.2. Reagents for Ablation Studies

1. Embryo medium (E3) *(20)*: 17.52 g of sodium chloride, 2.92 g of calcium chloride, 0.76 g of potassium chloride and 2.30 g of magnesium chloride into 1 L of distilled water to make a 60X stock. Make working solution of E3 in distilled water. (*see* **Note 1**).
2. 100X Phenylthiourea stock solution (PTU): 0.3% (w/v) 1-phenyl-2-thiourea (P7629 Sigma®) in distilled water. Use at 0.003% in E3 (E3/PTU). Note PTU does not dissolve completely at stock solution strength; therefore, shake well before use.
3. Phosphate buffered saline (PBS), pH 7.4 (Invitrogen/Gibco®).
4. Paraformaldehyde (PFA): 4% (w/v) paraformaldehyde (P6148, Sigma®) in 1X PBS.
5. Metronidazole (Met) (M3761, Sigma®). We store Met powder in the dark at 4°C.

2.3 .TUNEL Reaction Reagents

1. ApopTag Peroxidase In Situ Apoptosis Detection Kit (S7100, Millipore®).
2. PBT: 0.1% Tween in PBS.
3. Methanol.
4. Glacial Acetic Acid.
5. Proteinase K, 20 mg/mL (25530-049, Invitrogen®).
6. Goat Serum blocking solution (PBTS): 10% (v/v) normal Goat serum (0060-01, SouthernBiotech®) in PBT.
7. Anti-digoxigenin antibody-AP (11093274910, Roche®).
8. NTMT buffer: 0.1 M Tris–HCl, 50 mM magnesium chloride, 0.1 M sodium chloride, 0.1% Tween 20) pH 9.5.
9. NBT stock solution: nitroblue tetrazolium chloride (11383213001, Roche®) 100 mg/mL.
10. BCIP stock solution: 5-bromo-4-chloro-3-indolyl-phosphate (1383221, Roche®) 50 mg/mL.
11. Glycerol.

2.4. Stereomicroscopy

1. Stereomicroscopes: Stemi SV6 (Zeiss) and Stemi SV11 (Zeiss) or equivalent.
2. Digital camera: Axio Cam Color (Zeiss) and image handling software such as Axio Vision (Release 4.6.3) (Zeiss) or equivalent.
3. Tricaine stock solution: 3-amino benzoic acid (Sigma® A-5040) 400 mg/100 mL, pH adjusted to 7.0 with 1 M Tris–HCl (pH 9) *(20)*.
4. 3% methyl cellulose (Sigma® M0387) *(20)*.

2.5 Confocal Microscopy

1. Axiovert 200M microscope coupled to a LSM 5 Pascal system (Zeiss) or equivalent.

3. Methods

3.1. NTR-expressing Transgenic Zebrafish

3.1.1. Generation of NTR Expression Transgene

The gene *nfsB* can be easily cloned by PCR from DNA derived from any common laboratory strain of *E. coli*. Suitable primer sequences are shown below with coding regions capitalized. Exogenous sequence is in lower case and creates restriction sites for cloning. Restriction sites are underlined and endonucleases shown in parentheses.

Forward primer (*Nco*I) 5′ <u>ccATGG</u>ATATCATTTCT-GTCGCCTTA 3′

Reverse primer (*Cla*I) 5′ <u>atcgaT</u>TACACTTCGGTTAAG-GTGATGT 3′

In order to identify cells destined for ablation, observe cellular behavior during ablation, and detect any regeneration following transient prodrug treatment, it is desirable to simultaneously render cells susceptible to cell ablation and mark them with a fluorescent protein. Fusions consisting of either Cyan fluorescent protein (CFP) *(16)* or enhanced green fluorescent protein (eGFP) *(14)* at the N-terminus and either eGFP or mCherry at the C terminus *(17, 18)*, are functional both as fluorescent markers and catalysts of prodrug reduction. To generate a fusion protein with a C-terminal fluorescent protein the following reverse primer can be used to amplify a NTR open reading frame (ORF) without a stop codon.

Reverse primer no stop codon (*Cla*I) 5′ <u>atcgat</u>CACTTCG-GTTAAGGTGATGTTT 3′

Following polymerase chain reaction (PCR) amplification of the *nfsB* ORF (minus stop), sequence encoding fluorescent protein of choice (minus initial ATG) can be cloned in frame using the *Cla*I site. Once a suitable fusion expression cassette has been generated, it can be cloned downstream of a tissue specific promoter, or under the control of multimerized *Gal4*-responsive UAS *(18, 21, 22)*. It is highly recommended that the NTR expression cassette (consisting of regulatory sequence, *nfsB* gene and fluorescent protein encoding gene) be cloned into a Tol2 transposable element to facilitate transgenesis *(23, 24)*.

3.1.2. Gal4-VP16 Transactivation to Control Tissue-Specific NTR Expression

The Gal4/UAS system has been utilized extensively in *Drosophila* research *(25, 26)*. Once a Gal4 driver line has been created, it can be crossed to many different UAS responder lines to give spatially restricted transgene expression *(19, 26)*. To be able to direct NTR mediated cell ablation from existing or future zebrafish *Gal4* driver lines we generated a *UAS* responder line that when transactivated by *Gal4*-VP16, leads to expression of a NTR-mCherry fusion protein. To use this system the Gal4 driver line is simply crossed to the *UAS:nfsB-mCherry* responder line. Progeny inheriting both transgenes can be detected under a fluorescence microscope. These embryos are then used in ablation studies. (*see* **Note 2**)

3.2. Met Dependent Ablation

1. A working concentration of 10 mM Met is made in E3/PTU for ablation studies. This concentration of Met has no deleterious effect on wild-type embryos up to 7 dpf. Lower working concentrations of Met are noticeably less effective at removing all targeted cells. Met is difficult to dissolve and we do not recommend trying to make a higher concentration stock solution. Instead, the prodrug solution is made fresh just before use by adding Met powder into E3/PTU medium in a glass container followed by vigorous agitation until no particulate traces of Met are left (approx 2 min). (*see* **Note 3**).
2. In preparation for prodrug dependent ablation, dechorionate and transfer embryos to wells of a 24-well plate (5–10 embryos per well).
3. Remove as much E3/PTU as possible and replace with 10 mM Met in E3/PTU solution. Wrap plates in aluminum foil and incubate at 28°C for 24 h. (*see* **Note 4**).

   In this report, we illustrate the utility of the Gal4/UAS system by using *ptf1a:Gal4-VP16; UAS:nfsB-mCherry* transgenic fish (**Fig. 1**). These fish express the NTR-mCherry fusion protein in the *ptf1a* (*pancreas-specific transcription factor, 1a*) domain, which includes the exocrine pancreas. *Ptf1a* starts to be expressed in the exocrine pancreas around 32 hpf. By incubating in Met for 48 h starting at 24 hpf (i.e., before the onset of *ptf1a* expression in the pancreas) the formation of the exocrine pancreas in these transgenic embryos was completely prevented (**Fig. 1i, j**; *see* **Note 4**). Thus, incubating in Met prior to the expression of NTR can be used to ablate cells as they turn on the *nfsB* expression and prevent the formation of a whole tissue. (*see* **Note 5**).

3.3. Cell Death Detection

At the end of incubation, embryos are examined for the lack or disruption of fluorescence in the expression domain of interest (**Fig. 1d**). Because the fluorescence from the NTR-mCherry fusion protein lingers beyond the demise of the cell, this should not be the sole criteria for assessing cell death. In addition to lack or disruption of fluorescence, we have made use of Terminal deoxynucleotidyl Transferase Biotin-dUTP Nick End Labeling (TUNEL) to assess cell death. Examples of results are shown for the hindbrain (**Fig. 1f**) and eye (**Fig. 1h**) in Met treated *ptf1a:Gal4-VP16; UAS:nfsB-mCherry* embryos. Double transgenic embryos incubated in E3/PTU alone are used as negative controls (**Fig. 1e, g**). These and other examples of cell specific

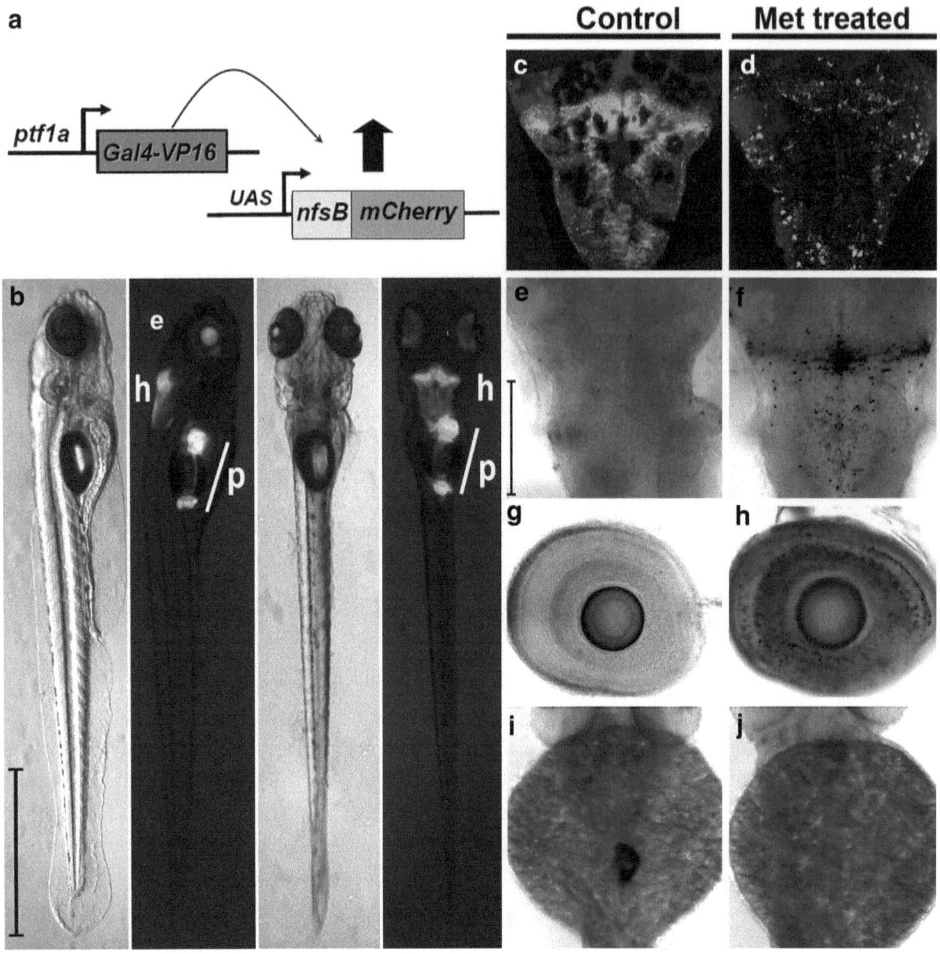

Fig. 1. Nitroreductase mediated cell ablation in *ptf1a:GAL4-VP16; UAS.nfsB-mCherry* double transgenic embryos. (**a**) Constructs used to make the bipartite GAL4-UAS transgenic system driving *nfsB* expression to the *ptf1a* domain. The *ptf1a:GAL4-VP16* construct is a recombineered BAC that expresses Gal4-VP16 in the expression pattern of endogenous *ptf1a*. The *UAS.nfsB-mCherry* construct is transactivated by Gal4-VP16 (*thin arrow*), leading to the production of NTR-mCherry fusion protein (*thick arrow*) in double transgenic fish. (**b**) Localization of NTR-mCherry in double transgenic embryos at 96 hpf. Lateral and Dorsal views under transmitted and fluorescent light (RFP filter). NTR-mCherry can be detected in the *ptf1a* domain which includes the hindbrain (h), the exocrine pancreas (p) and amacrine cells of the eyes (e). Scale bar in (**b**) = 1,000 μm. (**c–h**) Double transgenic embryos exhibit loss of fluorescence and apoptosis in a Met dependent fashion. Embryos were incubated in the absence (Control – **c**, **e**, and **g**) or presence (Met treated – **d**, **f**, and **h**) of 10 mM Met at 96 hpf for 24 h. At 120 hpf embryos were fixed and observed by confocal microscopy (**c** and **d**) or analyzed using TUNEL (**e–h**). The images (**c** and **d**) are 3D rendering of confocal sections using Zeiss LSM 5 Pascal software. Compared with the control (**c**), those exposed to Met (**d**) exhibited reduced fluorescence and disruption of tissue architecture. The loss of fluorescence corresponds to an increase of TUNEL positive nuclei in the hindbrain (**f**) and retina of the eyes (**h**) when compared to untreated controls (**e** and **g**). The number of cells undergoing apoptosis in the rhombic lip (seen as stripe across the midbrain/hindbrain boundary in (**f**) is particularly high and coincides with high expression of NTR-mCherry fusion protein in this tissue (see **b** and **c**). Scale bar in (**e**) = 200 μm. (**i** and **j**) Early cell ablation prevents formation of the exocrine pancreas. Embryos were incubated in the absence (**i**) or presence (**j**) of 10 mM Met at 24 hpf for 48 h. This regimen of ablation exposes the embryos to Met 8 h prior to the onset of *ptf1a* expression in the pancreatic tissue and maintains ongoing ablation until 72 hpf. Embryos at 72 hpf were processed for whole mount *in situ* hybridization for *trypsin*, and photographed from the ventral side through the yolk (**i** and **j**). Compared with controls (**i**), embryos treated with Met (**j**) exhibit a complete absence of *trypsin* expression. This confirms the failure in the exocrine pancreas formation.

Table 1
Transgenic Lines Expressing Nitroreductase (NTR) and Used in Metronidazole-(Met)-Mediated Cell Ablation Experiments

Transgenic fish	Tissues expressing NTR	Method of cell death detection	References
Tg(ptf1a:Gal4-VP16);Tg (UAS:nfsB-mCherry)	Hindbrain, amacrine cells of retina, and exocrine pancreas	TUNEL	
Tg(ins:nfsB-mCherry) and *Tg(ins:CFP-NTR)*^{s892}	Pancreatic β-cells	Nuclear fragmentation using confocal microscopy	*17*
		Activated caspase-3 immunostaining	*16*
Tg(Gal4-VP16;UAS:eGFP)^{c223}; *Tg(UAS:nfsB-mCherry)*	Floor plate	Reduction in cellular *mCherry* fluorescence and nuclear fragmentation using confocal microscopy	*18*
Tg(Gal4-VP16;UAS:eGFP)^{c230}; *Tg(UAS:nfsB-mCherry)*	Notochord	Reduction in cellular *mCherry* fluorescence and nuclear fragmentation using confocal microscopy	*18*
Tg(cmlc2:CFP-NTR)^{s890}	Cardiomyocytes	TUNEL and activated caspase-3 immunostaining	*16*
Tg(l-fabp:CFP-NTR)^{s891}	Hepatocytes	TUNEL	*16*
Tg(crestin:Gal4); *Tg(UAS:nfsB-mCherry)*	Dorsal peridermal cells	Cells failed to ablate	Yi-Yen Chen personal communication
Tg(Gal4-VP16;UAS:eGFP)^{c223}; *Tg (UAS:nfsB-mCherry)*	Pineal cells	Cells failed to ablate	*18*

ablation, using the NTR/Met system, are shown in **Table 1** *(16–18)*. Included in Table 1 are the methods used in each experiment to validate successful cell ablation.

The protocol used for TUNEL is described below and typical results are shown in **Fig. 1e–h**.

1. Fix embryos, dehydrate and then rehydrate in the same manner as carried out for whole mount *in situ* hybridization (WMISH).
2. Incubate embryos in proteinase K as per WMISH protocols. Proteinase K activity varies between batches and suitable treatment needs to be determined empirically. The **Table 2** below can be used as a guide to decide concentration for embryos/larvae of different ages. Digest for 25 min at room

Table 2

Age in hpf	Proteinase-K (μg/mL)
24	0
36	0.5
48	5
72	20

temperature and follow by three 5-min washes in PBT. Post fix in 4% PFA for 20 min at room temperature and then wash in PBT for 5 min, three times.

3. Post fix embryos by incubation in prechilled (–20°C) ethanol:glacial acetic acid (2:1) for 10 min at –20°C followed by three washes for 5 min in PBT at room temperature.
4. Incubate embryos for 1 h at room temperature in equilibration buffer. Remove as much equilibration buffer as possible and replace with working strength TdT enzyme mix. This mix is made by diluting TdT enzyme in reaction buffer (1:2) and then adding Triton X-100 to give a final concentration of 0.3%. Only very small volumes of this reaction mix need be made (e.g., 20 μL for five 48 hpf embryos in a 1.5-mL Eppendorf tube). Equilibration buffer, reaction buffer and TdT enzyme are supplied with the ApopTag Peroxidase *In Situ* Apoptosis Detection Kit. Incubate embryos overnight at 37°C.
5. Stop the reaction by washing the embryos in stop/wash buffer for 3 h at 37°C. Prepare this buffer by mixing 1 mL of stop/wash buffer (supplied with the ApopTag Peroxidase In Situ Apoptosis Detection Kit) with 17 mL of distilled water. Then wash embryos three times for 5 min each in PBT.
6. Block the embryos in goat serum blocking solution (PBTS) for 1 h at room temperature.
7. Incubate blocked embryos in 1:2,000 dilution anti-digoxigenin antibody-AP made in PBTS for 2 h at room temperature. Remove the antibody and wash for 15 min at room temperature in PBT. Repeat wash four times.
8. Wash embryos three times for 15-min each in freshly prepared NTMT buffer.
9. Make detection reagent fresh by mixing 3.375 μL of NBT stock solution with 3.5 μL of BCIP stock solution/mL of NTMT buffer.
10. Place embryos in the detection reagent. Perform the reaction in the dark at room temperature. Check progression of color reaction under a dissection stereomicroscope, after 15 min. Reaction time is normally fast taking only 15 min.

11. Stop the reaction by washing in PBT five times for 5-min each and then fix in 4% PFA overnight at 4°C. Remove the PFA by washing in PBT three times for 5-min each.
12. Clear the embryos by incubation in graded series of 25%, 50%, and 75% glycerol in distilled water. Finally, replace the 75% glycerol with 100% glycerol.
13. Image embryos in 100% glycerol (we use the ×1.6 objective of a Stemi SV11 [Zeiss] stereomicroscope). Collect digital images and manipulate using software such as Zeiss Axio Vision (Release 4.6.3). To create images of TUNEL positive nuclear present at different focal planes the extended focus applications of Axio Vision can be used to render a single image (**Fig. 1e, f**).

In our work on ablating the pancreatic β cells, we were unable to visualize TUNEL positive cells *(17)*. Apoptosis is a rapid process *(27)* and cells dying in this manner can only be detected by TUNEL transiently. In addition, many cells undergoing apoptosis do not undergo DNA fragmentation *(28)*. Hence, it can be difficult to observe apoptosis on small targeted cell populations. Alternative methods such as *in situ* detection of activated Caspase have been used successfully to detect the apoptotic process *(16, 29)*.

3.4. Stereomicroscopy

When necessary, anesthetize embryos/larvae by sequentially adding drops of tricaine stock into the E3 media until the embryos/larvae cease moving. As a reference, if embryos are in a 100-mm plate in 20 mL of E3, 3–5 drops (100–175 μL) of tricaine stock solution is adequate. If immobilization is required, suspend anesthetized subject in 3% methyl cellulose rather than E3. Collect and manipulate digital images using software such as Zeiss Axio Vision.

3.5. Confocal Microscopy

Visualize mounted specimens using a confocal microscope (e.g., Axiovert 200M microscope coupled to a Zeiss LSM 5 Pascal system). We use a Plan-Apochromat ×20 lens for widefield or Plan-NeoFluar ×40/1.3 Oil DIC objective for confocal sections. Red channel excitation uses a He/Ne laser and emission detection uses a LP 560 filter. Collection is controlled through the single channel-track mode in Zeiss AIM software.

4. Notes

1. We do not use methylene blue in the embryo medium since this dye increases the background fluorescence during image acquisitions.

2. It has been observed that *Gal4*-VP16 transactivation of *UAS* regulated genes can be variable and lead to mosaic expression *(18)*. Therefore, it is necessary to fluorescently tag any gene being expressed using the *Gal4/UAS* system.
3. Different transgenic systems may lead to different levels of NTR produced. We recommend titrating levels of Met to cause ablation in each new study, using the 10 mM concentration as a reference point.
4. Optimal ablation of cells deep within the embryo, like the pancreatic β cells, requires incubation in Met for at least 24 h. In our work on ablating the pancreatic β cells, the *ins:nfs-mCherry* embryos start to express NTR-mCherry fusion protein around 24 hpf *(17)*. By incubating these embryos in Met from 24 to 48 hpf, we were able to witness ongoing ablation throughout the period of this incubation *(17)*.
5. Not all cell types examined to date appear susceptible to NTR-mediated ablation (*see* Table 1). This resistance maybe due to cell specific metabolism or absorption of prodrug. It remains a possibility that other cell-types will also be resistant to NTR mediated ablation.

Acknowledgments

The authors wish to acknowledge Steve Leach for critical reading of the manuscript and Matthew Knabel for expert technical support. HP was supported by NIH NCRR T32 Grant 07002. MJP is supported in part by funding from the Juvenile Diabetes Research Foundation.

References

1. Yang, C. T., and Johnson, S. L. (2006). Small molecule-induced ablation and subsequent regeneration of larval zebrafish melanocytes, *Development* **133**, 3563–3573.
2. Borrelli, E., Heyman, R., Hsi, M., and Evans, R. M. (1988) Targeting of an inducible toxic phenotype in animal cells, *Proc Natl Acad Sci U S A* **85**, 7572–7576.
3. Clark, A. J., Iwobi, M., Cui, W., Crompton, M., Harold, G., Hobbs, S., Kamalati, T., Knox, R., Neil, C., Yull, F., and Gusterson, B. (1997). Selective cell ablation in transgenic mice expression E. *coli nitroreductase, Gene Ther* **4**, 101–110.
4. Drabek, D., Guy, J., Craig, R., and Grosveld, F. (1997). The expression of bacterial nitroreductase in transgenic mice results in specific cell killing by the prodrug CB1954, *Gene Ther* **4**, 93–100.
5. Smith, S. J., Kotecha, S., Towers, N., and Mohun, T. J. (2007). Targeted cell-ablation in Xenopus embryos using the conditional, *toxic viral protein M2(H37A), Dev Dyn* **236**, 2159–2171.
6. Poss, K. D., Keating, M. T., and Nechiporuk, A. (2003). Tales of regeneration in zebrafish, *Dev Dyn* **226**, 202–210.

7. Hu, L., Yu, C., Jiang, Y., Han, J., Li, Z., Browne, P., Race, P. R., Knox, R. J., Searle, P. F., and Hyde, E. I. (2003). Nitroaryl phosphoramides as novel prodrugs for E. coli nitroreductase activation in enzyme prodrug therapy, *J Med Chem* **46**, 4818–4821.
8. Johansson, E., Parkinson, G. N., Denny, W. A., and Neidle, S. (2003). Studies on the nitroreductase prodrug-activating system. Crystal structures of complexes with the inhibitor dicoumarol and dinitrobenzamide prodrugs and of the enzyme active form, *J Med Chem* **46**, 4009–4020.
9. Bridgewater, J. A., Springer, C. J., Knox, R. J., Minton, N. P., Michael, N. P., and Collins, M. K. (1995). Expression of the bacterial nitroreductase enzyme in mammalian cells renders them selectively sensitive to killing by the prodrug CB1954, *Eur J Cancer* **31A**, 2362–2370.
10. Cui, W., Gusterson, B., and Clark, A. J. (1999). Nitroreductase-mediated cell ablation is very rapid and mediated by a p53-independent apoptotic pathway, *Gene Ther* **6**, 764–770.
11. Cui, W., Allen, N. D., Skynner, M., Gusterson, B., and Clark, A. J. (2001). Inducible ablation of astrocytes shows that these cells are required for neuronal survival in the adult brain, *Glia* **34**, 272–282.
12. Isles, A. R., Ma, D., Milsom, C., Skynner, M. J., Cui, W., Clark, J., Keverne, E. B., and Allen, N. D. (2001). Conditional ablation of neurones in transgenic mice, *J Neurobiol* **47**, 183–193.
13. Felmer, R., Cui, W., and Clark, A. J. (2002). Inducible ablation of adipocytes in adult transgenic mice expressing the E. coli nitroreductase gene, *J Endocrinol* **175**, 487–498.
14. Kwak, S. P., Malberg, J. E., Howland, D. S., Cheng, K. Y., Su, J., She, Y., Fennell, M., and Ghavami, A. (2007). Ablation of central nervous system progenitor cells in transgenic rats using bacterial nitroreductase system, *J Neurosci Res* **85**, 1183–1193.
15. Wang, X. D., Shou, J., Wong, P., French, D. M., and Gao, W. Q. (2004). Notch1-expressing cells are indispensable for prostatic branching morphogenesis during development and re-growth following castration and androgen replacement, *J Biol Chem* **279**, 24733–24744.
16. Curado, S., Anderson, R. M., Jungblut, B., Mumm, J., Schroeter, E., and Stainier, D. Y. (2007). Conditional targeted cell ablation in zebrafish: a new tool for regeneration studies, *Dev Dyn* **236**, 1025–1035.
17. Pisharath, H., Rhee, J. M., Swanson, M. A., Leach, S. D., and Parsons, M. J. (2007). Targeted ablation of beta cells in the embryonic zebrafish pancreas using E. *coli nitroreductase, Mech Dev* **124**, 218–229.
18. Davison, J. M., Akitake, C. M., Goll, M. G., Rhee, J. M., Gosse, N., Baier, H., Halpern, M. E., Leach, S. D., and Parsons, M. J. (2007). Transactivation from Gal4-VP16 transgenic insertions for tissue-specific cell labeling and ablation in zebrafish, *Dev Biol* **304**, 811–824.
19. Fischer, J. A., Giniger, E., Maniatis, T., and Ptashne, M. (1988). GAL4 activates transcription in Drosophila, *Nature* **332**, 853–856.
20. Westerfield, M. (1993) The Zebrafish Book, University of Oregon, OR.
21. Koster, R. W., and Fraser, S. E. (2001). Tracing transgene expression in living zebrafish embryos, *Dev Biol* **233**, 329–346.
22. Scheer, N., and Campos-Ortega, J. A. (1999). Use of the Gal4-UAS technique for targeted gene expression in the zebrafish, *Mech Dev* **80**, 153–158.
23. Kawakami, K., and Shima, A. (1999). Identification of the Tol2 transposase of the medaka fish Oryzias latipes that catalyzes excision of a nonautonomous Tol2 element in zebrafish Danio rerio, *Gene* **240**, 239–244.
24. Kawakami, K., Shima, A., and Kawakami, N. (2000). Identification of a functional transposase of the Tol2 element, an Ac-like element from the Japanese medaka fish, and its transposition in the zebrafish germ lineage, *Proc Natl Acad Sci U S A* **97**, 11403–11408.
25. Brand, A. H., and Perrimon, N. (1993). Targeted gene expression as a means of altering cell fates and generating dominant phenotypes, *Development* **118**, 401–415.
26. Duffy, J. B. (2002). GAL4 system in Drosophila: a fly geneticist's Swiss army knife, *Genesis* **34**, 1–15.
27. O'Brien, B. A., Harmon, B. V., Cameron, D. P., and Allan, D. J. (1996). Beta-cell apoptosis is responsible for the development of IDDM in the multiple low-dose streptozotocin model, *J Pathol* **178**, 176–181.
28. Danial, N. N., and Korsmeyer, S. J. (2004). Cell death: critical control points, *Cell* **116**, 205–219.
29. Hughes, J., and Gobe, G. (2007). Identification and quantification of apoptosis in the kidney using morphology, biochemical and molecular markers, *Nephrology (Carlton)* **12**, 452–458.

Chapter 10

Focal Electroporation in Zebrafish Embryos and Larvae

Marcel Tawk, Isaac H. Bianco, and Jonathan D.W. Clarke

Summary

A method is described that allows the introduction by electroporation of either small dyes or larger RNA, DNA, or morpholino constructs into single cells or small groups of cells in zebrafish embryos or larvae. The dye or construct is delivered to cells via a patch-like microelectrode that also delivers the electroporation stimulus train. This technique allows the experimenter to target cells of their choice at a particular time of development and at a particular location in the embryo, and is useful for fate mapping, analysing neuronal organisation, ectopic expression and gene knockdown experiments.

Key words: Zebrafish, Electroporation, DNA, RNA, Fluorescent dextran, Morpholino.

1. Introduction

Manipulating gene function either by overexpression or by morpholino knockdown in zebrafish embryos has until recently been limited by the necessity to introduce constructs by microinjection into the very early embryo, often at the one or two cell stage. This results in essentially ubiquitous expression or knockdown from the outset of development. To obtain a more mosaic but still widespread distribution, individual cells at the 64–128 cell stages can be microinjected, but this approach does not allow the experimenter to target small patches of cells in particular locations at particular timepoints of development. Here we describe a method of focal electroporation that allows targeted delivery of DNA, RNA, charged morpholinos, or cell tracking dyes into very small numbers of cells or indeed single cells at any time from

Graham J. Lieschke et al. (eds.), *Zebrafish,* Methods in Molecular Biology, vol. 546
DOI: 10.1007/978-1-60327-977-2_10

12-h post-fertilization through at least 5 days post-fertilization. Targeting individual cells or small groups of cells is achieved by restricting both the construct or dye to be electroporated and the electric field required for electroporation to the 1–3 μm tip of a glass microelectrode. The principal uses of this technique are to allow experimenters to ectopically express genes or RNAs in precise locations and at particular times of development *(1)*, to examine cell morphology (neuronal morphology and axonal projections in particular), and to fate map the development of particular cell populations *(2)*. This technique was originally devised for use in *Xenopus* larvae and cultured tissue slices by the Cline lab in Cold Spring Harbor *(3)*. Very few minor modifications have been made for its application in zebrafish. This technique has been successfully used on embryos from the ten somite stage onwards.

2. Materials

2.1. Instrumentation

The physical arrangement of the experimental instrumentation is diagrammed in **Fig. 1**.

1. A fixed stage microscope (i.e., one that focuses by moving the objective lens rather than the specimen platform such as the Axioskop 2FS [Zeiss], with epifluorescent attachment). A ×10 or ×20 objective lens with a long working distance of about 0.5–1 cm is needed. Water immersion objectives are best, but dry lenses can also be used.
2. A fine resolution micromanipulator (e.g., MP-85 [Sutter Instruments Co.]). It is helpful to have one that has a fine axial drive on it (i.e., a drive that moves the electrode along the axis of the electrode).

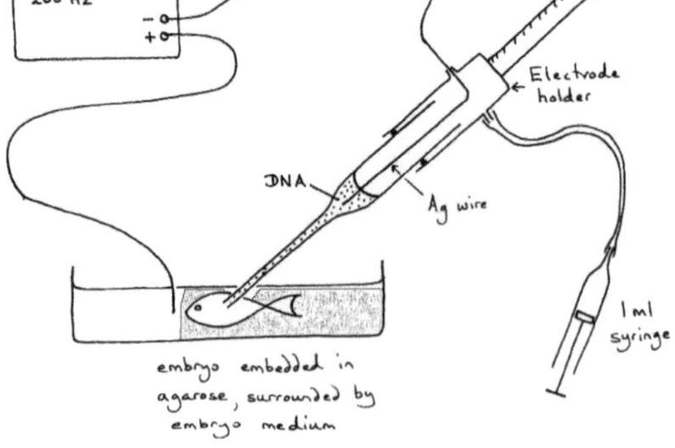

Fig. 1. Basic circuit diagram and experimental set-up for focal electroporation.

3. An anti-vibration table or equivalent anti-vibration device, for example a VFP or VFW (World Precision Instruments).
4. A electronic stimulator capable of delivering 2 ms pulses of between 1 and 50 V at a frequency of 200 Hz. We used a SD9 Square Pulse Stimulator (Grass Technologies).
5. Flaming/Brown micropipet puller P-97 (Sutter Instruments Co.).

2.2. Small Equipment

1. 1.2-mm diameter, thin-walled borosilicate glass pipets with internal filament GC120TF-10 (Harvard Apparatus).
2. A microelectrode holder with silver wire connector and optional side port for air pressure MEH7W (World Precision Instruments).
3. Sharp tungsten needle.
4. Silver wire.
5. 1-mL plastic syringe.
6. Plastic tubing.
7. Injection chamber slide (**Fig. 2**). This is most easily made from a standard glass microscope slide onto which a low solid rectangular wall of epoxy resin with approximately 1 × 2 cm dimensions has been laid. The agarose gel containing the embryo should occupy one end of the chamber and be in contact with the chamber walls to stop it floating off the glass base. The other end of the chamber should be free of agarose and serves as a space into which the bath electrode is placed.

2.3. Reagents and Solutions

1. E3 embryo medium, for 10 L of 60× (dilute in deionized water): 175 g NaCl (5 mM), 7.6 g KCl (0.17 mM), 29 g $CaCl_2 \times 6H_2O$ (0.33 mM), 49 g $MgSO_4 \times 7H_2O$ (0.33 mM).
2. Anaesthetic: tricaine methanesulfonate 0.03% *(w/v)* in E3 (cat. no. MS-222, Sigma).
3. Low melting point agarose.
4. DNA construct of interest containing fluorescent reporter gene (green flurescent protein [GFP], red fluorescent protein [RFP]) at concentration of 1 μg/μL.

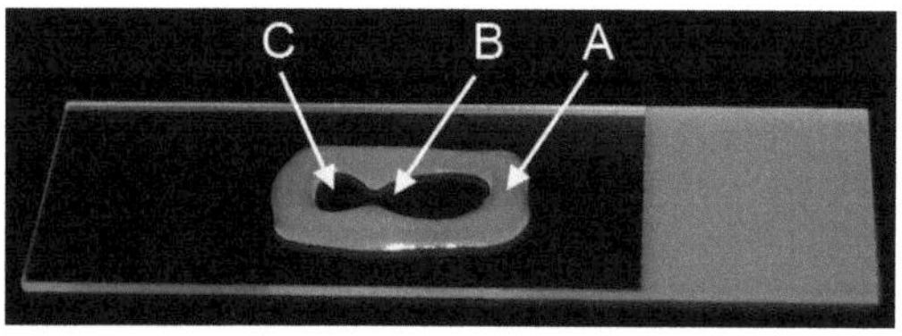

Fig. 2. The empty electroporation chamber slide.

5. Fluorescent tetramethylrhodamine and fluorescein dextran-coupled dyes diluted to 100 mg/mL in H_2O (cat no. D-3308 and D-3306, Molecular Probes, respectively).
6. Fluorescein-coupled anti-sense morpholino oligonucleotide of interest at concentration of 1 mM (GeneTools).

3. Methods

3.1. Preparation of Embryos or Larvae

1. Anaesthetize embryos and larvae in MS-222 and then embed in 1.5% low melting point agarose in electroporation chamber slide (*see* **Subheading 2.2**, **step 7**, **Fig. 2**, and **Note 1**).
2. Orient the embryo or larvae so that the tissue to be electroporated is most superficial in the agarose.
3. After agarose has set, add a small volume of embryo medium or clean tank water to the chamber until it just covers the agarose.
4. Carefully remove a small amount of the agarose immediately above the targeted tissue with a fine tungsten needle to allow easy access of the electroporation electrode to the tissue. If too much agarose is removed there is a danger the embryo will not be held sufficiently firmly by the agarose and is then likely to move when penetrated by the electrode.

3.2. Preparation of Microelectrodes

1. Pull a 1.2-mm diameter, thin-walled borosilicate glass with internal filament to form the electrode tip with a constant taper down to about 1–3 μm. The length of the taper of the electrode should be no more than 5 mm; the electrodes therefore look rather short and stubby at the business end (*see* **Note 2**).
2. Backfill the tip of the electrode with 1–2 μL of purified plasmid DNA (use a concentration of 1 μg/μL in distilled water) using a long thin microloader pipet tip (*see* **Note 3**).
3. Secure microelectrode into the electrode holder on the micromanipulator. Ensure the silver wire is in contact with the DNA solution. Keep this silver wire clean by gently scraping it with edge of a glass slide or emery paper. This should be performed at the end of every session to get rid of adherent debris.
4. Apply a tiny bit of positive air pressure to back of pipet via a 1-mL syringe attached to the side port of the microelectrode holder (move syringe plunger just 0.1 mL) (*see* **Note 4**).

5. Attach back of microelectrode to negative pole on stimulator. Set stimulator to deliver 2 ms square pulses of approximately 20–50 V (for DNA) or 2–5 V (for fluorescent dextran) at a rate of 200 Hz.
6. Attach the other pole of stimulator to another silver wire that is placed in the embryo medium in contact with the agarose embedded embryo (*see* **Note 5**).

3.3. Electroporation of Larva

1. Manipulate the DNA-containing electrode tip through the embryo skin and into the target tissue (*see* **Note 6**). Make sure the optics of your microscope are good enough to see precisely where the electrode is, so you can accurately target the cells of choice (*see* **Note 7**). Piercing the embryo skin with the electrode can be difficult as the skin is very elastic, especially at early stages. For this reason, and to minimize damage to the target tissue, it is best to penetrate the skin at some distance from the cells to be electroporated. Quick movements of the electrode will penetrate the skin better than slow ones. A sharp tungsten needle can be used to make a hole in the skin before trying to place the electrode into the tissue.
2. When confident that the electrode tip is next to the cells selected for electroporation, give two or three trains of pulses that last for about 1-s each. The pulse trains should run at 200 Hz, each pulse should be 2 ms long and be approximately 10–50 V to deliver DNA or RNA (*see* **Note 8**).
3. Observe the cells next to the electrode tip as you apply the pulses. One should see a "tip reaction" (i.e., as the cells react to the electric field they temporarily appear slightly more opaque and sometimes a small bolus of DNA solution can be seen leaving the tip of the electrode). If a tip reaction is not seen, it is likely the electric circuit has not been completed and all connections should be checked.
4. Wait a couple of minutes before carefully withdrawing the electrode from the cells and tissue. If the electrode is moved too quickly after the electroporation, very often cells will adhere to the electrode tip and be removed from the tissue, but this is much less likely to happen if the electrode is left in place for a few minutes.
5. Carefully remove embryo or larva using fine forceps to cut through the agarose and allow it to recover in fresh media or tank water. Following electroporation of DNA, it takes a few hours before RNA or protein is expressed by the cells (**Fig. 3**) (*see* **Note 7**).

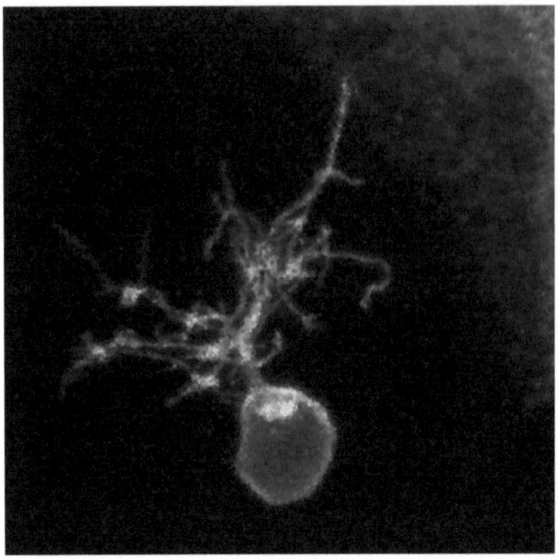

Fig. 3. Single forebrain neuron expressing membrane GFP following electroporation 10 h earlier.

4. Notes

1. Electroporating DNA or RNA in embryos of 24 h or older should give a success rate of about 50–70%. For embryos younger than 24 h we get a lower success rate. We do not fully understand why the technique is less efficient in younger embryos but guess that the lower density of cell packing may allow DNA or RNA constructs to move more easily through the extracellular space rather than entering cells.
2. Use a new electrode at least every three embryos. Cell debris accumulates on the electrode tip and this inhibits movement of the electrode through the tissue and injection of solutions through the tip. The best electrode is a new one. Poor electrode design and shape are the most likely reason for lack of success. It is worth measuring the tip size accurately. To do so, a dry ×40 objective lens will be needed. Larger electrode tips (say 3 μm) will electroporate more cells (~5–10 cells). It is possible to move the electrode around in the tissue and try to hit more cells by doing multiple electroporations, but this is not the strength of this particular technique.
3. By mixing two plasmids into the electrode it is possible to introduce two independent constructs into the same cells.
4. This ensures the DNA is right at the electrode tip, but if the electrode holder does not have a side port, do not worry, as this does not appear to be a crucial part of the protocol.

5. The best way to do this is to move the bath electrode around using a second micromanipulator, but one can just hook this bath electrode into the slide chamber if this can be done in a secure way. It may help to arrange the bath electrode so that the target cells will lie directly between the glass electrode and the bath electrode. This will align the electric fields so they pass through the target cells.
6. Vibration must be eliminated. Make sure the microscope and manipulator are solidly attached to the base-plate and that the manipulator only moves when intended to.
7. Although this protocol suggests using a fixed stage compound microscope with relatively high magnification objectives, if the precise location of the electroporation is not so important, then this technique can also be used with a dissecting microscope. In this case, one is unlikely to see the exact location of the electrode tip or be able to monitor the tip reaction at the time of stimulation, but successful electroporations are none the less possible.
8. This method can also be used to label cells with a fluorescent dye such as dextran labeled with tetramethylrhodamine or fluorescein. Use concentrations of 100 mg/mL in distilled water. These dyes will require a much lower voltage than DNA or RNA, requiring only between 2 and 5 V. The same stimulus regime applies. If the electroporation is performed on a fluorescence microscope, the success of the cell labeling can be monitored immediately. Some fluorescent dextrans will carry the opposite electric charge to DNA and so will require the opposite polarity of pulses from the stimulator. This technique will also load charged morpholinos (available from GeneTools) into small groups of cells and is thus theoretically useful for localized knockdown experiments. We have not pursued this approach beyond establishing that fluorescent morpholinos can enter cells using this technique; one experimental difficulty of this approach would be estimating the concentration of morpholino received by the cells.

References

1. Concha, M.L., Russell, C., Regan, J.C., Tawk, M., Sidi, S., Gilmour, D.T., Kapsimali, M., Sumoy, L., Goldstone, K., Amaya, E., Kimelman, D., Nicolson, T., Grunder, S., Gomperts, M., Clarke, J.D.W. and Wilson, S.W. (2003). Local tissue interactions across the dorsal midline of the forebrain establish CNS laterality. *Neuron* **39**, 423–438.
2. Crump, J.G., Swartz, M.E., Eberhart, J.K. and Kimmel, C.B. (2006). Moz-dependent Hox expression controls segment-specific fate maps of skeletal precursors in the face. *Development* **133**, 2661–2699.
3. Haas, K., Sin, W-C., Javaherian, A., Zheng, L. and Cline, H.T. (2001). Single cell electroporation for gene transfer in vivo. *Neuron* **29**, 583–591.

Chapter 11

Tissue Micromanipulation in Zebrafish Embryos

Alexander Picker, Daniela Roellig, Olivier Pourquié, Andrew C. Oates, and Michael Brand

Summary

Although a common approach in large vertebrate embryos such as chick or frog, manipulation at the tissue level is only rarely applied to zebrafish embryos. Despite its relatively small size, the zebrafish embryo can be readily used for micromanipulations such as tissue and organ primordium transplantation, explantation, and microbead implantation, to study inductive tissue interactions and tissue autonomy of pleiotropic, mutant phenotypes or to isolate tissue for organotypic and primary cell culture or RNA isolation. Since this requires special handling techniques, tools, and tricks, which are rarely published and thus difficult to apply without hands-on demonstration, this article provides detailed instructions and protocols on tissue micromanipulation. The goal is to introduce a broader scientific audience to these surgical techniques, which can be applied to a wide range of questions and used as a starting point for many downstream applications in the genetically tractable zebrafish embryo.

Key words: Zebrafish, *Danio rerio*, Transplantation, Bead implantation, Presomitic mesoderm, Optic vesicle, Dissection, Micromanipulation.

1. Introduction

The classical experimental approach to study inductive cellular interactions in the vertebrate embryo is to transplant tissue. If a tissue comprising cells with strong inductive potency is transplanted from a donor embryo to an ectopic site in a host embryo, creating a chimera composed of tissue from two individuals, it can, depending on the competence state of the host and developmental timing of the transplantation, change the fate of neighboring cells and tissue. The ectopic transplantation of dorsal blastopore lip tissue and secondary axis induction in the amphibian

Graham J. Lieschke et al. (eds.), *Zebrafish,* Methods in Molecular Biology, vol. 546
DOI: 10.1007/978-1-60327-977-2_11

embryo by Hans Spemann and Hilde Mangold in the 1920s is the famous prototype of these experiments *(1)*. Transplantations can also serve to test other fundamental properties of developing tissue, e.g., commitment and competence. Can transplanted tissue autonomously develop according to the position and fate it had in the donor, or is it recruited into the developing host and subsequently changes its fate in a host-dependent respecification process? A similar, but even further reaching question asks what degree of self-organization does morphogenesis and differentiation of a tissue or whole organ primordium show upon explantation from its normal developmental site. What are the extrinsic and intrinsic contributions to these processes and what is their timing? All these questions have classically been addressed by tissue micromanipulation in the embryo.

With the advances in biochemical, molecular biological, and genetic approaches to animal development, a vast canon of methods to study inductive tissue properties have emerged. Nonetheless, experimental manipulation of embryonic tissue is not an outdated approach but a powerful technical complement to genetic and molecular approaches: the protein microbead implantation procedure described in this article is a technique at the interface between biochemistry and tissue micromanipulation. On one hand, it requires purified proteins to be tested for their activity; on the other, it applies experimental techniques that were developed long before the discovery of inductive molecules. Bead implantations have proven to be a powerful tool to study the inductive nature of proteins, dosage effects, tissue responsiveness, and timing *(2–6)*. The optic vesicle transplantation technique described in this article was originally developed to dissect the tissue autonomy of pleiotropic phenotypes of zebrafish mutants isolated in large-scale, forward genetic screens *(3, 7–11)*.

Although performed routinely in chicken and frog embryos, tissue micromanipulation is rarely done in the zebrafish embryo, probably due to its relatively small size. Nevertheless, features such as the complete transparency and the easy access to large numbers of embryos and genetic mutants available make the zebrafish an ideal model for tissue manipulation. The focus of this article is *manual* micromanipulation techniques that can be applied to zebrafish embryos despite their small size: the surgical methods described here are all performed freehanded, under a conventional stereomicroscope with transillumination. Although they require very steady hands, they can be performed without sophisticated micromanipulators, which clearly enhances throughput. Protocols describing cell transplantation and injection with the aid of micromanipulators have been previously published *(12–14)*. Importantly, the concrete experimental procedures outlined here exemplify the principles and key techniques of tissue micromanipulation in zebrafish embryos in such a way that enable scientists to vary and adopt the protocols to their particular need and application.

2. Materials

2.1. Media and Reagents

1	Ringer medium	116 mM NaCl, 3 mM KCl, 4 mM $CaCl_2$ anhydrous, 1 mM $MgCl_2$, 5 mM HEPES.
2	E3 medium	10 L of 60 × E3 (dilute in deionized water): 175-g NaCl (5 mM), 7.6-g KCl (0.17 mM), 29-g $CaCl_2 \cdot 6H_2O$ (0.33 mM), 49-g $MgSO_4 \cdot 7H_2O$ (0.33 mM).
3	L15 medium	Leibovitz's L-15 medium (1×), liquid – with L-glutamine, without phenol red (Invitrogen Cat. No. 21083-027).
4	Trizol	TRIZOL® reagent, 100 mL (Invitrogen).
5	AMP-PNP, 40 mM in water	AMP-PNP adenosine 5′-(β,γ-imido) triphosphate tetralithium salt hydrate, ~95% (HPLC), powder (Sigma-Aldrich Cat. No. A2647).
6	4% Paraformaldehyde	Paraformaldehyde, reagent grade, crystalline (Sigma-Aldrich Cat. No. P6148); dilute to a working concentration of 4% in PBST.
7	Microbeads	Polybead Polystyrene 45.0-μm microspheres, 2.51% solids-latex (Polysciences, Inc. Cat. No. 07314) (*see* **Note 1**).
8	Fgf8 protein (*see* **Note 2**)	Recombinant Mouse FGF-8b, CF (R&D Systems Cat. No. 423-F8-025/CF).
9	Low melting point (LMP) agarose	LMP-agarose (Invitrogen).
10	Mineral oil	Mineral oil, mouse embryo tested, light oil (neat) (Sigma Cat. No. M8410) (*see* **Note 3**).
11	0.5 M NaCl in water.	
12	PBS.	
13	Ethanol.	
14	Electrophoresis-grade agarose.	

2.2. Instrumentation

1	Micropipette puller	e.g., Flaming/Brown micropipette puller, Model P-97 (Sutter Instruments Co.)
2	Water bath, 42°C	
3	Standard laboratory power supply with cables and alligator clip or a 9-V block battery	
4	Benchtop centrifuge	
5	Heating block, 50°C	
6	Bunsen burner	

2.3. Other Materials

1	Borosilicate glass tubing	World Precision Instruments Cat. No. TW100F-3
2	Microelectrode holder with a male Luer port	World Precision Instruments Cat. No. MPH3
3	Handle for microelectrode holder	World Precision Instruments Cat. No. 2505
4	Pasteur pipettes, glass	
5	Plastic pipettes	
6	Micropipette set	
7	0.5-mL reaction tubes, siliconized	
8	1.5-mL reaction tubes	
9	15-mL Falcon tubes	
10	Diamond glass knife	
11	Superglue	
12	Dumont forceps	
13	Tungsten wire	e.g., a set of 5, 20, 100, and 200-μm diameter wires (*see* **Note 4**)
14	Blunt needle on holder	e.g., a sewing needle
15	27G needles	
16	Petri dishes, 35 mm	
17	Petri dishes, 35 mm, coated with silicone elastomer (*see* **Note 5**)	
18	Scalpel, pointed blade	(World Precision Instruments)
19	Plastic tubing (*see* **Note 6**)	(World Precision Instruments)

3. Methods

3.1. Immobilizing Embryos in Agarose

For microbead implantation and tissue manipulation at late stages of development it is mandatory to rigidly mount the embryos (*see* **Note 7**). A protocol was developed to mount embryos at somitogenesis stages for bead implantation and manipulation of the optic vesicle.

1. Melt 1.2% Ringer–LMP agarose in a 15-mL Falcon tube using a microwave oven (the tube has to be almost full). Place the tube into a water bath at 42°C in reach of your stereomicroscope and let it "cool" there for at least 30 min.
2. Dechorionate embryos using Dumont forceps in a Petri dish (coated with 1–2% E3 agarose) containing E3 medium.
3. Place the dish with the dechorionated embryos next to your stereomicroscope.
4. Open the Falcon with LMP-agarose and place it back into the water bath.
5. Draw an embryo into a fire-polished Pasteur pipette with as little medium as possible (*see* **Notes 8** and **9**).
6. Pick up the LMP-agarose tube from the water bath; carefully pipette the embryo into it and quickly draw the embryo with agarose back into it **(Fig. 1a)**.
7. Pipette the embryo in an agarose drop to the center of a 35-mm Petri dish cover (*see* **Note 10**, **Fig. 1b**).
8. Rapidly orientate the embryo with a blunt needle and hold the embryo until the agarose solidifies.
9. Let the agarose solidify for another 2–5 min at room temperature (*see* **Note 11**).
10. Use a pointed scalpel to cut a wedge pointing toward the embryo into the agarose (*see* **Note 12**, **Fig. 1b**).
11. Follow the contours of the wedge with a blunt tungsten needle up to the embryo surface to loosen it (*see* **Note 13**, **Fig. 1c**, top).
12. Fill the dish with Ringer medium until the agarose block is fully covered.
13. Use the blunt tungsten needle to remove the agarose wedge: insert the needle under the agarose wedge and flap it upward as one piece (**Fig. 1c**, bottom).
14. Check if the embryo is fully exposed in the wedge – if not, use a blunt tungsten needle to remove any remaining agarose pieces. Now the embryo is ready for further manipulation (*see* **Note 14**).

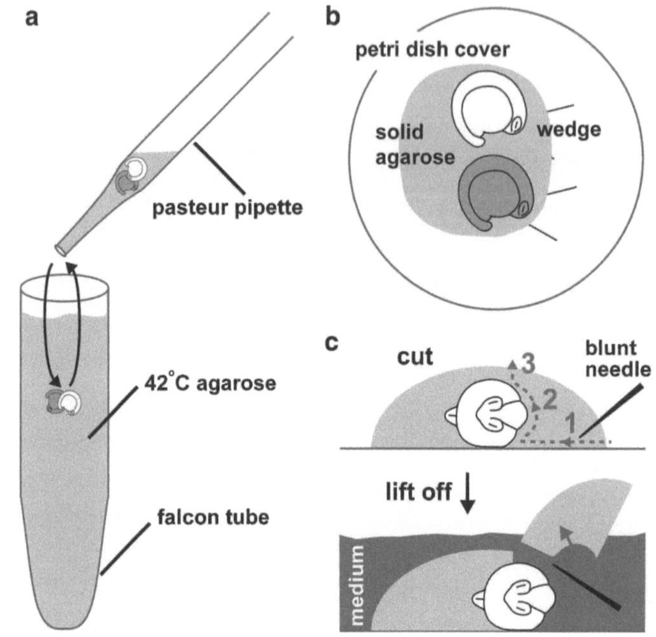

Fig. 1. Immobilizing embryos in agarose. (**a**) Transfer dechorionated donor (*red*) and host (*white*) embryos with a fire-polished Pasteur pipette into 42°C LMP agarose. (**b**) Place and orient embryos side-by-side in a Petri dish cover until agarose solidifies and cut a wedge shape into the agarose at the side of the embryos with a scalpel (*top view*). (**c**) Follow the wedge contours with a blunt tungsten needle to loosen the agarose block; cover the preparation with medium and lift off the agarose (*side view*).

This protocol can also be used for embedding embryos for in vivo imaging purposes: mount embryos (depending on developmental stage) in agarose between 0.5 and 1.5% in a dish, fill the dish with a medium, and image with a dipping lens. Try to spread the agarose in a way that only a thin film covers the embryo.

3.2. Opening the Embryonic Epidermis

To access deep manipulation sites at late stages of development, it is necessary to locally remove the primitive embryonic epidermis. The method of choice to achieve this, without destroying the embryo as a whole, is to locally destroy and remove the epidermis with mineral oil.

1. Embed the embryo in agarose (*see* **Subheading 3.1**).
2. Use the handheld micropipette (*see* **Subheading 3.6.2** and **Fig. 5**) to apply a droplet of mineral oil onto the epidermis at the site that you want to open (*see* **Notes 15** and **16** for oil application prior to optic vesicle transplantation).
3. Incubate the embryos for 3–5 min at room temperature – during that time the oil will destroy the epidermis underneath the drop (*see* **Note 17**).
4. Using a fine tungsten needle, carefully lift off the destroyed patch of epidermis (*see* **Note 18**). Now the embryo is ready for further manipulation (*see* **Note 19**).

3.3. Microbead Implantation

This section covers a method of microbead implantation into zebrafish embryos. It has proven to be a very useful technique, e.g., to deliver inductive proteins, with high spatial and temporal resolution and control over concentration. The procedure differs for embryonic stages before and after gastrulation – therefore, two different protocols are outlined. The method is exemplified by the implantation of Fgf8-protein beads (a) at the animal pole of embryos at the 50% epiboly stage (b) and into the eye primordium at the 15-somite stage.

3.3.1. Implantation into Embryos at Blastula Stages

1. Dechorionate embryos using Dumont forceps in a Petri dish (coated with 2% E3 agarose) containing E3 medium.
2. Transfer embryos into a premade mold of 2% E3 agarose (**Fig. 2a**, *see* **Note 20**) using a fire-polished Pasteur pipette (*see* **Note 21**).
3. Once in the mold, orient the embryos with the desired implantation site facing upward, using a blunt needle.

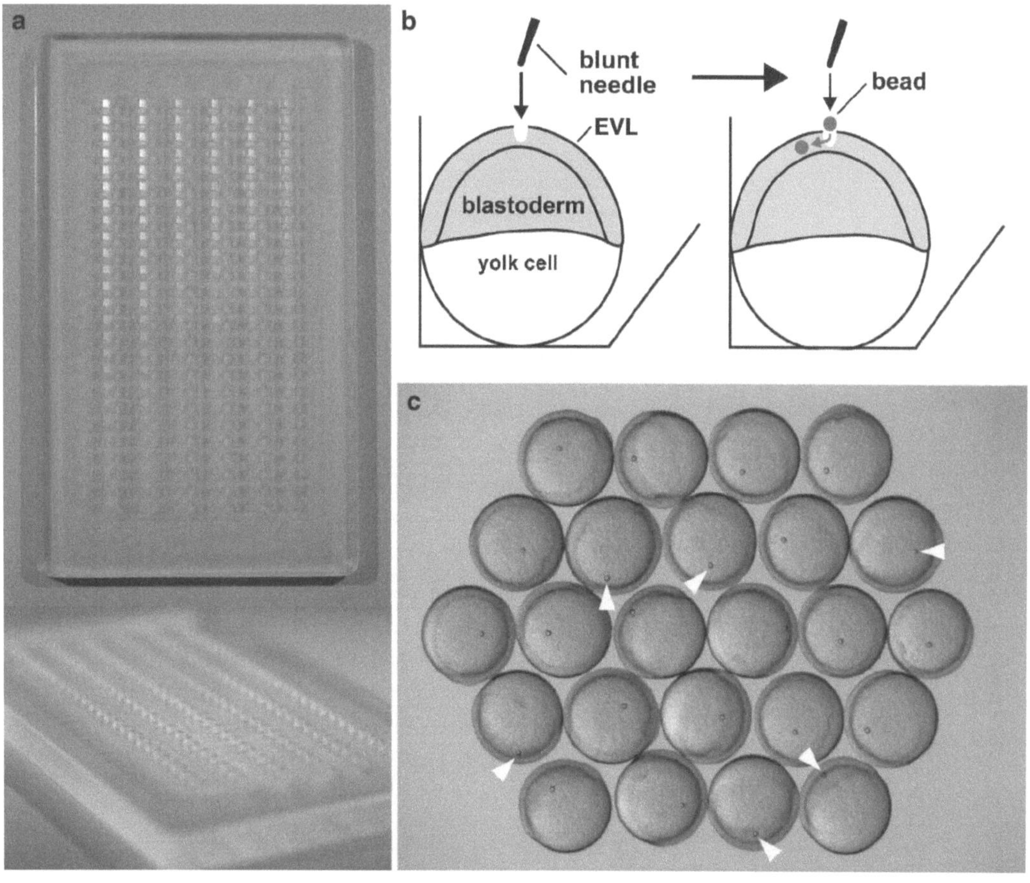

Fig. 2. Microbead implantation into blastula-stage embryos. (**a**) Plastic form for casting agarose molds [*see (11)* for details]. (**b**) Opening of the embryos with a blunt tungsten needle and bead implantation. (**c**) Live embryos, 1 h after bead implantation at the 50% epiboly stage.

4. Resuspend beads in the tube by carefully pipetting up and down. Transfer 0.5 μL of bead suspension (*see* **Subheading 3.3.3**) slowly into the dish and let the beads sink onto the agarose bed before you continue (*see* **Note 22**).
5. Open the first embryo by gently poking it with a blunt tungsten needle. The hole has to be at least the size of the bead. If too small, try to make it bigger by gentle sideward movements with the needle inserted in the embryo (*see* **Note 23**).
6. Approach the beads lying on the agarose bed with the needle and move one bead onto the opened implantation site (*see* **Note 24**).
7. Once the bead lies on top of the opening in the embryo, gently push it in with the needle (*see* **Note 25** and **Fig. 2b, c**).
8. After implantation, transfer the embryos with a fire-polished Pasteur pipette into a fresh, agarose-coated Petri dish with E3 medium; place it in a 28.5°C incubator and let the embryos heal for 30–60 min.
9. Now, the embryos can be handled like normal, dechorionated embryos.

3.3.2. Implantation into Embryos at Somitogenesis Stages

Bead implantation into somitogenesis stage embryos requires prior agarose mounting and opening of the epidermis. The protocol was developed for implantation of Fgf8-protein-loaded microbeads into the optic vesicle between the 5- and 15-somite stages.

1. Embed embryos in agarose (*see* **Subheading 3.1**).
2. Open the embryos (*see* **Subheading 3.2** and **Note 26**).
3. Continue with **step 4** of **Subheading 3.3.1**.
4. Insert a fine tungsten needle through the hole in the epidermis and gently push the cells at the implantation site apart. This loosens the tissue connectivity and allows implanting the bead deep into the tissue.
5. Continue with **steps 6** and 7 of **Subheading 3.3.1**.
6. Continue with **steps 10** and **11** of **Subheading 3.4.1**.

3.3.3. Loading Microbeads with Protein

Microbeads can serve as carriers for various substances. Different bead diameters and materials exist. The protocol describes the principles of this method for 45-μm polystyrene microbeads (*see* **Note 27**) loaded with recombinant mouse Fgf8 protein:

1. Resuspend beads in the container that they are delivered in and transfer one drop (*see* **Note 28**) of bead suspension to 1-mL tube.
2. Add 1 mL of 100% ethanol and mix by inverting the tube a few times.
3. Split into four ~250 μL samples by transferring into new, silicon-coated (*see* **Note 29**) 500-μL tubes. Carefully resuspend the

beads before splitting to achieve an equal distribution over the four samples.

4. Pellet the beads by 5-min centrifugation in a standard bench-top centrifuge at 13,000 rpm and remove supernatant.
5. Dry the bead pellets in opened tubes for ~20 min on a 50°C heating block. All ethanol needs to evaporate before the next step. Test if the beads are dry: they should fall as white powder to the bottom of the tube, when you gently flick it.
6. Add protein solution at the desired concentration (e.g., 250 μg/mL Fgf8 in PBS, prepared according to manufacturer's guide) to the beads (*see* **Note 30**).
7. Suspend the bead pellet by pipetting up and down (*see* **Note 31**).
8. Incubate at 4°C for 24 h to let the protein solution completely soak the beads.
9. Control beads are loaded accordingly with, e.g., BSA or PBS.

3.4. Tissue Transplantation

This protocol can also be used as a starting point for explant culture or isolation of embryonic cells or as a source for tissue-specific RNA preparations (*see* **Subheading 3.5**).

3.4.1. Optic Vesicle Transplantation

The following protocol was developed for explantation/transplantation of the optic vesicle between the 5- and 20-somite stages but can be easily adapted to other developmental tissues and stages.

1. Mount two embryos (one donor and one host for the transplantation) at the 5–10-somite stages in 1.2% LMP-agarose in Ringer medium (*see* **Subheading 3.1**) (**Fig. 1**). Mount the embryos, lying on their side (right body half facing up), close to each other, with their heads down.
2. Locally remove the agarose to expose the optic vesicle to be transplanted (*see* **Subheading 3.1**), and then open the host embryo with mineral oil (*see* **Subheading 3.2**): apply only a small oil drop to the epidermis posterior to the optic vesicle.
3. Lift off the destroyed patch of epidermis with a fine tungsten needle.
4. Using a tungsten hook (*see* **Subheading 3.6.1** and **Fig. 4c**), increase the size of the opening in the epidermis if necessary: the opening needs to be a little bigger than the grafted tissue block.
5. Use the tungsten hook to cut the optic vesicle loose off the forebrain primordium (*see* **Note 32**).
6. Once the optic vesicle tissue lies loose in the embryo use the tungsten loop (*see* **Subheading 3.6.1** and **Fig. 4b**) like a spoon to take it out. Remnant host tissue can be removed by scraping the created cavity with the loop.

7. Repeat **steps 2–6** for the donor embryo (*see* **Note 33**).
8. Transfer the donor optic vesicle and place it onto the opening in the host embryo using the tungsten loop (*see* **Note 34**).
9. Gently push the graft into the cavity in the host embryo using a blunt tungsten needle.
10. Incubate the embedded embryos in Ringer medium at 28.5°C until the hole in the epidermis has completely healed.
11. Carefully remove donor and host embryos from the agarose (by cutting the embedding blocks apart with forceps) and transfer them to a new agarose-coated dish with E3 medium.

The success of tissue transplantation can be followed using fluorescently labeled embryos as donors. This allows monitoring the amount, extent, and integrity of the graft at different time points after transplantation *(7)*. An alternative protocol for optic vesicle transplantation has been published before *(11)*.

3.5. Flat-Mount Tissue Dissection

Removing an optic vesicle, as described earlier (**Subheading 3.4**), takes advantage of the natural tissue boundaries in the embryo. However, if a subsection of a tissue without naturally occurring internal boundaries or structures is required, a more aggressive technique must be used to obtain it. As an example in this section we describe the dissection of one side of the posterior part of the embryonic presomitic mesoderm (PSM). This micromanipulation was not practical using the whole embryo mounting technique described earlier, but rather uses a live flat-mounting technique after yolk cell removal *(15)*. The protocol combines the advantages of high precision in dissecting the same part of the PSM of similarly staged embryos, a significant reduction of contamination by other tissues, e.g., lateral plate mesoderm (LPM), and the ability to exert increased force upon the immobilized embryonic tissue. It can be simply adopted to dissect other parts of the embryonic body, especially mesenchymal tissue. The dissected tissue can be used for subsequent RNA isolation to study stage- and tissue-specific gene expression *(16)*.

3.5.1. "Deyolking" and Flat Mounting of Live Embryos for Dissection

To be able to flatten the embryo on a stable silicone surface the yolk cell needs to be removed. When cutting with a fine tungsten tool, this should yield just enough to stabilize the embryo and to not break the tools.

1. Transfer three to four 12–13-somite stage embryos (*see* **Note 35**) to L15 medium-covered silicone plates and dechorionate using Dumont forceps.
2. Position one embryo in a small dell in the silicone surface to stabilize it (*see* **Note 36**).
3. Inject an 8-nL bolus of AMP-PNP solution (*see* **Note 37**) into the yolk cell with a micropipette attached to a plastic tube for mouth aspirating, or the handheld micropipette (without oil)

(*see* **Subheading 3.6.2** and **Note 38**). Yolk will start to float out of the yolk cell.

4. After waiting 30–60 s for the yolk cell cortex to paralyze, extrude most of the yolk by gently streaking along the ventral side with a blunt tungsten needle (*see* **Subheading 3.6.1**).
5. Cut open the ventral yolk cell epidermis (between head and tail) by holding the embryo with a thicker tungsten needle and cutting with a fine, pointed scalpel.
6. By holding the embryo still via the head, pull off the yolk cell enveloping layer and epidermis with a slightly hooked tungsten needle (*see* **Note 39**).
7. Cut off the head of the embryo with a fine, pointed scalpel and remove the periderm from the caudal part of the embryo (*see* **Note 40**).
8. Remove the LPM using a sharp tungsten needle (**Fig. 3a**; *see* **Note 41**).

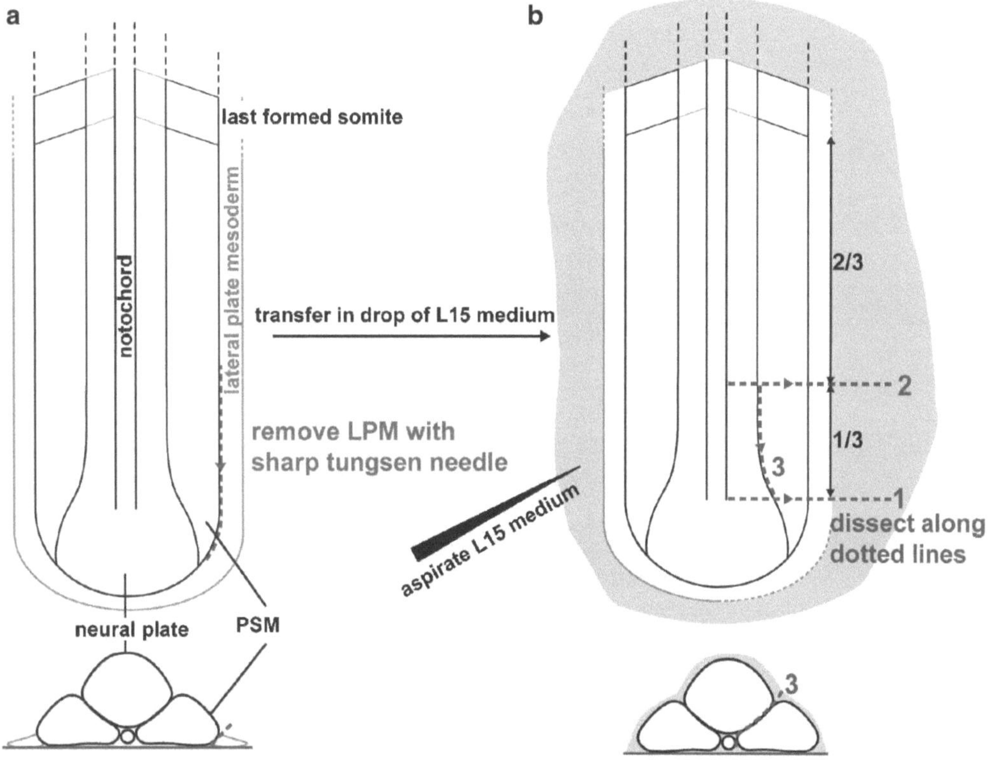

Fig. 3. Flat-mount presomitic mesoderm (PSM) dissection. (**a**) Embryo is shown in a dorsal and a cross-sectional view. In a L15 medium-containing, silicone-coated Petri dish, the deyolked and decapitated embryo is held ventral side down, and the lateral plate mesoderm is removed locally by cutting along its border with the PSM. (**b**) The embryo is transferred to a new silicone-coated Petri dish in a drop of L15 medium. While holding the embryo ventral side down with a blunt tungsten needle, the medium is aspirated until the embryo is trapped between the silicone surface and the liquid layer, as seen in cross-sectional view (*bottom*). Dissection can now be completed along the *red dotted* lines.

9. Using a fire-polished glass pipette, transfer the embryo to a fresh silicone plate in a drop of L15 medium and position the ventral side down by holding it with a tungsten needle in the anterior trunk area.
10. Aspirate the L15 medium with the handheld micropipette (without oil) (*see* **Subheading 3.6.2**) or mouth aspirate using a micropipette attached to a plastic tube until the embryo is trapped between the silicone surface and the water surface that immobilizes it (Alexander Aulehla, personal communication; **Fig. 3b**; *see* **Note 42**).

3.5.2. Dissection and Transfer of PSM and Embryo

1. With a very fine sharp Tungsten needle, make an incision in the PSM perpendicular to the end of the notochord (**Fig. 3b**; *see* **Note 43**).
2. Make a second cut at a distance two-thirds from the anterior end of the PSM (*see* **Note 44**).
3. Complete the first cut perpendicular to the notochord end.
4. Make a third cut along the PSM–neural tube boundary (*see* **Note 45**).
5. Draw the dissected PSM piece into a fresh micropipette with the handheld micropipette with as little medium as possible and expel it into a prepared 1.5-mL reaction tube containing ice-cold 500-μL Trizol.
6. Immediately transfer the tube onto dry ice or immerse in liquid nitrogen.
7. To replenish the dissection area, carefully apply L15 medium next to the embryo using the handheld micropipette or by mouth pipetting to lift it from the silicon surface (*see* **Note 46**).
8. Transfer the embryo to a prepared 0.5-mL reaction tube containing ice-cold 4% paraformaldehyde for fixation.

The embryo is then used for standard whole-mount in situ hybridization *(14)*. The PSM piece can now be used for RNA isolation and subsequent DNA microarray analysis.

3.6. Custom-Made Tools for Micromanipulation

The earlier protocols require a few custom-made dissection and manipulation tools. Using standard workshop and laboratory materials they can be easily built.

3.6.1. Tungsten Tools for Microdissection

Tools made of tungsten wire can be sharpened to very fine-pointed needles and bent into different shapes ideal for microdissection (*see* **Note 47**). These protocols describe how to make them.

Electrolytic Sharpening:

1. Cut a 2-cm piece off the tungsten wire.
2. Attach two cables to a standard laboratory power supply (*see* **Note 48**) that you place next to a stereoscope. Clamp the piece of tungsten wire into an alligator clip attached to the anode (+) of the power supply.

3. Fill a small glass beaker with 0.5 M NaOH (*see* **Note 49**).
4. Turn the power supply to 5–10 V.
5. Insert the cathode (–) into the NaOH solution.
6. Repeatedly dip the tungsten wire in and out of the solution by hand. The solution will bubble with ongoing electrolysis. Check the progress of sharpening under the stereomicroscope. You can determine the degree of sharpening and shape of the needle tip by adjusting the voltage and dipping rate.

Fire-Polishing

A less sophisticated way to sharpen tungsten tools is to fire-polish them. For thinner tungsten wires, that works well. Since it is a slow process, it leaves time to adjust the thickness and sharpness precisely.

1. Attach a 2-cm piece of tungsten wire to a needle and handle (*see* **Subheading "Attaching to Holders and Shaping"**) and hold it into a Bunsen burner flame until it glows orange for 30–60 s.
2. Check the progress of sharpening under a stereomicroscope.

Attaching to Holders and Shaping

For holding the tungsten needles a simple and cheap design can be used (**Fig. 4a**):

1. Cut the cannula of a 27G needle with pliers to a length of ~0.5 cm.
2. Use flat nose pliers to carefully reopen the now flat cannula tip by squeezing on the short edges.
3. Insert the piece of tungsten wire a few millimeters into the cut cannula.
4. Fix the tungsten wire in the cannula by strongly clamping the tip with pliers.

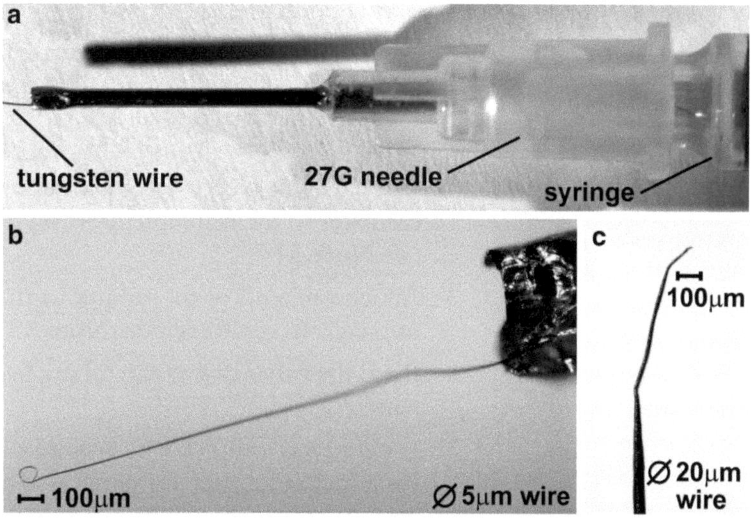

Fig. 4. Tungsten tools for microdissection and transplantation. (**a**) Holder for tungsten tools, consisting of a 1-mL syringe and a cut 27G needle. (**b**) Tungsten loop for optic vesicle transplantation. (**c**) Tungsten "hook" for microdissection.

5. If the tungsten is not completely fixed (this will be the case with small-diameter wire), apply a drop of superglue.
6. The tungsten needle can now be attached via its female Luer lock port to many holders (e.g., use a 1-mL syringe as a very cheap but good handle).
7. To bend the tip of the needle to different tool shapes (e.g., a sharp hook or a loop) attach the holder to a micromanipulator and place the needle tip into the field of view. Use a pair of Dumont forceps to bend the needle tip to the desired shape (*see* **Note 50**).
8. To electrolytically sharpen a needle after use, take it off the handle and hold the cannula with the alligator clip (*see* **Subheading "Electrolytic Sharpening"**).
9. Store needles in the cannula on a piece of modeling clay in a Petri dish.
10. Used tungsten needles can be conveniently cleaned and sterilized by flaming with a lighter.

3.6.2. Handheld Micropipette

To apply mineral oil as small droplets onto the surface of the embryo a custom-made, handheld micropipette, which is simple to put together, is ideal. It can also be used to transfer small pieces of explanted tissue in a small liquid volume (*see* **Subheading 3.5.2**).

Parts

The device consists of three parts **(Fig. 5a)**:

- Glass needle, pulled from borosilicate glass tubing with internal filament, pulled to the shape of a standard injection micropipette on a micropipette puller and broken to a controlled tip diameter.
- Microelectrode holder with a male Luer port.
- Handle (World Precision Instruments, 2505).

Assembly and Operation

1. Attach the microelectrode holder to the handle.
2. Cut a 15-cm piece of flexible tubing that fits with its inner diameter to the male Luer port on the microelectrode holder (*see* **Note 51**).
3. Attach one end of the tubing to the male Luer port on the microelectrode holder (*see* **Note 52**).
4. Bend the other end of the tubing back toward the end of the handle.
5. Loop a piece of thick wire around the tubing and handle and twist-lock it in place using pliers. This air-seals the tubing.
6. Use flexible tape to loosely glue the tubing at a mid level to the handle.
7. Fill a micropipette by attaching it (e.g., using a piece of modeling clay as holder) to an oil-filled reaction tube with the

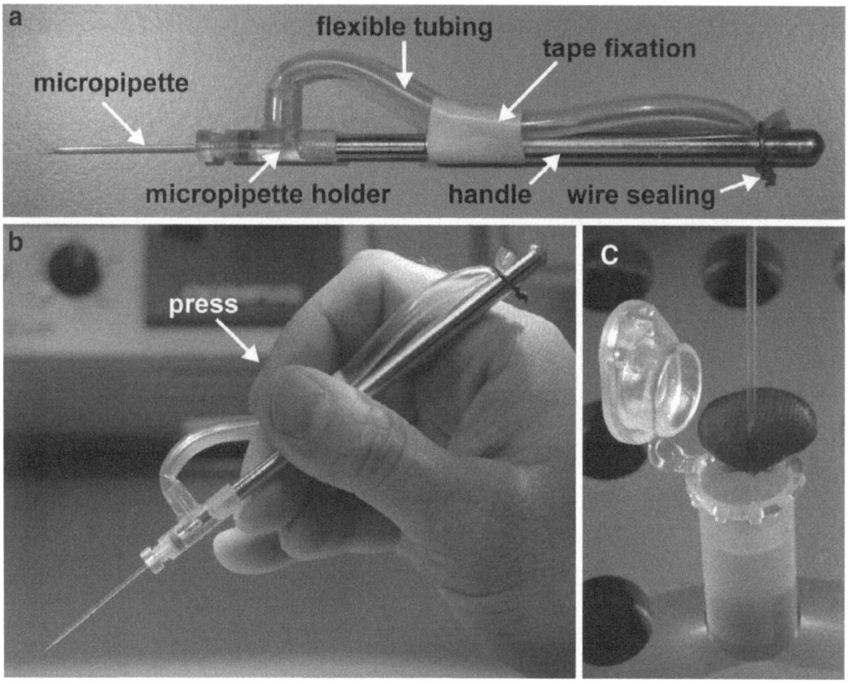

Fig. 5. Handheld micropipette. (**a**) A commercial micropipette holder is attached to a handle and an airtight piece of plastic tubing serves to expel and aspirate liquids through a pulled, glass micropipette. (**b**) The device is held like a pen and operated by compressing the plastic tubing. (**c**) For oil application, glass micropipettes (with internal filament) are filled by capillary force, by inserting into an oil-filled reaction tube, attached via a piece of modeling clay.

tip inserted – the internal filament will draw the oil into the needle. This can take a few hours (**Fig. 5c**).

8. Fit the oil-filled micropipette into the microelectrode holder.
9. Hold the device like a pen using the index finger to compress the tubing on the handle to expel small oil droplets (*see* **Note 53** and **Fig. 5b**). Test the oil gun by expelling small droplets into a medium-filled Petri dish under a stereomicroscope. The size of the droplet can be regulated by breaking the tip to different diameters.

4. Notes

1. Polysciences, Inc. offers a broad range selection of microbeads. Polystyrene beads with a diameter down to 25 μm have been successfully used for manual implantation.
2. Or any other protein/substance that you want to test by local, time-controlled application through microbeads.
3. Important: use exactly this oil – not every mineral oil from you lab stock will work.

4. Tungsten wire of different lengths and thicknesses can be bought from various suppliers, e.g., World Precision Instruments (WPI), Clark Electromedical Instruments/Harvard Apparatus, or California Fine Wire.
5. The Petri dish should be coated up to 1.5 mm under the rim, making it easier to reach the embryo with dissection tools.
6. For example, Tygon tubing, ID 1/32, OD 3/32, 50′ is distributed by VWR.
7. Only by so doing can you exert force against the embryo, which is necessary if you want to explant or implant.
8. As soon as you manage to rapidly orientate embryos in agarose you can do parallel embeddings with two or more embryos in one drop – this enhances throughput. Regarding fire-polishing of Pasteur pipettes, *see* **Note 21**.
9. It is important to avoid transfer of excess medium into the agarose to prevent gradual dilution. Excess medium at ambient temperature also tends to clog the melted agarose in the Falcon tube. Hold the tube against the light to find embryos.
10. Use the cover, not the dish's bottom, because the high rims of the bottom hinder manual access with the dissection tools during manipulation.
11. Embedded embryos can be stored in a "moist chamber" (e.g., a parafilm-sealed Petri dish with water-soaked pieces of tissue).
12. The embryo will be exposed in this wedge. The size and width of the wedge depend on the subsequent surgery. For eye manipulation, cut a wedge that forms a 45–60 angle – this leaves the embryo embedded in enough agarose to not become loose during surgery.
13. This needs to be done before adding medium – otherwise, the contours cut into the agarose will not be visible and cannot be followed with the needle.
14. Embryos will grow into the wedge during further incubation and thus should be manipulated rapidly after completed mounting. If you want to store mounted embryos, do so before cutting the wedges.
15. Gently press the tip of the oil-filled micropipette to the embryo and then press the handheld micropipette until a droplet sticks to the epidermis. Alternatively, first nick the epidermis with a sharp, fine-pointed tungsten needle and then place the droplet.
16. For all manipulation steps that require very precise manual operation, let the working hand rest on the stereomicroscope stage. Support the working hand using the other hand.
17. After this step the epidermis detaches from the embryo as a necrotic "blister."

18. If it does not detach easily, incubate for longer after application of the oil.
19. Healing happens rapidly – thus, directly proceed with manipulation after this step, before the wound closes.
20. It is best if the mold size is a little smaller than the diameter of the embryos. Thus "squeezed-in," embryos do not roll during implantation.
21. For fire-polishing, nick the pipette with a diamond knife at the shank position that corresponds to the desired tip diameter. Then, gently break off the tip. Pull the pipette a few times through the flame of a Bunsen burner to polish it – unpolished pipettes can damage dechorionated embryos easily.
22. After drawing beads into the pipette tip, clean outside of the tip by dipping it once into a Petri dish containing embryo medium – this removes beads from the outside of the tip, which otherwise would float on the surface of the medium in your manipulation dish.
23. Use a fairly blunt needle – this ensures that the cells are pushed apart rather than destroying them.
24. For this, use a kind of "underwater hockey" technique, generating swirls that move the bead ahead.
25. It is sometimes favorable to push the bead a little sideward after inserting it, which assures that during healing it does not slip out. This might require making the initial opening with a little spatial offset to the desired, final implantation site.
26. For late bead implantations, create only a small hole by applying a smaller oil droplet. Beads tend to slip out of the tissue if the hole is too big.
27. Often, heparin-coated beads (e.g., heparin–acrylic beads, Cat. No. H5263, Sigma) are used. But these beads are variable in size and shape and therefore get loaded with unpredictable amounts of protein. Comparison of implantation results of two beads from the same preparation is not possible. Size-controlled polystyrene beads have very reproducible effects.
28. The beads are delivered in drop dispensers (one drop corresponds to approximately 35 μL).
29. Protein and protein-loaded beads can stick to the wall of uncoated plastic ware, decreasing effective protein concentration during loading and poor retrieval of beads.
30. Depending on how much protein solution is added, the amount of beads/volume is controlled. Following the instructions of this protocol, adding volumes as small as 10 μL/tube is fine.
31. Do not introduce bubbles – otherwise, beads will stick to the wall of the reaction tube.

32. Do so by following the contours of the optic vesicle in a stepwise fashion, first superficially and then increasingly deeper in the tissue. This step needs to be done with extreme care to avoid disruption of the optic vesicle. Do not disrupt the epidermis above the optic vesicle: once the host vesicle is removed it will serve as a "pocket" for the donor vesicle. Without this epidermal pocket, the graft will not heal in. This also applies to other tissue grafts – always place the tissue under an intact piece of epidermis; otherwise, it will fall off and disintegrate.
33. For the donor embryo, less care to preserve the epidermis above the optic vesicle is necessary. Rather, the tissue integrity of the optic vesicle to be grafted is more important, which is sometimes better achieved if larger parts of the head epidermis are removed by applying several small oil droplets.
34. **Steps 8** and **9** have to be done rapidly to avoid the disintegration of the graft.
35. This stage was picked because the PSM is compact enough to dissect it as a whole piece and contamination by endodermal cells is almost negligible since they occur as a scattered, single-cell layer ventral to the PSM.
36. The dell does not need to be deep, the embryo should fit in half way – it is only needed to facilitate injection. Make the dell with a nonpulled micropipette.
37. AMP-PNP is needed to inhibit contractions of the yolk cell by paralyzing the actomyosin skeleton. This reduces curling of the embryo and facilitates surgery. AMP-PNP is membrane impermeable and will not diffuse into the embryonic cells.
38. Mouth pipetting was always used since it works more accurately than the handheld micropipette. Mouth pipetting is not allowed in every country, so consult your local regulations.
39. It can often be taken off conveniently as one piece.
40. The periderm is very flexible and difficult to cut with a tungsten wire and this makes it difficult to dissect the PSM properly. That is already apparent when cutting off the head. The advantage is that it can be pulled off in one piece and it will already detach from the embryo trunk when cutting off the head. When trying to culture the embryo, the periderm needs to remain in place, otherwise the embryo will lose its normal morphology.
41. The border between PSM and LPM is clearly visible at this stage from dorsal or ventral. The embryo was positioned with the ventral side down and was stabilized with the other hand.
42. Positioning the embryo always in this way facilitates dissection and increases precision. Also, do not aspirate too much

fluid, otherwise the preparation will dry out by the heat produced by the microscope lamp.

43. Do not cut all the way through the tissue because it will rapidly split open causing a morphological change of the tissue, which makes it difficult to make a precise second cut.
44. The position for the second cut was defined using an eyepiece reticule. Position the embryo with its midline parallel to the reticule and the perpendicular notochord incision line at the null position of the reticule. Based on test dissections and subsequent ISH stainings the mark on the reticule had been defined that would demarcate the two-third PSM piece for that orientation at the given magnification.
45. Try to follow that line all the way through to the ventral side, ending next to the notochord, since in the posterior end of the PSM the neural tube covers the PSM on the dorsal site.
46. This is a critical step because if working too slowly the surface will dry off, which will damage the tissue. Also, when adding L15 to the embryo, it can stick to the water/air interface, which destroys the embryo.
47. Fine tungsten needles can also be used for dissection and thick sectioning of fixed embryos, e.g., for imaging of whole-mount stainings. A good set of tungsten needles is very precious and can be used for many years.
48. Alternatively use a 9-V block battery.
49. The NaCl solution can be recycled many times.
50. The loop for optic vesicle transplantation is made of 5-μm diameter tungsten wire and bent to a diameter slightly bigger than the tissue block to be transplanted (90 μm). The sharp hook for tissue loosening and dissection is made of sharpened 20-μm diameter tungsten wire.
51. If the tubing is too soft it will fold and close. To prevent this, you can insert a piece of more rigid, small-diameter tubing into the soft tubing to support it.
52. Silicon grease helps to make the connection airtight.
53. If the oil does not expel, the system is either not airtight or the tip of the glass needle is too small.

Acknowledgments

This work was supported by grants to MB from the Deutsche Forschungsgemeinschaft (SFB 655), and the EU (ZF-Models and Endotrack) and by the Max Planck Society. The authors would like to thank Mary-Lee Dequeant, Aurelie Krol, Alexander Aulehla, and Karen Echeverri for advice on PSM dissection and technical support, the fish facilities of the MPI-CBG and Biotechnology Center, Dresden and Stowers Institute, Kansas City, MO, USA, and Katrin Bergman (CRTD) for photography.

References

1. Spemann, H. and Mangold, H. (1924). Über Induktion von Embryonanlagen durch Implantation artfremder Organisatoren. *Arch. Mikr. Anat. Entw. Mech.* **100**, 599–638.
2. Reifers, F., Walsh, E.C., Léger, S., Stainier, D.Y. and Brand, M. (2000). Induction and differentiation of the zebrafish heart requires fibroblast growth factor 8 (fgf8/acerebellar). *Development* **127**, 225–235.
3. Picker, A. and Brand, M. (2005). Fgf signals from a novel signaling center determine axial patterning of the prospective neural retina. *Development* **132**, 4951–4962.
4. Jászai, J., Reifers, F., Picker, A., Langenberg, T. and Brand, M. (2003). Isthmus-to-midbrain transformation in the absence of midbrain–hindbrain organizer activity. *Development* **130**, 6611–6623.
5. Scholpp, S. and Brand, M. (2004). Endocytosis controls spreading and effective signaling range of Fgf8 protein. *Curr Biol.* **14**, 1834–1841.
6. von der Hardt, S., Bakkers, J., Inbal, A., Carvalho, L., Solnica-Krezel, L., Heisenberg, C.P., et al. (2007). The Bmp gradient of the zebrafish gastrula guides migrating lateral cells by regulating cell-cell adhesion. *Curr Biol.* **17**, 475–487.
7. Picker, A., Brennan, C., Reifers, F., Clarke, J.D., Holder, N. and Brand, M. (1999). Requirement for the zebrafish mid-hindbrain boundary in midbrain polarisation,mapping and confinement of the retinotectal projection. *Development* **126**, 2967–2978.
8. Wagle, M., Grunewald, B., Subburaju, S., Barzaghi, C., le Guyader, S., Chan, J., et al. (2004). EphrinB2a in the zebrafish retinotectal system. *J. Neurobiol.* **59**, 57–65.
9. Fricke, C., Lee, J.S., Geiger-Rudolph, S., Bonhoeffer, F. and Chien, C.B. (2001). Astray, a zebrafish roundabout homolog required for retinal axon guidance. *Science* **292**, 507–510.
10. Hutson, L.D., Campbell, D.S. and Chien, C.B. (2004). Analyzing axon guidance in the zebrafish retinotectal system. *Methods Cell Biol* **76**, 13–35.
11. Hutson, L.D., Campbell, D.S. and Chien, C.B. (2004). Analyzing axon guidance in the zebrafish retinotectal system, in Methods in Cell Biology, Volume 76, The Zebrafish: cellular and Developmental Biology (Detrich, H.W., III., Westerfield, M., Zon, L.I., eds.), Elsevier Academic Press, London, pp. 13–32.
12. Kane, D.A. and Kishimoto, Y. (2002). Cell labelling and transplantation techniques, in Zebrafish: A Practical Approach (Nuesslein-Volhard, C., Dahm, R., eds.), Oxford University Press, Oxford, pp. 95–119).
13. Detrich, H.W., III., Westerfield, M. and Zon, L.I. (eds.) (2004). Methods in Cell Biology, Volume 76, The Zebrafish: Cellular and Developmental Biology, Elsevier Academic Press, London.
14. Westerfield, M. (2000). The zebrafish book. A guide for the laboratory use of zebrafish (Danio rerio). 4th ed., University of Oregon Press, Eugene.
15. Langenberg, T., Brand, M. and Cooper, M.S. (2003). Imaging brain development and organogenesis in zebrafish using immobilized embryonic explants. *Dev Dyn.* **228**, 464–474.
16. Déquéant, M.L., Glynn, E., Gaudenz, K., Wahl, M., Chen, J., Mushegian, A., et al. (2006). A complex oscillating network of signaling genes underlies the mouse segmentation clock. *Science* **314**, 1595–1598.

Part IV

Analyzing Gene Expression

Chapter 12

Zebrafish Spotted-Microarray for Genome-Wide Expression Profiling Experiments. Part I: Array Printing and Hybridization

Siew Hong Lam, Sinnakarupan Mathavan, and Zhiyuan Gong

Summary

The availability of microarray technology for zebrafish research has enabled the expression of tens of thousands of genes to be studied simultaneously in one experiment. The experiment usually involves measuring and comparing the relative abundance of tens of thousands of mRNA species in experimental samples obtained from mutant versus wild-type embryos, disease versus normal tissues, embryos/fish of different developmental stages, physiologic states, or from multiple treatments and/or time-points. A microarray experiment comprised of several stages can be divided into two distinct parts (i.e., the "wet-lab" and the "dry-lab"). The success of a microarray experiment hinges on the "wet-lab" procedures, which include technology that allows for generation of arrays with very high-density DNA where tens of thousands of genes are represented in an area smaller than a standard glass microscope slide, and procedures that enable extraction of high-quality RNA, efficient fluorescent labeling of nucleic acids, as well as specific hybridization of fluorescent labeled-samples with arrayed probes. This chapter describes these "wet-lab" procedures. "Dry-lab" procedures are described in the next chapter.

Key words: Zebrafish spotted-microarray, Gene expression profiling, Array printing, Hybridization, Transcriptome.

1. Introduction

Microarrays are based on nucleic acid complementary hybridization principles, which have long been applied to identify, quantify, and characterize nucleic acids in biologic samples (e.g., dot blots, colony hybridizations, Southern blots, and Northern blots). In microarray procedures, the complex nucleic acid sample is

Graham J. Lieschke et al. (eds.), *Zebrafish,* Methods in Molecular Biology, vol. 546
DOI: 10.1007/978-1-60327-977-2_12

labeled and is probed by tens of thousands of specific nucleic acid sequences immobilized in known positions on a solid substrate. Thus, instead of probing and studying one gene at a time—as in most conventional hybridization experiments—a single microarray will allow researchers to probe and study thousands of genes at the genome-wide level simultaneously. Although having this high-throughput genome-wide probing capability is the main feature and advantage of the microarray technique, the large representation of genes in itself poses challenges in terms of microarray fabrication and hybridization as well as microarray data acquisition and analysis.

Presently, there are two popular microarray platforms, spotted arrays and the Affymetrix GeneChip® (Affymetrix; CA), available for zebrafish. Because Affymetrix GeneChip® has its own proprietary standardized protocols for array fabrication and hybridization (http://www.affymetrix.com/support/technical/manual/expression_manual.affx), this communication will only describe procedures for spotted arrays. Each array can be made with up to 30,000 unique DNA probes spotted orderly onto a standard glass microscope slide. They can be spotted in-house using polymerase chain reaction (PCR) products (usually 100–500 bp) generated from cDNA libraries (*1, 2*), or oligonucleotide probes (40–70 mers) that are available commercially (.Operon, CA). Alternatively, the oligonucleotide probes can be designed in-house and synthesized by a company. In-house spotted arrays are printed using a robotic spotter/arrayer mounted with several array printing tips that can carry small quantities of probes, each representing a specific gene, from microtiter plates to a designated position on the surface of a glass slide, thus enabling high precision and high density spotting of probes. Spotted arrays are also available commercially (Agilent, CA; DNAmicroarray Inc., CA).

Spotted arrays are usually used in genome-wide expression profiling (transcriptomics) experiments by measuring and comparing relative abundance of mRNA species in two or more biological samples of interest (e.g., mutant versus wild-type [*3*], disease versus normal [*4*], different developmental [*5*] or physiologic state [*6*], multiple treatments and/or time-points [*7*]). To maximize the reliability of the quantification of the differences in the abundance of each RNA species, spotted arrays use a dual-label system in which two RNA samples are separately reverse-transcribed, labeled with distinct fluorescent dyes, and mixed together for simultaneous hybridization to one array. The ratio of the fluorescent intensities of the two dyes at one hybridized probe-spot on the array represents the relative abundance of an RNA species in the two samples. When using dual-label arrays, however, one must decide on a design for pairing and labeling samples. Several designs have been proposed and discussed (*8–10*). We recommend the use of common reference design (i.e., using a common reference sample labeled with the same dye for each of the hybridizations required in an entire experiment). We and others find the common reference design to be straightforward and compatible with most experimental designs and statistical analyses, in particular those involving multiple-group comparisons (*4, 5, 7, 8*).

Planning of the experiment is especially important because microarray experiments are costly and required substantial downstream work for data analysis and validation of the findings. Several key decisions need to be addressed prior to performing the actual experiment: (a) the aim of the experiment; (b) the experimental and control groups available/required for comparisons in order to fulfill experimental aim; (c) the desirable statistical and downstream data analyses; (d) the sample size, replicates (pooling/non-pooling) and the design for pairing of samples during hybridization; (e) validation methods for the microarray results/findings. Thus, a microarray experiment can be viewed as one comprised of multiple stages—from the initial experimental planning to the "wet-lab" procedures such as array printing, sample preparation and array hybridization and followed by "dry-lab procedures" involving image/data acquisition and data normalization, to statistical filtering and data analysis (**Fig. 1**).

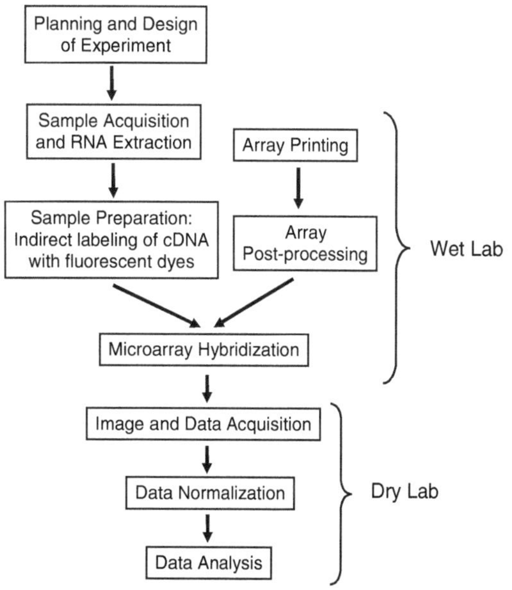

Fig. 1. An overview of the multiple stages in a microarray experiment.

2. Materials

2.1. Array Printing and Post-processing

2.1.1. Cleaning and Poly-L-Lysine Coating of Glass Slides for Array Printing

1. Microscope slides (Gold Seal; Erie Scientific Company, Portsmouth, NH).
2. Microscope slide cleaning solution (450 mL/dish). Dissolve 65 g NaOH in 130 mL MilliQ water in a 1-L glass beaker with a magnetic stirrer in a fume hood. While gently stirring the NaOH solution, slowly and carefully add 320 mL ethanol (**WARNING**: This is an exothermic reaction and ethanol will react violently with NaOH). Continue the gentle stirring and slowly add water to top up the solution to 450 mL. The solution should be clear after adding water. This solution is freshly made each time it is used.
3. Lysine solution: Add 45 mL poly-L-lysine solution (Sigma; St. Louis, MO) to 360 mL MilliQ water followed by 45 mL phosphate buffered saline (PBS) (137 mM NaCl, 2.7 mM KCl, 10 mM Na_2HPO_4, 1.8 mM KH_2PO_4) If necessary, adjust to pH 7.4 with HCl with continuous stirring to ensure solution is well mixed.
4. Stainless steel slide racks (30 slides) (Shandon Lipshaw Inc.; Pittsburgh, PA).
5. Plastic dishes/containers (size 2 and 5 from Komax Industrial Co. Ltd.; Seoul, Korea).
6. Centrifuge with microtiter plate holder/carrier (Sorvall Legend RT; Ramsey, MS).

2.1.2. Preparation of Oligo-Probes and Printing

1. Robotic spotter/arrayer (Genemachines™; Genomic Solutions, Ann Arbor, MI).
2. Tweezers (Canemco & Marivac; Quebec, Canada).
3. 3X SSC (printing buffer): 0.15 M NaCl and 0.015 M sodium citrate adjusted to pH 7.0 (with of 1N HCl). The solution is sterilized by autoclaving and is prepared freshly each time it is required.
4. PCR products at approximately 0.1 μg/μL or oligo-probes at 20 μmol/μL in 3× SSC.

2.1.3. Array Post-processing

1. Ultraviolet (UV) cross linker (Stratagene; La Jolla, CA).
2. Diamond edging pen (Samco; UK).
3. Blocking solution. Dissolve 7.7 g of succinic anhydride (Sigma) in 430 mL of 1-methyl-2-pyrrolidinone (Sigma) in a dry beaker with continuous stirring using a magnetic stirrer. After succinic anhydride is dissolved, immediately add 20 mL of 1 M boric acid, pH 8.0, and allow the solution to mix thoroughly. The blocking solution is to be freshly prepared each time it is required.

2.2. Sample Acquisition and Extraction

1. TRIZOL® reagent (Invitrogen; Carlsbad, CA).
2. Chloroform (Sigma).
3. Isopropanol (Sigma).
4. 75% *(v/v)* Ethanol in RNase-free water. Store at room temperature.
5. RNase-free water can be prepared by adding 0.01% *(v/v)* diethylpyrocarbonate (DEPC) (Sigma) to water contained in RNase-free glass bottles. Let it stand overnight in room temperature and autoclave. Store at room temperature.

2.3. Sample Preparation

2.3.1. Aminoallyl-Label cDNA Synthesis

1. Oligo-dT 20-mer, which can be synthesized by any local company that does PCR primer synthesis. Resuspend the oligonucleotide at a concentration of 0.5 μg/μL in RNase-free water. Store in small aliquots at –20°C.
2. SUPERSCRIPT II RNase H^- reverse transcriptase kit (Invitrogen) supplied with 5× first strand buffer and 0.1 M dithiothreitol (DTT).
3. RNaseOUT ribonuclease inhibitor (5000 units) (Invitrogen).
4. 100 mM dNTP sets (Invitrogen).
5. aa-dUTP (aminoallyl-dUTP) (Sigma). Dissolve the aa-dUTP in RNase-free water to make 100 mM of solution and store at –20°C.
6. 20X low-dT/aa-dUTP mix is prepared by adding 10 μL of each 100 mM dATP, dCTP, dGTP, 2 μL of 100 mM dTTP, 8 μL of 100 mM aa-dUTP and top it up with 60 μL of RNase-free water to make 100 μL of the solution. Store at –20°C.
7. 500 mM ethylenediamine tetraacetic acid (EDTA) in RNase-free water. Store at room temperature.
8. 1 M NaOH in RNase-free water. Store at room temperature.
9. 1 M Tris—Hcl, pH 7.5, in RNase-free water. Store at room temperature.
10. Microcon-YM30 column (Millipore; Billerica, MA).

2.3.2. Coupling with N-Hydroxysuccinimidyl (NHS)-Ester Cy3 and Cy5 Dyes

1. Mono-functional NHS-ester Cy3 and Cy5 dyes (Amersham Biosciences; Piscataway, NJ). Dissolve dry pellet in 20 μL dimethyl sulfoxide (DMSO). Aliquot 2 μL into each of ten single-use tubes. Dry the aliquots by SpeedVac rotary evaporator, without heat and store desiccated at 4°C in the dark. NHS-ester conjugated Cy dye is rapidly hydrolyzed in water, therefore it is important to store in a dried state and preferably with desiccant.

2. 500 mM $NaHCO_3$ (pH 9) in sterile water. Freshly made.
3. 4 M hydroxylamine in sterile water. Store at room temperature.
4. 100 mM sodium acetate, (pH 5.2, in sterile water. Store at room temperature.
5. QIAquick PCR purification kit (Qiagen; Venlo, The Netherlands) supplied with QIAquick columns, PB and PE buffers.
6. Herring Sperm DNA (10 μg/μL) (Invitrogen).
7. DIG Easy Hyb (Roche Applied Science; Mannheim, Germany)

2.4. Array Hybridization and Post-hybridization Washes

1. Hybrislip (Grace Bio-Labs; Bend, OR). 20% *(w/v)* Sodium dodecyl sulfate (SDS) in sterile water. Store at room temperature.
2. 20X SSC is made up of 3.0 M NaCl and 0.3 M sodium citrate adjusted to pH 7.0 (with a few drops of 14N HCl). The solution is sterilized by autoclaving.
3. Post-hybridization washing solutions can be prepared with the following components:

Washing solutions (200 mL)	MilliQ water	20X SSC	SDS (20% *[w/v]*)
2X SSC + 0.1%SDS	179 mL	20 mL	1 mL
2X SSC	180 mL	20 mL	–
0.2X SSC	198 mL	2 mL	–
0.02X SSC	200 mL	0.2 mL	–

3. Methods

3.1. Array Printing and Post-processing

3.1.1. Cleaning and Poly-L-Lysine Coating of Glass Slides for Array Printing

1. Select good quality microscope slides (Gold Seal™ microscope slides) that are free from any obvious surface/edge irregularities (scratches, cracks or chipped edges) and place them in a steel slide rack (Shandon Lipshaw) for cleaning (**Fig. 2a**). One rack can carry up to 30 slides and a total of four racks (120 slides) can be processed simultaneously. Always use powder-free gloves during the slide preparation.
2. Prepare the slide cleaning solution in the fume hood (*see* **Note 1**).

Fig. 2. Selected instruments and materials used in microarray wet-lab. (**a**) Stainless steel slide rack used during cleaning and poly-L-lysine coating of glass slides. (**b**) A robotic spotter or arrayer placed in a printing chamber to provide a dust-free environment. (**c**) Solid steel printing pins spotting probes on a glass slide. (**d**) A microarray slide with visible printed spots, as a result of the dried salt deposition on the glass surface, prior to post-processing. The probes are spotted orderly in grids with high precision and density (Inset). (**e**) The MAUI™ hybridization system, which allows four slides to be hybridized simultaneously in a controlled environment. (**f**) A centrifuge with a swing-out rotor and microtiter plate holders that can accommodate staining rack is required for rapid drying of microarray slides.

3. Place the slides into plastic dishes (size 2; Komax) and add the cleaning solution until it covers the slides. Allow gentle mixing of solution with the slides in the plastic dishes by shaking slowly on a 2-dimension shaker for 2 h. Ensure sufficient cleaning solution is added to avoid edges of slides being exposed to air or dried during shaking.
4. Transfer the slide rack to a plastic dish (size 5, Komax) filled with MilliQ water (3/4 of the dish) and plunge the slides vigorously for 10 s. Transfer to another plastic dish filled with fresh MilliQ water and plunge the slides vigorously for 30 s. Transfer to a plastic dish filled with fresh MilliQ water and repeat the process of washing the slides vigorously 3–4 times (30 s each time) with fresh MillliQ water each time to ensure that residual NaOH is removed prior to lysine coating.
5. Add about 450 mL of lysine solution into a clean plastic dish (size 2, Komax). Quickly transfer the slides from MilliQ water and plunge vigorously in the lysine solution for 10 s. Shake the plastic dish with the slides gently on a shaker for 45 min to allow mixing of lysine solution with the slides. The slides should be completely covered with lysine solution during shaking.
6. Wash the slides by plunging into MilliQ water in plastic dish (size 5, Komax) vigorously for 10 s, and let it stay for few minutes. Remove the rack with slides from the water and dry the slides in the rack by centrifuging at 100*g* for 4 min at room temperature (*see* **Note 2**).
7. Lysine coated slides should be transferred to a clean slide box and maintained in desiccators or vacuum chamber until use (*see* **Note 3**). Allow slides to age for at least 1 wk before printing as this is the time needed to acquire the optimal hydrophobic character necessary for printing small and uniform spots.

3.1.2. Preparation of Oligo-Probes and Printing

1. The oligo-probes (40–60 mer) received from the synthesis company are usually in lyophilized powder form in 96- or 384-well plates. They are resuspended in 3X SSC (the printing buffer) at a concentration of 20 μmol/μL for printing. For PCR products generated from a cDNA library, they are re-suspended in 3× SSC at a concentration of approximately 0.1 μg/μL (*see* **Note 4**).
2. Arrange the lysine-coated slides according to the designated alignment on the printing platform of the robotic spotter/arrayer (**Fig. 2b, c**); *see* **Note 5**).
3. Set the program for printing and run a test print by using the printing buffer (3X SSC) alone on 10–20 slides to test the quality of the printed spots and ensure correct operation of the printing process. The slides can be examined under a

dissecting microscope to check the uniformity (i.e. intensity, alignment, shape, and size) of the printed spots (**Fig. 2d**; *see* **Note 6**).

4. If the test print is good, proceed with the actual printing using the oligo-probes with a preset program. Ensure that the microtiter plates containing the oligo-probes are stacked in the proper order during printing. Although the printing process is fully automated and will take several hours, it is advisable to monitor the printing progress closely.
5. After printing, use sharp tweezers to remove the slides from the printing platform and carefully examine the slides for uniformity of the printed spots under a dissecting microscope. Slides with irregular spots (e.g., missing spots and varying spot intensities, shape or size) are to be discarded.
6. All the printed slides are kept in labeled slide boxes and placed in same orientation with the printed side facing the same direction. The printed slides should be stored at room temperature in a humidity-controlled cabinet (desiccators) for several days or kept in a drying oven at 55°C for overnight prior to post processing. The printed slides can be kept for at least 1 mo prior to post processing.

3.1.3. Array Post-processing

1. Before post-processing, carefully edge the borders of the printed area and number each slide on the reverse side of the printed surface using a diamond pen (*see* **Note 7**).
2. Place the printed slides in the chamber of a UV Cross-linker with the printed probes facing up towards the UV light source. Expose the slides to UV at 600 μJ/cm^2 for 1 min. Once UV cross-linking is complete, transfer the slides to a slide rack and keep it in a humidity-controlled cabinet (desiccator) to ensure that slides are dry.
3. Prior to blocking, transfer the slides into a drying oven at 55°C for 10–20 min to ensure the slides are completely dry.
4. Pour the freshly prepared blocking solution in a plastic dish (size 2; Komax). Remove the slides from the oven and vigorously plunge the printed slides in the blocking solution for 30 s.
5. Keep the slides in the blocking solution and shake them on an orbital shaker for 15 min. Ensure the slides are not exposed to air during shaking.
6. Rinse the slides in the rack with MilliQ water in a plastic dish (size 2; Komax) for about 10 s. (If cDNA probes are used, the printed slides are immersed in almost boiling hot water for 2 min to denature the cDNA. If oligo-probes are used, the hot water is not required, simply rinse the slides a second time in MilliQ water).

7. Plunge the rack with the slides in 95% ethanol contained in a plastic dish (size 2; Komax) for about 10 s.
8. Dry the slides in the rack by centrifuging at 100 *g* for 4 min at room temperature (*see* **Note 2**).
9. Transfer the slides immediately to a clean slide box. The slides are ready for hybridization. Post-processed printed slides can be stored for about 12 months in humidity-controlled cabinet (desiccator) without any significant deterioration of quality.

3.2. Sample Acquisition and RNA Extraction

The success of an expression microarray is highly dependent on the quality of the RNA samples and therefore all standard precautionary steps while working with RNA need to be taken to avoid degradation and contamination of the RNA samples. Depending on experimental set-up and sampling protocols, samples can be immediately processed for RNA extraction or snap-frozen in liquid nitrogen followed by storage in −80°C. Alternatively samples are stored in RNA*later*® (Ambion; Austin, TX) until RNA extraction is ready to be carried out (*see* **Note 8**). The RNA extraction protocol described here follows the TRIZOL® Reagent (Invitrogen) protocol although many other compatible RNA extraction methods will also be suitable. Reference RNA can be obtained by pooling a predetermined amount of total RNA extracted from embryos, specific tissues and/or adult zebrafish following the same procedure described below (*see* **Note 9**).

1. Homogenize tissue samples (<100 mg) in 1 mL of TRIZOL® Reagent using a glass or power homogenizer. As a general guide, the sample volume should not exceed 10% of the volume of TRIZOL® Reagent used for homogenization. Incubate the homogenized samples for 5 min at 15–30°C to permit the complete dissociation of nucleoprotein complexes.
2. Add 0.2 mL of chloroform per 1 mL of TRIZOL® reagent. Cap sample tubes securely and shake tubes vigorously by hand for 15 s, incubate them at 15–30°C for 5 min.
3. Centrifuge the samples at 12,000*g* for 15 min at 2–8°C. Following centrifugation, the mixture separates into a lower red phenol-chloroform phase, an interphase, and a colorless upper aqueous phase. Transfer the aqueous phase, which exclusively contains RNA to a fresh tube.
4. Add 0.5 mL of isopropanol into the transferred aqueous phase and incubate the mixture at 15–30°C for 10 min to precipitate the RNA.
5. Centrifuge at 12,000*g* for 10 min at 2–8°C. The RNA precipitate will form a gel-like pellet on the side and bottom of the tube.
6. Remove the supernatant. Wash the RNA pellet once with 75% ethanol by adding 1 mL of 75% ethanol, vortexing the pellet sample and centrifuge at 7500*g* for 5 min at 2–8°C.

At this stage, the RNA precipitate can be stored in 75% ethanol at 2–8°C for at least 1 wk, or at least 1 year at −20°C (*see* **Note 10**).

7. Air- or vacuum-dry the RNA pellet for 5–10 min. It is important not to let the RNA pellet dry completely as this will greatly decrease its solubility.
8. Dissolve RNA in RNase-free water by passing the solution a few times through a pipet tip, and incubating for 10 min at 55–60°C.
9. The quality/purity of the extract can be evaluated by UV spectrophotometry at 230, 260, and 280 nm (*see* **Note 11**). The purity (if contaminated by DNA) and integrity of RNA samples can also be verified by denaturing gel electrophoresis (*see* **Note 12**). Alternatively, the quality and quantity of the total RNA can be determined using Agilent 2100 Bioanalyzer. Adjust RNA concentration to 1–2 μg/μL with RNase-free water.

3.3. Sample Preparation: Aminoallyl-Label cDNA Synthesis and Coupling with Dyes

Indirect labeling with aminoallyl-labeled dUTPs during cDNA synthesis followed by coupling with reactive esterified-fluorescent dyes are described here. Although this procedure requires more steps and take a longer time, its added advantages of greater labeling efficacy (higher dye:nucleotide ratio), lack of labeling bias, and lower cost (fluorescent-labeled dNTPs are expensive) make the additional efforts worthwhile.

3.3.1. Aminoallyl-Label cDNA Synthesis

1. Add 2 μL of oligo-dT (20 mer at 0.5 μg/μL) and 20 μg of total RNA to an RNase-free microfuge tube (or PCR strip tubes) and top up with RNase-free water to a final volume of 23 μL to prepare an RNA reaction mixture for each reverse transcription reaction (*see* **Note 13**).
2. Incubate the RNA reaction mixture at 65°C for 5 min and then transfer the RNA reaction mixture on ice.
3. Assemble the following reverse transcription mixture (if a master mix is made, it is recommended to make an extra reaction on top of the required number of reactions):

Component	1 reaction	11 reactions
5× first strand buffer	8 μL	88 μL
0.1 M DTT	4 μL	44 μL
20× low-dT/aa-dUTP mix	2 μL	22 μL
SUPERSCRIPT II reverse transcriptase	2 μL	22 μL
RNaseOUT ribonuclease inhibitor	1 μL	11 μL

4. Add 17 μL of the reverse transcription master mix to each RNA reaction mixture and incubate at 42°C for 90 min.
5. Add 5 μL of 500 mM EDTA to the reaction mixture to stop the reaction.
6. Add 10 μL of 1 M NaOH to the reaction mixture and incubate at 65°C for 45 min to hydrolyze residual RNA.
7. Add 25 μL of 1 M Tris–HCl (pH 7.5) to neutralize the reaction mixture. The neutralized reaction mixture is ready to be cleaned up and concentrated.
8. Add 400 μL of sterile water to the neutralized reaction mixture and transfer to Microcon YM-30 column. The Microcon YM-30 column is to be pre-wetted and tested (*see* **Note 14**) before this step is performed.
9. Centrifuge at 13,000*g* for 5 min. About 10–40 μL of solution will remain in the column. Discard the flow-through.
10. Add about 500 μL of sterile water (or until the bottom first line of the column) and repeat **step 9**.
11. Add about 500 μL of sterile water (or until the bottom first line of the column) and centrifuge at 13,000*g* until the volume of the remaining solution in the column is less than 9 μL. This can be measured using a 10 μL volume micropipet and will require several short centrifugation at 13,000*g* followed by measurement of the remaining volume in the column (*see* **Note 15**).
12. Column is inverted and placed in a clean tube and centrifuge at 13,000*g* for 2 min to collect the concentrated aminoallyl-labeled cDNA samples.
13. Top up the sample volume to exactly 9 μL with sterile water. Aminoallyl-labeled cDNA samples are ready to be coupled with NHS-esterified fluorescent dyes or should be stored frozen in −20°C if not used immediately. This is a good step to stop before continuing with the remaining steps on another day.

3.3.2. Coupling with NHS-Ester Cy3 and Cy5 Dyes

1. Add 1 μL of freshly made 500 mM $NaHCO_3$, pH 9.0. to each of the 9 μL volumes of aminoallyl-labeled cDNA sample.
2. Transfer the 10 μL bicarbonate-buffered cDNA solution to the appropriate pre-prepared dried NHS-ester Cy3 and Cy5 dyes (*see* **Note 16**). Re-suspend the dye pellet with the bicarbonate-buffered cDNA solution by vigorously pipeting the mixture up and down.
3. Transfer the coupling reaction mixture back to the original labeled tube where the 9 μL volume of cDNA was first collected, and let the mixture incubate for 1 h at room temperature in the dark. (During the 1 h-incubation time, it

is recommended to proceed to prehybridization preparation as described in **Subheading 3.4.1.**)

4. Add 4.5 μL of 4 M hydroxylamine to each of the reaction mixture and incubate 15 min at room temperature in the dark to allow for quenching of uncoupled/free esterified dye.
5. Add 35 μL of 100 mM sodium acetate, pH 5.2, to each reaction mixture.
6. Combine the predetermined pairing of the Cy3- and Cy5-labeled samples (*see* **Note 17**). The combine paired Cy3- and Cy5-labeled samples is now ready for removal of uncoupled and quenched Cy dyes using the QIAquick PCR purification kit.
7. Add 500 μL of PB buffer to each of the combine Cy3- and Cy5-labeled sample.
8. Transfer each of the sample mixture to a QIAquick column and centrifuge at 10,000*g* for 60 s. Discard flow-through.
9. Add 750 μL of PE buffer (added with appropriate volume of ethanol as described in manufacturer protocol) to the column and spin at 10,000*g* for 30 s. Discard flow-through.
10. Add 400 μL of PE buffer (with ethanol) and centrifuge at 10,000*g* for 30 s. Discard flow-through.
11. Centrifuge the column at 10,000*g* for 1 min to remove any remaining solution trapped in the column. Place column onto a fresh 1.5-mL tube and let it sit for 1 min to allow any excess ethanol to evaporate.
12. Add 30 μL of sterile water to the center of the silica membrane in the column and let it sit 1 min at room temperature to prepare for elution.
13. Centrifuge at 10,000*g* for 1 min. Do not discard the eluted sample in the tube.
14. Repeat **steps 25** and **26** with another 30 μL of sterile water and collect the eluted sample in the same tube.
15. Transfer eluted sample to a fresh Microcon YM-30 column [The Microcon YM-30 column is to be pre-wetted and tested (*see* **Note 14**) before this step is performed].
16. Add 400 μL of sterile water (or until the bottom first line of the column) and centrifuge at 13,000*g* for 5 min. Discard flow-through.
17. Add 500 μL of sterile water (or until the bottom first line of the column) and centrifuge at 13,000*g* until the volume of the remaining solution in the column is less than 16 μL, similar to **step 11** (*see* **Note 15**).

18. Column is invert-placed in a fresh tube and centrifuge at 13,000*g* for 2 min to recover the concentrated combine Cy3- and Cy5-labeled sample.
19. Adjust the sample volume to exactly 16 μL with sterile water.
20. Add 4 μL of Herring Sperm DNA (10 μg/μL) and 20 μL of DIG Easy Hyb™ to each combine Cy3- and Cy5-labeled sample, which gives a final volume of 40 μL (*see* **Note 18**).
21. Incubate the labeled samples at 65°C for 5 min in the dark to denature the cDNA for hybridization.
22. Cool the samples on ice and briefly centrifuge to gather down solution to the bottom of each sample tube. Store in the dark at 42°C until ready to hybridize. (It is recommended that samples by hybridized within 1 h).

3.4. Array Hybridization and Post-hybridization Washes

The goal of array hybridization is to achieve high signal-to-background ratio with minimum cross-hybridization (resulting from small regions of homology) between labeled targets and arrayed probes by carrying out hybridization and washes under moderate to high stringency. This is effected by the use of high ionic strength hybridization buffers to reduce electrostatic repulsion and to promote complementary pairing between labeled targets with arrayed probes, blocking agents and detergents to reduce background noise, and moderately elevated temperature. In order to provide constant environment, the hybridization needs to be performed in a hybridization chamber, which can be improvised from standard laboratory material or manufactured commercially (**Fig. 2e**; *see* **Note 19**).

3.4.1. Array Prehybridization

1. Pre-warm DIG Easy Hyb™ at 42°C.
2. Add 35 μL of sterile water to each end-well of the hybridization chamber to maintain humidity during hybridization.
3. Place 50 μL of pre-warmed DIG Easy Hyb™ on the surface of each array slide. Carefully lower the coverslip (Hybri-slip™) on top of the solution, ensuring that the solution spread evenly over the surface of the array and avoiding bubble accumulation beneath the coverslip (*see* **Note 20**).
4. Place the arrays in the hybridization chamber and close it firmly, followed by incubation at 42°C for at least 1 h.

3.4.2. Array Hybridization

1. Carefully remove arrays from the hybridization chamber and immerse array-side-down in a staining rack and dish filled with MilliQ water.
2. Plunge the staining rack gently until the coverslip falls away from the arrays. Remove the coverslip gently to avoid scratching the array surface (*see* **Note 21**).

3. Transfer the staining rack with the arrays into a second staining dish filled with MilliQ water and wash gently for 1 min,
4. Transfer the staining rack with the arrays into a third staining dish filled with isopropanol and wash gently for 1 min.
5. Transfer the staining rack with the arrays into a centrifuge with a swing-out rotor and microtiter plate holders (**Fig. 2f**). Place the staining rack with the arrays on paper towels (to absorb remaining solution) on the plate holder and make sure that the rotor is properly balanced. Centrifuge at 100*g* for 4 min to dry the slide.
6. Place the 40 μL of hybridization sample onto the surface of each array as described in **step 3** of **Subheading 3.4.1** (*see* **Note 19**). Incubate in the dark at 42°C for 14–16 h (or overnight) using a hybridization chamber.

3.4.3. Post-hybridization Washes

1. Carefully remove arrays from the chamber and quickly (to avoid evaporation) immerse array-side-down in a staining rack and dish filled with 2X SSC and 0.1% SDS. Plunge gently until coverslips fall away from the slide (*see* **Note 21**).
2. Transfer the staining rack with the arrays into a second staining dish filled with 1X SSC and plunge gently for 1 min.
3. Transfer the staining rack with the arrays into a third staining dish filled with 0.2X SSC and plunge gently for 30 s.
4. Transfer the staining rack with the arrays into a fourth staining dish filled with 0.02X SSC and plunge gently for 30 s.
5. Transfer the staining rack with the arrays into a centrifuge with a swing-out rotor and plate holders. Place the staining rack with the arrays on paper towels (to absorb remaining wash buffer) on the microtiter plate holder and make sure that the rotor is properly balanced. Centrifuge at 100*g* for 4 min at room temperature to dry the arrays.
6. Proceed to acquire the fluorescent signals on each array by scanning with an appropriate array scanner as soon as possible (within the same day). The arrays are to be stored in the dark until scanning. This step ends the "wet-lab" procedures.

4. Notes

1. This process involves explosive exothermic reaction and will generate excessive heat. Care should be taken during the process (*see also* **item 2** under **Subheading 2.1.1**).

2. To ensure rapid drying and clean slide surfaces, ensure centrifuge chamber is dry and relatively clean prior to centrifuging the slides. Fold 2 or 3 low-lint paper towels and place neatly in bottom of each microtiter plate holder to absorb excess water from the wet slide racks; change these paper towels between spins as they will quickly become saturated with water. If a vacuum oven is available, place racks in oven at 40°C for 5–10 min in a vacuum. The Lysine solution can be refrigerated and reused for up to 7 days.
3. It is important to keep the slides in low humidity (we recommend storing in a dessicator cabinet at <35% relative humidity) and slides can be used for printing 1 wk after lysine coating. An alternative to the 1-week aging process is to dry the freshly-coated slides in an oven at 55°C for more than 4 h or overnight to achieve the optimal hydrophobic characteristics. To ensure high quality prints, do not use lysine-coated slides that have been aged for more than 4 months. It is very important to avoid touching the surface of the glass where the array will be printed – not even with gloved fingers. Handle the slides via edges and extreme corners only.
4. The oligo-probes can be aliquoted as "stock plates" and "working (printing) plates". The oligo-probes in the stock plate can be lyophilized and stored at −80°C. It is important to note the volume in the wells prior to lyophilization as this would be the volume required to reconstitute the oligo-probes when required for printing in the future. After printing, the oligo-probes in the printing plate can be maintained at 4°C (if the next printing is to be done within 2 wk). Otherwise, the oligo-probes in the "working plate" can also be lyophilized and stored at −80°C till they are reconstituted for the next printing.
5. Care should be taken to avoid introducing dust or touching the surface of the slides during the process of slide arrangement and printing and slide collection after printing. Use powder-free gloves. Avoid contact with printed areas in the slide. The printing capacity depends on the printing machine (from 40 to 220 slides for each print). The number of printing pins varies from 8 to 48 pins depending on the models. The model "Gene machine™" has 48 printing pins and can print about 220 slides for each print.
6. If the intensity, shape, and size of the printed spots appear not to be uniform, check, and clean the printing pins/tips and make sure they are place properly on the holder. If the spots appear not to be properly aligned, check the alignment of the print-tip holder and the slides on the printing platform.

7. The boundary of the printed area can be seen easily because the printed spots, as a result of the dried salt deposition on the glass surface, are visible to the naked eye prior to post-processing. The edging of the printed slide is important to indicate the printed side of the slide and to locate the area/boundary of the printed probes which are essential during hybridization because after post-processing, the printed spots and area are no longer visible to the eye.
8. If samples required shipping, we recommend storage and shipment in RNA*later*® in dry ice, or ship precipitated total RNA in 75% ethanol in dry ice (after **step 8** of **Subheading 3.3**).
9. The reference RNA need not be related to the samples being examined. The purpose of the reference RNA is that it should ideally provide a hybridization signal (and thus a non-zero denominator for the hybridization ratio) at each probe on the array so that the relative abundance of the RNA species can be calculated. A convenient approximation of this ideal reference RNA is an equal mixture of RNA/tissue material from each of the experimental samples. In such a mixture, any gene represented in one of the samples will be represented in the reference. Using the same reference RNA for all experiments in a laboratory makes it possible to compare expression among different experiments (although other sources of variation may make this difficult). Therefore, a laboratory can decide on a suitable reference (such as a pre-determined ratio of total RNA from different stages of developing embryos, male and female adult fish) if it is planning to do a lot of microarray experiments. This is to maximize hybridization/signal coverage (without saturation) of the probes on the array while providing a common reference for comparison between groups/experiments. Even so, it is important to prepare excess reference RNA for all the hybridizations required for each set of experiment, as variation will still occur for each different batch of similar reference RNA. The reference RNA should be stored in single use aliquots (e.g., 100–200 μL of 1 μg/μL reference RNA which is sufficient for 10 hybridizations) to avoid repeated thawing and freezing.
10. If samples are not going to be processed for microarray hybridization within the next few days, it is better to store precipitated RNA samples in 75% ethanol at −20°C to ensure stability. If there are excess amount of RNA samples after hybridization, it is also advisable to reprecipitate the RNA samples with 3 volumes of ethanol and store at −20°C for future validation assays. Recover RNA by centrifugation.

11. The ratio of the readings at 260 nm and 280 nm (A_{260}:A_{280}) provides an estimate of the purity of RNA with respect to contaminants that absorb in the UV. The A_{260}:A_{280} ratio of RNA should be 1.9–2.1. Significant absorption at 280 nm indicates contamination of protein (particularly, aromatic amino acids). Significant absorption at 230 nm indicates contamination of phenolic and organic compounds (in the case of pure samples, the ratio A_{260}:A_{230} should be approximately 2.2). Quantify RNA yield by spectrophotometric analysis using the convention that 1 absorbance unit at 260 nm equals 40 μg/mL RNA [RNA concentration (μg/mL) = (OD_{260}) × (dilution factor) × (40 μg RNA/mL)/(1 OD_{260} unit)].
12. A denaturing gel system is suggested because most RNA forms extensive secondary structure via intramolecular base pairing, and the addition of formaldehyde to the agarose gel will help to maintain the RNA in its linear (denatured) form that will allow it to migrate strictly according to its size. Nevertheless, we find that native agarose gel electrophoresis may be sufficient to gauge the integrity and overall quality of a total RNA preparation by inspection of the 28S and 18S rRNA bands. The gels should show prominent bands of 28S and 18S rRNA and low molecular weight RNAs. The upper 28S band should stain about twice more intensely than the lower 18S band. A heavy smear or intensely staining of low-molecular-weight RNA band are indicators of RNA degradation or DNA contamination (the latter may include intense high-molecular-weight bands and staining in wells). In case of DNA contamination, the RNA sample can be re-extracted using TRIZOL® reagent.
13. The amount of total RNA required for synthesis of aminoallyl-labeled cDNA can be dependent on the hybridization system used in **Subheading 3.4** (*see* **Note 19**). If we are using conventional hybridization chambers such as a self-improvised slide box, pipette tip box or commercially available hybridization cassette (GeneMachines™), the amount of total RNA required to obtain strong signals is 20 μg. However, using the MAUI™ hybridization system (BioMicro Systems, Salt Lake City, Utah), we have obtained comparable strong signals with 8–10 μg of total RNA.
14. On rare occasions, we have experienced Microcon-YM30 columns with faulty membrane. Therefore, it is important to check and make sure that the Microcon-YM30 columns are not faulty before using them for cleaning up the aminoallyl-labelled cDNA samples. To do so, add 500 μL of sterile water to each Microcon-YM30 column that is planned to be used for cleaning up each reaction mixture and centrifuge

for 5 min at 13,000*g*. If the column membrane is in good condition, a small amount (10–40 μL) of water will remain in the column, which can be removed by inverting and tapping the column on a paper towel before using it for cleaning up. However, if all the water passes through the column after centrifugation, the column membrane is likely to be faulty and the column is to be discarded.

15. During measurement of the remaining volume in the column, care needs to be taken not to touch or puncture the delicate column membrane with the pipette tip. Use the orange gasket around the perimeter in the column as a touch point for the pipet tip.

16. Depending on the experimental design, the labeling of samples with Cy5 or Cy3 dyes has to be predetermined during experimental planning. For convenience and for ease of comparison with future experiments in our laboratory, we have adopted the common reference design (*see* **Fig. 1a**) and decided to label all experimental samples (control and test) with Cy5 and all reference RNA samples with Cy3.

17. As in **Note 3** above, the pairing of samples has to be predetermined during experimental planning. With the common reference design, we can pair any of our Cy5-labeled experimental samples with any of the Cy3-labeled reference samples.

18. The 40 μL hybridization volume for the sample mixture solution is suited for a 22 × 60 mm coverslip. As for a 22 × 40 mm and 22 × 22 mm coverslips, the hybridization volume for the sample mixture solution should be adjusted to 30 μL and 20 μL, respectively, while maintaining the volume ratio of the components.

19. In order to prevent dehydration, the array slide is incubated in a hybridization chamber that provides a sealed humid environment. Satisfactory results can be obtained using a plastic slide storage box or an empty pipette tip box containing filter paper or paper towel soaked in hybridization buffer (DIG Easy Hyb™). The array slide can be placed on top of another microscope slides to elevate it slightly from the filter paper in the slide box or can be placed flat on the perforated empty pipette tip rack with paper towels below the tip rack. The lid of the box can be sealed with Parafilm™ before placing in an incubator/oven at 42°C. Alternatively there are commercially available hybridization chambers (GeneMachines). We have found the MAUI hybridization system to be useful as it allows mixing of the labeled samples on the array surface and this leads to brighter signal and less of a gradient in intensity across the surface of the array. By using this system, greater

sensitivity is attained due to the larger signal to noise ratio, and less input total RNA is required for cDNA synthesis (*see* **Note 13**).

20. The placement of the hybridization solution and coverslip is one of the more technically challenging steps. Briefly, we here describe three procedures, but the choice between them is a matter of personal preference because once mastered, they yield fairly reproducible results. In the first procedure, the hybridization solution is placed in three large droplets linearly along the length and midline of the array. The coverslip is lowered until it almost touches the droplets before dropping gently onto the droplets so that the solution disperses radially (outward) from each droplet. In the second procedure, the hybridization solution is placed on the middle of a coverslip laid on a clean and level surface (e.g., on a clean microscope slide). The array is inverted with the probe-spotted surface facing the hybridization solution on the coverslip. The array is slowly lowered until it makes contact with the hybridization solution and the coverslip is transferred to the array slide and the array slide can be quickly inverted so that the coverslip is now on top of the array. In the third procedure, the hybridization solution is placed at one of the long ends of the array slide. Hold the coverslip at an angle so that the lower edge of the coverslip is resting on the surface of the array slide behind the hybridization solution. Slowly drag the coverslip forward so that the edge meets the hybridization solution. Now, slowly transfer support of the elevated end of the coverslip to the tip of a small-gauge needle and gently lower the coverslip until it is almost parallel and resting on the array slide surface before slipping the needle out from under the coverslip. The hybridization solution will spread between the coverslip and array as the coverslip is lowered onto the array surface. The investigator should practice on clean dummy slides and coverslips until confident with one procedure (i.e., hybridization solution spread evenly with little accumulation of bubbles under the coverslip).
21. Do not let array slides dry out at any stage during the washing procedure. Do not attempt to manually remove the coverslip. Allow the array slide with coverslip to immerse in wash solution (2X SSC and 0.1% SDS) and wait a while (~30 s) for coverslip to detach from the array slide. If the coverslip does appear to be detaching, gently plunge the array slide at an angle into wash solution (2X SSC and 0.1% SDS) to ensure coverslip falls away from array without scratching it. During plunging, watch out for coverslips that have already detached and have been carried by

the current off the bottom of the dish and up towards the surface of the wash solution where they may scratch your array. It is therefore better to remove the coverslip after it has detached from the array slide.

Acknowledgments

This work is supported by Biomedical Research Council (BMRC) and Environment and Water Industry (EWI) Development Council, Singapore.

References

1. Ton, C., Stamatiou, D., Dzau, V.J., and Liew, C.C. (2002). Construction of a zebrafish cDNA microarray: gene expression profiling of the zebrafish during development. *Biochem. Biophys. Res. Commun.* **296**, 1134–1142.
2. Lo, J., Lee, S., Xu, M., Liu, F., Ruan, H., Eun, A., He, Y., Ma, W., Wang, W., Wen, Z., and Peng, J. (2003). 15000 unique zebrafish EST clusters and their future use in microarray for profiling gene expression patterns during embryogenesis. *Genome Res.* **13**, 455–466.
3. Sumanas, S., Jorniak, T., and Lin, S. (2005). Identification of novel vascular endothelial-specific genes by the microarray analysis of the zebrafish cloche mutants. *Blood* **106**, 534–541.
4. Lam, S.H., Wu, Y.L., Vega, V.B., Miller, L.D., Spitsbergen, J., Tong, Y., Zhan, H., Govindarajan, K.R., Lee, S., Mathavan, S., Murthy, K.R., Buhler, D.R., Liu, E.T. and Gong, Z. (2006). Conservation of gene expression signatures between zebrafish and human liver tumors and tumor progression. *Nat. Biotechnol.* **24**, 73–75.
5. Mathavan, S., Lee, S.G., Mak, A., Miller, L.D., Murthy, K.R., Govindarajan, K.R., Tong, Y., Wu, Y.L., Lam, S.H., Yang, H., Ruan, Y., Korzh, V., Gong, Z., Liu, E.T. and Lufkin, T. (2005). Transcriptome analysis of zebrafish embryogenesis using microarrays. *PLoS Genet.* **1**, 260–276.
6. van der Meer, D.L., van den Thillart, G.E., Witte, F., de Bakker, M.A., Besser, J., Richardson, M.K., Spaink, H.P., Leito, J.T., and Bagowski, C.P. (2005). Gene expression profiling of the long-term adaptive response to hypoxia in the gills of adult zebrafish. *Am. J. Physiol. Regul. Integr. Comp. Physiol.* **289**, R1512–R1529.
7. Lam, S.H., Winata, C.L., Tong, Y., Korzh, S., Lim, W.S., Korzh, V., Spitsbergen, J., Mathavan, S., Miller, L.D., Liu, E.T., and Gong, Z. (2006). Transcriptome kinetics of arsenic-induced adaptive response in zebrafish liver. *Physiol. Genomics* **27**, 51–361.
8. Simon, R.M. and Dobbin, K. (2003). Experimental design of DNA microarray experiments. *Biotechniques* **34**, S16–S21.
9. Yang, Y.H. and Speed, T. (2002). Design issues for cDNA microarray experiments. *Nat. Rev. Genet.* **3**, 579–588.
10. Churchill GA. (2002). Fundamentals of experimental design for cDNA microarrays. *Nat. Genet.* **32** Suppl, 490–495.

Chapter 13

Zebrafish Spotted-Microarray for Genome-Wide Expression Profiling Experiments: Data Acquisition and Analysis

Siew Hong Lam, R. Krishna Murthy Karuturi, and Zhiyuan Gong

Summary

In expression microarray experiments, post-hybridization procedures involving data acquisition and analysis are mainly computational and hence are described as "dry-lab procedures." The aim of these procedures is to acquire the fluorescent signals representing the relative abundance of tens of thousands of RNA species in samples with biological interest, and transform them into meaningful data with biological significance. These procedures begin with image and data acquisition followed by data normalization and processing, and thereafter primary and secondary data analyses. Although the actual data analysis is carried out after the wet-lab procedures, the strategy for data analysis should be planned prior to wet lab procedures when designing the experiment, to ensure effective downstream data handling and interpretation. Principal steps and advisory notes for acquiring, processing, and analyzing data that are commonly associated with various zebrafish expression profiling experiments are described in this chapter.

Key words: Zebrafish spotted-microarray, Transcriptome, Gene expression profiling, Data acquisition, Data processing, Data analysis, Data mining.

1. Introduction

The primary goal of an expression microarray experiment is to determine the relative representation of RNA species found in two or more samples and thereafter to make sense and make use of the data in a biologic context. The relative representation of specific RNA species in the samples is determined by measuring the ratio of the fluorescent intensities of the two dyes used to label the samples that compete to hybridize to specific probes spotted on the array. Fluorescent images of the array are acquired

Graham J. Lieschke et al. (eds.), *Zebrafish,* Methods in Molecular Biology, vol. 546
DOI: 10.1007/978-1-60327-977-2_13

for both fluorescent dyes using a laser scanning confocal microscope (scanner) where fluorescent intensity of each dye on each spotted probe is captured. The fluorescent image of the array is analyzed and processed into a raw digital data file using software that can assign coordinates to each of the spotted-probes located within grids, segment the image pixels into foreground and background, and calculate and extract the foreground and background pixels for each spot on the array (*1*). The general workflow of the major downstream procedures involved from this step onwards and described in this chapter is summarized in **Fig. 1**.

The raw gene expression data is then extracted, usually having uploaded to a database compliant to the MIAME [Minimal Information About a Microarray Experiment (*2*)] standard, for further processing. Background correction and normalization procedures are applied to the data of each array to account for any variations or differences that are not caused by the condition(s) being studied. Most normalization algorithms operate based on the assumption that majority of the transcripts on the array do not change between different samples and therefore certain distribution patterns are assumed when the total spots on the array are accounted (*3*). The normalized data is then transformed to $\log_2$ ratio and formatted (individual sample datasets are compiled into one large dataset) for primary data analysis to identify differentially-expressed genes associated with the condition(s) of interest. Depending on the experimental goal and design, appropriate statistical and visualization procedures have to be used to obtain a list of genes that are significantly associated with the condition of interest. Regardless of the statistical procedures used, a multiple hypotheses testing correction is usually incorporated to account for the high number of statistical tests conducted due to the simultaneous analysis of thousands of genes, which by chance alone could generate significant associations (e.g., false-positives) (*4, 5*). Most software incorporate options for this type of correction, such as Bonferroni, False Discovery Rate (FDR) or permutation corrections, into their statistical analysis. Once a list of genes of interest is obtained, secondary data analyses and data mining are performed on these genes to extract biological significance and insights.

As it is not within the scope of this chapter to cover all aspects of microarray data analysis, we have described statistical and data analysis procedures based on the type of experimental designs commonly associated with various zebrafish experiments. The principal steps for carrying out some of these procedures are described to help researchers to understand the available options, motivation, and effects of these procedures during experimental planning and subsequent data analysis, and when necessary, to facilitate further discussion with a statistician. In addition, hyperlinks for freely available software/tools are provided as starting

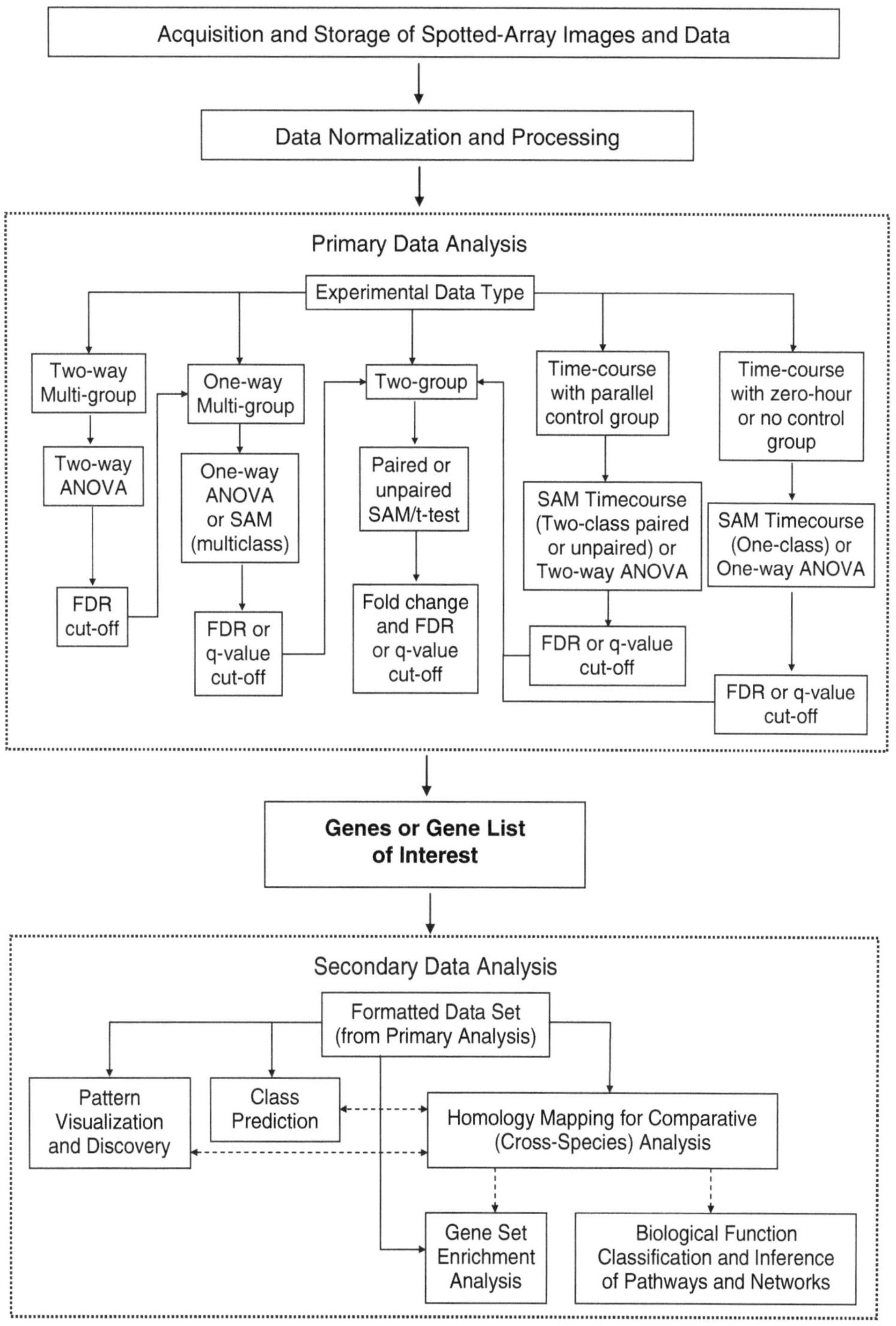

Fig. 1. A general work-flow of the microarray "dry-lab" procedures. The procedures begin with image and data acquisition, followed by data normalization and processing, and finally primary and secondary data analysis. The data analysis work-flow is merely a guide (*solid arrow* indicates direction of work-flow), as different analysis strategies can be applied depending on the experimental design and aim. In secondary data analysis, the dashed arrows indicate possible involvement of cross-species data which requires homology mapping and the work-flow can be one way or both ways.

points to enable the researcher to explore the many options available for data analysis. The researcher is encouraged to refer to the software manuals for detailed step-by-step operation. We have also included references for zebrafish-related microarray publications demonstrating some of the statistical and computational procedures described here.

2. Materials

1. Axon Scanner (Molecular Devices; CA)
2. Computer with good configuration [Intel "Core 2" processor 2.0 GHz or faster, Windows XP or Vista 32-bit edition operating system, 2 GB RAM or more, Dedicated Video Card 128 MB or greater, 120 GB hard drive (for image and data storage)]
3. Image Analysis Software: Freely available TM4 Spotfinder [The Institute for Genomic Research (TIGR); http://www.tm4.org/spotfinder.html] or commercially available GenePix Pro® (Molecular Devices, CA).
4. Data Storage Software: Freely available TM4 MADAM (TIGR; http://www.tm4.org/madam.html), BASE (Lund University; http://base.thep.lu.se/) or commercially available GeneTraffic® (Iobion Informatics LLC, CA).
5. Data Normalization and Processing Software: Freely available BASE, TM4 MIDAS (TIGR; http://www.tm4.org/midas.html) and Express Converter (TIGR; http://www.tm4.org/getprogram.cgi?program=expcnvt), R (R Foundation; http://www.r-project.org/), SNOMAD (Johns Hopkins School of Medicine; http://pevsnerlab.kennedykrieger.org/snomad.htm), or commercially available GeneSpring™ (Agilent Technologies; CA), Matlab™ (The MathWorks Inc, MA), GeneTraffic®.
6. Primary Data Analysis Software: Freely available TM4 MeV (TIGR; http://www.tm4.org/mev.html), SAM (Standford University; http://www-stat.stanford.edu/~tibs/SAM/), R, or commercially available Genespring™, Matlab®, Excel® with Analysis ToolPak (Microsoft Corporation, WA).
7. Secondary Data Analysis Software: Freely available TM4 MeV, Eisen Cluster and Treeview (University of California, Berkeley; http://rana.lbl.gov/EisenSoftware.htm), GSEA (Broad Institute; http://www.broad.mit.edu/gsea/), GOTM (Vanderbilt University; http://bioinfo.vanderbilt.edu/gotm/), DAVID/EASE (National Institute of Allergy and Infectious

Diseases; http://david.abcc.ncifcrf.gov/), ExpressionWave (Genome Institute of Singapore; http://giscompute.gis.a-star.edu.sg/~karu/ExpressionWave.html) or the commercially available GeneSpring™, Ingenuity® (Ingenuity Systems Inc., CA), Metacore™ (GeneGO Inc., CA) and Pathway Studio® (Ariadne Genomics Inc., MD).

3. Methods

3.1. Acquisition and Storage of Spotted-Array Images and Data

The purpose of this step is to transform fluorescent images and signals to digital images and data in order to store and analyze them computationally. A scanning process is performed to capture and measure the fluorescent intensities emitted by the fluorochrome-labeled targets hybridized to individual probes spotted on the arrays. The scanning process usually involves the use of a laser to excite the fluorochrome and the emitted fluorescence is captured through a photomultiplier tube (PMT). Hence, the captured signal intensity are influenced by both laser strength and PMT settings, and these parameters need to be adjusted for every array as a result of variations in hybridization efficiencies, fluorochrome incorporation and decay. Although the procedures described here are based on Axon™ scanner, they are generally applicable to other scanners.

3.1.1. Image Acquisition

1. Switch on the scanner (**Fig. 2a**) to warm up (about 15 min) before scanning.
2. Make sure that the array slide is dry and load the slide appropriately in the scanner with the array spotted-probes facing the laser scanning and in the correct orientation so that each of the spotted probe matches the assigned identity in the Gene Array List (GAL) file (a file containing the identities and positions of all the probes spotted on the array).
3. Set the image balance at a minimum intensity of 500 and a maximum intensity of 65,530. This determines the range of intensity that will be captured by the scanner (*see* **Note 1**).
4. Set the PMT voltage between 500 and 700 for each wavelength channel (635 and 532 nm) and perform an initial "Preview Scan" at a low resolution (40 μm per pixel). While scanning, adjust the PMT voltage in each channel until the pixel ratio (or count ratio 635 nm/532 nm) is between 0.8 and 1.2 and the histogram images depicting the cumulative frequency/count for the captured signal intensity (from 500 to 65,530) for 635 nm and 532 nm channels overlaps each other closely (**Fig. 2b**). This will help set the scan area and

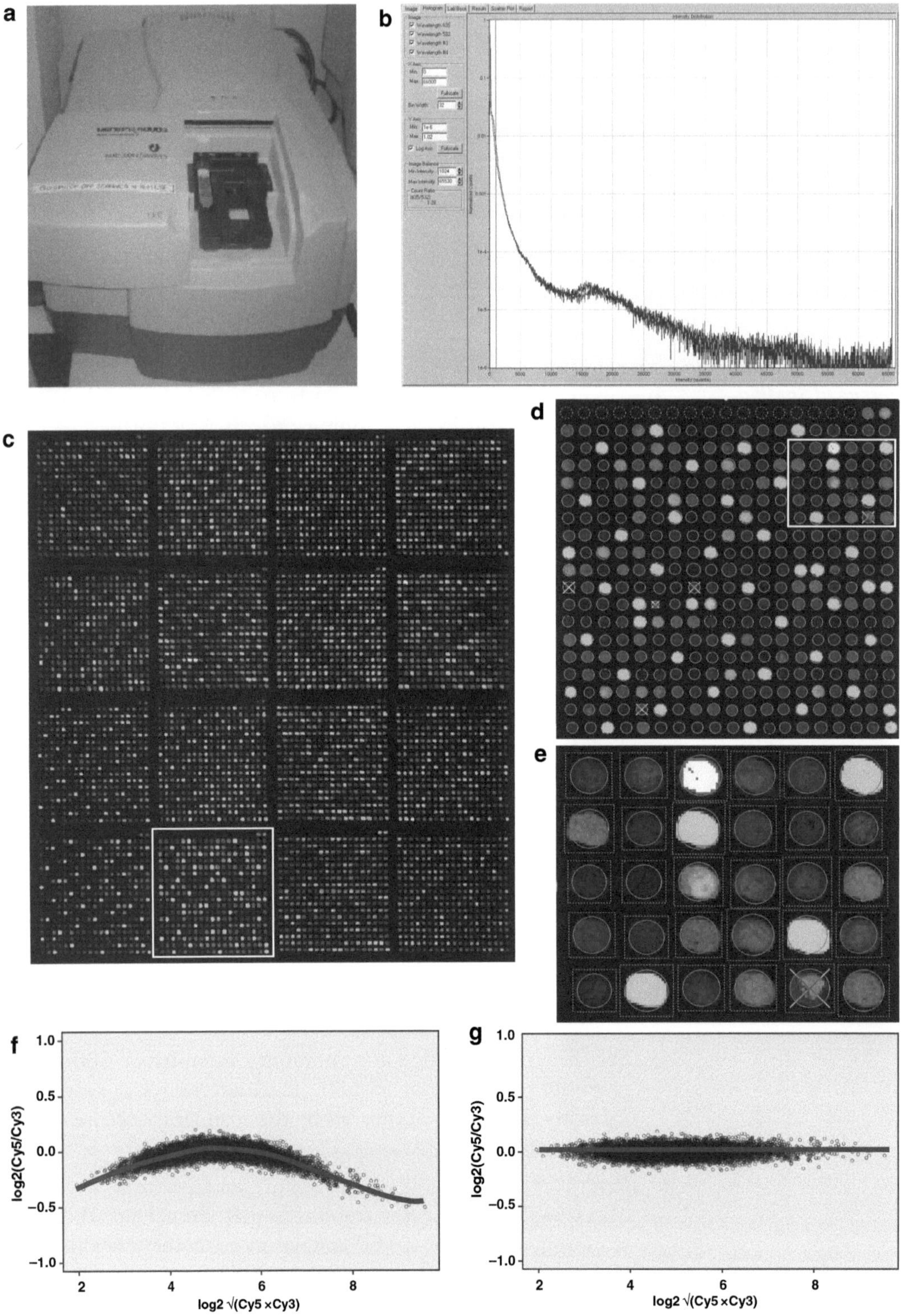
a
b
c
d
e
f
g
log2(Cy5/Cy3)
log2 √(Cy5 ×Cy3)
1.0
0.5
0.0
−0.5
−1.0
2
4
6
8

provide some idea of whether the initial PMT settings are appropriate (*see* **Note 2**).

5. Once confident of the PMT settings, ensure that the scan area covers all the spotted-probes on the array and that the image resolution has been set to 5–10 μm per pixel.
6. Perform a full scan. Do not change any scanning settings while the full scan is ongoing.
7. Once the full scan is complete, checked if the pixel ratio (count ratio 635 nm/532 nm) is within 0.8–1.2 and the histogram images for the cumulative captured signal intensity for each channel of the wavelength overlaps each other closely. If required, rescan the array with minor adjustment on the PMT settings until these criteria are achieved before saving the raw image data for analysis.
8. The image data can be saved as multi-image or single-image TIFF files, with single-image TIFF files being the recommended format since it is accessible from most image data analysis programs. Raw data should not be saved in the JPEG format, since this is not suitable for image data analysis.

3.1.2. Image Analysis

Spotted-array images consist of spots arranged in a regular grid-like pattern and each spot can be localized on the array by specifying its location in terms of column and row within sub-grids (a group of spots that are printed by the same print tip are localized together in sub-grids). Because it is known in advance the number of spots and their location and identity on the array, image analysis software can pre-configure a meta-grid cell mask and designate the identity of the spots according to the distribution pattern with which they are printed (GAL-File). The purpose of image analysis can be generally described as three-fold (*1*):

Fig. 2. Microarray image-data acquisition, analysis, and normalization. (**a**) Axon™ scanner used for acquiring fluorescent image of spotted-microarray. (**b**) Closely overlapping histogram images of both green and red channels depicting the cumulative frequency/count (*y*-axis) across the captured signal intensity from less than 500 to 65,530 (*x*-axis). (**c**) Scanned image of a spotted-microarray with 16 subgrids shown. A subgrid (*white-boxed*) can consist up to 360–400 spotted probes and a close-up view of the subgrid is shown in (**d**). A close-up view of a few selected spots of the subgrid is shown in (**e**). Pixels that fall within the *white-dashed circle* around a spot are considered to be foreground signals while pixels that are found outside the *white-dashed circle* but within the *white-dashed square* area around a spot are considered to be local background. Spots with questionable quality (bad spots or artifacts) are flagged (*crossed*). Here, experimental sample was labeled with Cy5 (*red*) whereas the reference was labeled with Cy3 (*green*). *Yellow spots* indicate equal abundant of transcripts in both experimental sample and reference, while shades of *red* or *green* spots indicate higher abundant of a transcript in the experimental sample or reference, respectively. A *bright white spot* indicate intensity saturation in at least one channel (in this case is the red channel only). (**f**) Before LOWESS normalization, intensity-dependent variation (a skewed decentralize distribution of data points along the *blue-line*) is observed when $\log_2$(Cy5/Cy3) (*y*-axis) for individual arrayed probes are plotted against the $\log_2\cdot$(Cy5×Cy3) (*x*-axis) on a scatter plot. (**g**) After LOWESS normalization, most of the data-points are distributed centrally along the *blue-line* (*see Color Plates*).

(a) to assign coordinates of the exact location and the identity of each spot on the array based on pre-defined grids and the active GAL-file; (b) to segment or partition each of the spot images according to its size and/or shape, and classify the pixels as foreground signal (within a spot) or background; (c) to calculate the foreground and background pixel intensities for each of the fluorescent dyes and estimate the signal intensity of a spot. In addition, most image analysis software such as TM4 Spotfinder or GenePix Pro® provide the ability to mark or *flag* spots that are considered to be "bad" due to poor printing or hybridization artifacts, in order to exclude them from subsequent analysis. The final goal of the image processing is to compute values that hopefully are directly proportional with the quantity of specific mRNA present in the samples that hybridized to each probe spotted on the array.

1. Open the scanned image file of the array to be analyzed using the image analysis software (**Fig. 2c**).
2. Load the pre-defined grid and/or GAL files. Superimpose the preconfigured meta-grid cells onto the scanned image of the array, making sure that the array spots at the edge (left, right, top bottom) of the scanned image are within the meta-grid.
3. Align the meta-grid cells with the array spots of the scanned image so that the each grid cell is centered on a spot (**Fig. 2d**). Although the alignment procedure is algorithm-automated, manual intervention by the user is required to ensure the reliability of the grid alignment with the array spots. The user has to make sure that the rows and columns of the array spots within each grid and between the grids are well separated and correctly aligned, because each spot will be assigned an identity based on the GAL file.
4. Define the boundary of each spot based on the geometry (size and shape) of the spot by adjusting the size and aligning the grid cell that is centered on a spot. The pixels that fall within a spot are considered to be foreground signals while pixels that are found outside a spot (local area around a spot) are considered to be background. Flag spots that cannot be distinguished from the background (empty/weak spots) or have questionable quality (bad spots or artifact) (**Fig. 2e**) to exclude them from subsequent analysis. Again, this segmentation procedure is automated, but manual intervention and inspection by the user is required to ensure the reliability of the process.
5. Once all the spots have been qualitatively assessed (**steps 3** and **4**), quantify the foreground signal and background pixels for all the spots using the image analysis software. Examples of some values commonly computed for individual spots for both the channels (635 and 532 nm) are: total signal intensity, mean signal intensity, median signal intensity, mode signal intensity,

background intensity, intensity ratio, standard deviation and the regression ratio across the two channels.

6. The data is saved in text delimited ASCII file format and usually stored in a database before being extracted for further processing and normalization.

3.1.3. Data Storage and Extraction (MIAME-Compliance)

Storing and managing array data effectively requires development and maintenance of a database that allows intelligent tracking of experimental parameters and results that are essential for the interpretation of expression results in downstream analyses. The MIAME set of protocols (*2*), developed by the Microarray Gene Expression Database Group (MGED: http://www.mged.org/) requires researchers to provide sufficient information needed for interpretation of data quality submitted for publication and to allow others to repeat published studies (*see* **Note 3**). Selected software programs designed to manage microarray databases compliant to MIAME standards are TM4 MADAM, BASE, and GeneTraffic®. These software programs will prompt the user to enter all MIAME-compliant information in the required format for storage of both the image and data files in the database. Submission of data file to public repositories such as ArrayExpress (http://www.ebi.ac.uk/arrayexpress/) at the EBI (UK), GEO (http://www.ncbi.nlm.nih.gov/geo/) at NCBI (US) and CIBEX (http://cibex.nig.ac.jp/) at DDBJ (Japan) which are designed to accept, hold, and distribute MIAME compliant microarray data, are required by some journals.

3.2. Data Normalization and Processing

The purpose of this step is to enhance and extract meaningful data characteristics so that any results obtained represent relevant biologic factors and not irrelevant systematic variation. Because of the multiple stages taken to produce spotted array data, it is common to observe systematic variation in the data as a result of print-tip variation, different quantities of starting RNA material, different efficiencies of dye labeling, disproportionate distribution of labeled sample during hybridization, topographical variation of slide, and/or scanning bias between the two channels. Most of these factors can be corrected/normalized by using one or more algorithms built into specialized microarray analysis software such as freely available BASE, TM4 MIDAS, R, and SNOMAD, or commercially available GeneSpring™, Matlab®, GeneTraffic®. While there is no consensus to a prescription for data normalization, this crucial step can be visualized as a pipeline with raw data entering at one end followed by a series of commonly applied data transformation steps as described below, before coming out as normalized/processed data at the other end.

1. Format raw data file according to the requirement of the software. If using MIDAS, TIGR Express Converter (http://

www.tm4.org/programs/ExprConvtl_9.zip) can transform a variety of files into the format that can be read by MIDAS.

2. Load formatted files. Most software can read multiple files in a batch analysis. Both channel A and B intensities will be read for individual spots.
3. Preliminary filtering is applied to filter out spots that have very low intensities (e.g., less than 200) in one or both channels or flagged during image analysis.
4. Perform background correction on the raw data. The background intensities of both channels are read and subtracted from the total intensity for each spot. A signal to noise ratio is also calculated for each spot and if it is less than the threshold set by the user (e.g., less than 1.5), the spot can be filtered out (*see* **Note 4**).
5. Perform total intensity normalization on the data (*see* **Note 5**). A normalization factor F is computed from the ratio (total intensity for channel B for all the arrayed probes)/(total intensity for channel A for all the arrayed probes). The normalizing factor is then used to rescale the intensities of one appropriate channel (non-reference) for all the arrayed probes.
6. Calculate the ratio of mean or median intensities of the two channels (Cy5/Cy3 or Cy3/Cy5) for each arrayed probes (most software that is used for normalization would have computed log-transformed ratios of mean or median of the two channels). Ratio of median intensity is preferred as it is lesser affected by outliers compared to ratio of mean intensity. Apply a $\log_2$ transformation on the calculated intensity ratio for each spot.
7. Perform LOWESS (LOcally Weighted polynomial regreS-Sion) Normalization on the data. The purpose of LOWESS normalization is to correct for intensity-dependent variations of ratios (*3*). This is done by estimating the variations according to a best-fit curve generated across intensity [$\log_2\sqrt{(Cy5/Cy3)}$] and subtracting it from the experimentally observed ratio for each data point (*see* **Note 6**; **Fig. 2f**, **g**). Only a certain percentage (user-defined smoothing parameter) of locally neighboring data points are used for estimating points along the best-fit curve and subsequently "smoothed" to fit the curve. The greater the percentage of data points used (as defined by the user) or the higher the smoothing parameter is set, the more adjustment is performed on the data and the smoother the fit. Typically, the smoothing parameter value ranges 20–30%.
8. Combine the $\log_2$(ratio) values of replicated probes (spot replicates) within a slide to calculate a mean $\log_2$(ratio) value for the gene.

9. Compile the $\log_2$(ratio) of individual arrayed-probes for all arrays into one file (usually in text delimited or Excel format) for primary data analysis. It is important to make sure that the file contains the correct unique identifier, GenBank accession number with all its corresponding $\log_2$(ratio) values for each arrayed probe/gene.

3.3. Primary Data Analysis

The purpose of primary data analysis is to identify genes (to simplify, all arrayed-probes are referred to as genes from this section onwards) significantly associated (up- or down-regulated) with an experimental condition. Here, the normalized data is subjected to appropriate statistic procedure to identify differentially expressed genes in accordance with the experimental condition (*see* **Note 7**). The choice of statistical procedure depends on the underlying hypothesis and design of the experiment. The generic procedures for various types of analyses commonly associated with zebrafish studies are described: (a) two-group analysis; (b) one-way multi-group analysis; (c) two-way multi-group analysis; and (d) time-course analysis (with parallel control, with zero-hour control or no control) (**Fig. 1**). All analysis procedures described here assume a common reference design (i.e., all arrays have a common reference within an experiment). Many freely available software tools such as TM4 MeV, R, SAM or commercial products such as GeneSpring™, Matlab®, including Excel® (preferably with Analysis ToolPak) can perform statistical analysis on gene expression data.

3.3.1. Two-Group Analysis

Such experiment involves two types of biologic samples and the resultant data is used to identify genes differentially expressed in one group compared to the other. There are two types of two-group experiments whose distinction is important from both statistical as well as biological perspectives: (a) unpaired (e.g., mutant vs wild type; treated vs control) and (b) paired (e.g., disease and normal tissue from the same individual fish with each replicate obtained from different individuals; treated and control samples from the same batch of embryos/fish with each replicate sampled from different batches). The procedure to analyze a two-group experiment is as follows:

1. Load the datasets into data analysis software such as SAM, MeV, or Excel and choose the appropriate comparison options (i.e., paired or unpaired). Label the sample/array datasets in each group according to the software requirements. In paired comparison, the datasets from the same individual/batch of fish in both groups have to be kept in the same order. MeV requires the *p*-value threshold to be set before executing the analysis (set corrected p-value < 0.05; *see* **steps 3** and **4** below).

2. Run the software. A *p*-value (or *q*-value in SAM) and mean $\log_2$-fold change between the two groups will be computed for each gene.
3. Correct the *p*-value for False Discovery Rate (FDR; *see* **Note 8**). Most software would include options for False Discovery correction. If not, *p*-value correction for FDR can be done via http://giscompute.gis.a-star.edu.sg/~vega/jstools/pval2fdr.html. The *q*-value reported by SAM has factored in the FDR and therefore needs no further correction.
4. Set a threshold for the FDR or *q*-value cut-off (e.g., <0.05 or 5%) for selecting genes that are considered to be significantly differentially expressed. In addition, a fold-change threshold (e.g., >1.5-fold change) in both directions (up- or down-regulated) can be included as selection criteria (*see* **Note 9**).
5. Save the selected list of differentially-expressed genes with its corresponding $\log_2$-fold change value in a text delimited or Excel format for secondary analysis.

3.3.2. One-Way Multi-Group Analysis

This type of experiment involves different sets of samples that are treated differently or of different biologic origin such as different (levels of) treatments/conditions, different strains/mutants or tissues. The aim of such experiments is to identify genes that are differentially expressed in at least one such treatment/condition compared to the others (or control group). This can be performed using ANOVA or SAM-multi-class followed by two-group analysis on each group with its respective control (**Fig. 1**).

1. Load the datasets into a data analysis software (if using SAM, choose the "Multiclass" option while for MeV or Excel Analysis ToolPak™, choose One-way ANOVA). Label each sample/array dataset according to the requirement of the software. This is usually done by assigning a group number/alphabetic character to each treatment group and all of the replicates from a treatment group are given same number/alphabetic character, which is different from the replicates of the other groups.
2. Run the software and a *p*-value (corrected for FDR as in **Subheading 3.3.1**) or *q*-value (for SAM) will be computed for each gene. Using a predefined cut-off (e.g., FDR or *q*-value < 5%), select the differentially-expressed genes.
3. On the selected differentially-expressed genes, use the above two-group analysis procedures (*see* **Subheading 3.3.1**) to identify the affected genes between each group with respect to its control group. Alternatively, especially in cases where there are multiple groups with no control group (e.g., comparing different segments/parts of the same organ such as brain,

intestine, or liver), use pattern visualization and discovery algorithms (*see* **Subheading 3.4.1**) on the selected differentially-expressed genes to identify peak expression or discover differential expression patterns among the groups.

3.3.3. Two-Way Multi-Group Analysis

Two-way multi-group experiment involves different sets of samples treated with different combinations of two independent factors (A and B) such as a drug treatment (factor A) with a gene knockdown or an over-expression of a gene (factor B) or, different concentration levels of a treatment (factor A) on different strains/sexes (factor B). The aim of such experiments is to identify differentially-expressed genes that are: (a) affected by factor A independently; (b) affected by factor B independently; and, (c) affected by interaction of factor A and B (may or may not be affected by the individual factors). This can be done using two-way ANOVA, and depending on experimental design and aim, it can be followed by one-way multi-group analysis on each factor and/or thereafter by two-group analysis with its respective control (**Fig. 1**).

1. Load data into a data analysis software such as MeV, Excel Analysis ToolPak™, or Genespring™. Label each sample/array dataset according to the requirement of the software.
2. Run two-way ANOVA and three *p*-values (corrected for FDR as in **Subheading 3.3.1**) will be generated for each gene, for factor A, factor B, and interaction of factors A and B, respectively.
3. Using a predefined cut-off (e.g., FDR <5%), three categories of significantly affected genes can be listed.
4. Depending on experimental design and aim, the significantly affected genes can be used for pattern discovery using clustering algorithm (*see* **Subheading 3.4.1**). Alternatively, if the experiment consists of multiple levels within factor A (e.g., different concentration levels 0, 1, 2) and/or within factor B (e.g. strain 1 and 2), the above One-way Multi-group analysis (*see* **Subheading 3.3.2**) can be performed on the selected differentially-expressed genes for factor A (concentration) to identify the genes significantly affected by the change of concentration (0, 1, 2) at certain level of factor B (e.g., for strain 1). The selected genes are then subjected to Two-group analysis (*see* **Subheading 3.3.1**) between concentrations (0, 1, 2) leading to finding differentially-expressed genes relative to different concentrations or relative to a control group (if concentration 0 is considered to be the control group) for strain 1. This analysis can be repeated at different levels of factor B (strain 2) or, if required, for factor B itself (at different levels of factor A).

3.3.4. Time-Course Analysis

Time-course experiments are useful to identify the kinetics of gene expression and/or identify differentially-expressed genes at different treatment durations or developmental stages (including aging). The length of time-course may vary from two to more than ten time-points. There are three common designs for such experiments (**Fig. 1**): (a) *Time-course with parallel control group* has both test and control groups sampled for each time-point (e.g., mutant and wild-type groups sampled at similar developmental time-points or, treated and control groups sampled at multiple time-points). Such experimental data can be analyzed using SAM (use Time-course Two-class unpaired or paired options) or using a two-way multi-group analysis as in **Subheading 3.3.3** (where "time" is treated as a factor); (b) *Time-course with zero-hour control group* has only one large control group sampled at the onset (zero-hour) of the experiment followed by multiple time-point sampling. Such experimental data can be analyzed using SAM (select time-course One-class) or using one-way multi-group analysis as in **Subheading 3.3.2** (where each time-point, including the zero-hour control, is treated as individual group) followed by Two-group analysis between each sampling time-point with the zero-hour control group; and (c) *Time-course with no control group* such as in normal developmental/aging study. Such experimental data can be analyzed using SAM (use time-course one-class) or using one-way multi-group analysis followed by two-group analysis. The procedure for analyzing time-course experiments is as follows:

1. Load the datasets into data analysis software such as SAM, MeV, or Excel Analysis ToolPak™ and choose the appropriate comparison options. Label the sample datasets in each group according to the software requirements.
2. If using SAM, the data can be analyzed by a *slope* or a *signed area* statistics depending on the expression profile pattern. The *slope* statistic is useful for finding genes with a consistent increase or decrease of expression level over time whereas *signed area* statistics is useful for finding genes with expression that increase or decrease and then level off or return to their baseline (but the expression level should not fluctuate between positive values and negative values). If the predominant patterns are not known for the dataset, then use SAM pattern discovery option (*see* **Subheading 3.4.1**). Alternatively, (One- or Two-way) ANOVA can be used and "time" is treated as a factor and each time-point as an individual group.
3. Run the software and a q-value (for SAM) or a p-value (for ANOVA and corrected for FDR as in **Subheading 3.3.1**) will be generated for each gene.
4. Using a predefined cut-off (e.g., FDR or q-value <5%), select the differentially-expressed genes. Pattern visualization and

discovery algorithms (*see* **Subheading 3.4.1**) on the selected differentially-expressed genes can also be used to identify peak expression or discover differential expression patterns among the groups.

3.4. Secondary Data Analysis

The purpose of secondary data analysis is to make use of the differentially-expressed genes identified in the primary analysis in order to identify biologic significance, discover novel insights, develop potential application, or propose a new hypothesis. The procedures usually involve, but are not limited to, the following: (a) pattern visualization and discovery; (b) class prediction; (c) homology mapping for comparative data mining and analysis; (d) gene set enrichment analysis; and (e) biologic function classification and inference of pathways and networks (**Fig. 1**). These procedures can be performed using a combination of data analysis software such as the freely available Eisen Cluster and Treeview, ExpressionWave, GOTM, MeV, DAVID, GSEA, or the commercially available GeneSpring™, Ingenuity®, MetaCore™, and Pathway Studio®.

3.4.1. Pattern Visualization and Discovery

The purpose of this approach is to obtain an overview and capture distinct patterns/structures in the data by either reducing the complexity (dimensionality) of the data or reorganizing and regrouping the data based on similarity of expression pattern changes. This approach can be applied to a large dataset as an initial explorative analysis without using any knowledge of the sample dataset grouping scheme (hence considered an "unsupervised" approach) or it can be applied to a statistically selected differentially-expressed gene set by using knowledge of the sample dataset grouping scheme as obtained from primary data analysis (hence a "supervised" approach). These procedures are useful to visualize similarities and differences between the samples and/or gene expression profiles. They can be done using the following: (a) principal component analysis; (b) clustering analysis; and (b) peak-finding analysis.

Principal Component Analysis

The purpose of principal component analysis (PCA) is to reduce complex multidimensional data to 2–3 dimensions in order to capture distinct patterns in the data. A large (multidimensional) dataset projected into a new reduced space will generally trend toward a certain direction. PCA identifies this direction and draws a solid line that best fits the direction (i.e., the first principle component [PC]). The remaining variation that is not captured by the first PC is captured by the subsequent orthogonal (hence statistically independent) PCs. Therefore, each successive PC accounts for less of the data variability than the previous one resulting in the reduction of the variability of the data to several components. The first three PCs could themselves act as Cartesian axes (x, y, z) where the data can be plotted according to these axes providing a two- or three-dimensional visual representation of a multidimensional data that may reveal valuable insight/trend. PCA has been performed on zebrafish datasets in order to capture major gene expression patterns in ovary and testes (*6*), as well as in embryos treated with different chemical compounds (*7*). PCA is available in MeV, SAM (known as pattern discovery), Eisen Cluster, or GeneSpring™.

1. Format the dataset according to the requirement of the software.
2. Load the formatted dataset onto the software and select the settings. In Mev, missing values are imputed using the κ-nearest neighbors (kNN) algorithm (default setting is 10). In MeV and Eisen Cluster, all three PCs are analyzed by default. In SAM or GeneSpring™ the user specifies the number of PCs (also refer as eigengenes), usually starting with 1, then 2 or 3 until the generated FDRs are too high.
3. Run the software and it will compute the requested/default setting PC(s), and finds the variables that have high correlation with it.
4. View the analysis. In MeV, the first three principal components are used to map each element into a 3-D or 2-D viewer **(Fig. 3a)**. GeneSpring™ provides viewing of the PCA in cluster sets and dendrogram. If using SAM or Eisen Cluster, the user should make a scatterplot of the PCs and the significant genes in Excel and study its shape. In general, some expression trends of the dataset may be found to correlate to the PCs, although the interpretation can be somewhat subjective and therefore requires further analysis.

Clustering Analysis

The purpose of clustering analysis is to organize and group multivariate data according to how they behave in the experiments, thus highlighting structure similarities and differences in the data **(Fig. 3b–e)**. The procedures are useful for inferring gene function from clusters of genes similarly expressed across many samples or for classifying new samples. There are several clustering algorithms available such as: (a) *hierarchical clustering*

Fig. 3. Gene expression profile of normal developing zebrafish at progressive developmental time-points viewed using different pattern visualization and discovery algorithms. The dataset is obtained from Mathavan et al. *(8)* and contains the expression of 3657 developmentally regulated genes from egg (unfertilized) to 48 h hour post-fertilization. (**a**) Principal component analysis generated by MeV showing distinct shift and distribution pattern of samples at progressive developmental time-points in 3-D space. (**b**) Unclustered gene expression profile (identity of genes not shown), (**c**) Hierarchical clustered gene expression profile (with dendograms), (**d**) *K*-means clustered gene expression profile ($K = 10$ discrete clusters), (**e**) Self-organizing map (10 discrete clusters) and (**f**) Peak-finding analyses visualized as heatmaps using Treeview software. Cells that are *red*, *green*, and *black* represent increase, decrease, and unchanged transcript abundance when compared to the reference sample, respectively (*see Color Plates*).

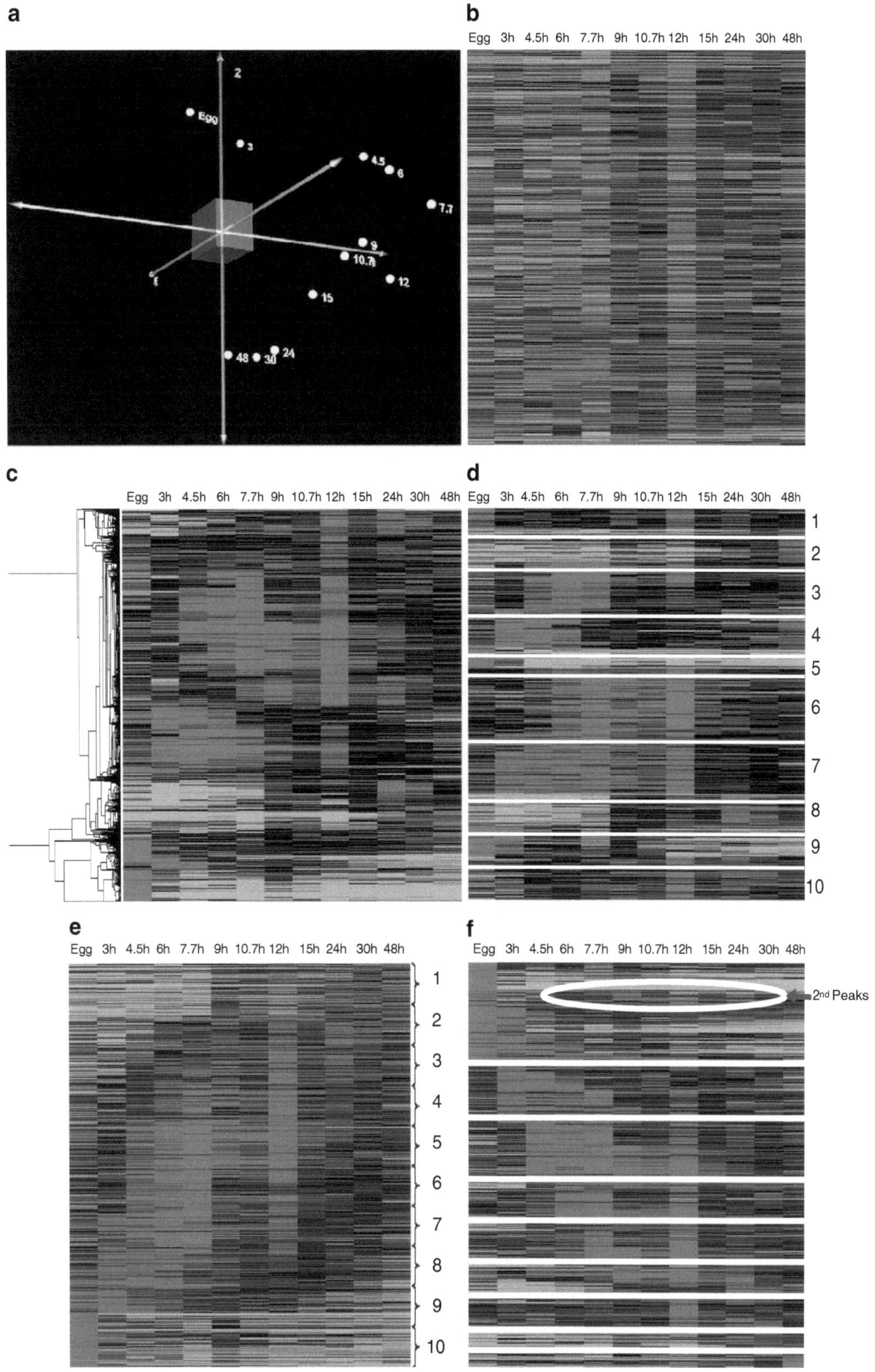
a
Egg
3
4.5
6
7.7
9
10.7
12
15
24
30
48
b
Egg 3h 4.5h 6h 7.7h 9h 10.7h 12h 15h 24h 30h 48h
c
Egg 3h 4.5h 6h 7.7h 9h 10.7h 12h 15h 24h 30h 48h
d
Egg 3h 4.5h 6h 7.7h 9h 10.7h 12h 15h 24h 30h 48h
1
2
3
4
5
6
7
8
9
10
e
Egg 3h 4.5h 6h 7.7h 9h 10.7h 12h 15h 24h 30h 48h
1
2
3
4
5
6
7
8
9
10
f
Egg 3h 4.5h 6h 7.7h 9h 10.7h 12h 15h 24h 30h 48h
2nd Peaks

(HC) which produces a dendrogram (hierarchical tree organization) of genes/samples by grouping the most similar expression profiles together. The items are joined by very short branches if they are very similar to each other, and by increasingly longer branches as their similarity decreases **(Fig. 3c)**; (b) *K-means clustering* (*KMC*) partitions data into a user-defined number (*K*) of discrete clusters, where expression profiles within a cluster are similar and across other clusters dissimilar **(Fig. 3d)**. *K* is the parameter to be chosen by trial and error by the user and it ranges from 2 to the number of genes/samples; and, (c) *self-organizing map* (*SOM*) which is similar to KMC except that the clusters are arranged in certain topological order so that clusters with greater similarities are neighbors, and dissimilar clusters are placed far apart each other **(Fig. 3e)**. When clustering a large dataset, it is useful to apply KMC or SOM (which is less memory intensive and therefore faster) to obtain discrete clusters and thereafter to apply HC to the smaller cluster(s) to visualize finer relationships within the cluster(s). The procedures are available in MeV, Eisen Cluster, and GeneSpring™ software. Clustering algorithms have been used to cluster different types/classes of toxicant-treated zebrafish embryos (*7*), visualize expression profiles during zebrafish development (*8*), between normal liver and liver tumors in zebrafish (*9*), between liver tumor progression in zebrafish and human (*10*), and between zebrafish *mind bomb* (*mib*) mutants and wild type embryos (*11*).

1. Format the dataset according to the requirement of the software.
2. Load the formatted dataset onto the software and choose the appropriate clustering method (HC, KMM or SOM) and the respective parameters [Distance Metrics (*see* **Note 10**) and Linkage Method (*see* **Note 11**)].
3. Run the clustering algorithm. Both MeV and GeneSpring™ generate clustering outputs that can be viewed immediately within the software. Eisen cluster will generate three files (filename.cdt, filename.gtr, and filename.atr).
4. View the analysis. MeV and GeneSpring™ provide viewing of the clustering in multiple formats (cluster sets of graphs, dendograms, and heat-maps) depending on the clustering method and software employed. For viewing Eisen cluster output, load the cdt file onto Eisen Treeview software for visualizing the heatmaps **(Fig. 3c–e)**. Specific expression trends will be revealed for different clusters of genes across different samples.

Peak-Finding Analysis

Peak-finding or expression wave analysis allows the researchers to visualize and demarcate the wave-like pattern of expression changes in a time-course study such as during zebrafish development.

With such an analysis, different subgroups of genes activated at different stages of development or at different time-points can be identified; from onset of activity to peak/highest activity to offset of activity, and this time course can be used to infer their biologic meaning. Definitions of onset and offset for such analysis can be based on expression level in which onset of a peak is the time at which gene expression crosses above certain threshold level and offset is when expression goes below that threshold (the threshold expression level can be either the average of the whole time-course or the average of certain time-points). The online peak-finding analysis tool ExpressionWave is available at http://giscompute.gis.a-star.edu.sg/~karu/ExpressionWave.html. The procedure has been used to visualize and determine expression peaks during zebrafish development (*8*).

1. Format the dataset according to the software requirement. Make sure that the missing values are replaced by entering "NA" (non-available).
2. Load the data into ExpressionWave and keep default parameters for initial analysis (fine-tuning of parameters can be applied for subsequent analysis).
3. Run the software and it will generate a file containing peak parameters (onset time, peak time and offset time) for each gene. The analysis can be visualized as a heatmap **(Fig. 3f)**.
4. The genes with similar group of peaks or expression wave can be further interpreted using pathway analysis as in **Subheadings 3.4.3–3.4.5**.

3.4.2. Class Prediction

The purpose of class prediction is to predict/classify samples into known/predefined groups using available data. The procedure includes a training phase in which the algorithm learns from selected datasets whose grouping schemes are already known (i.e., supervised) hence building a model, and a testing phase, in which the model (trained algorithm) generalizes the rules learnt from the training data to predict the classification of other unknown samples. In the training phase, a presumptive classification (supplied knowledge based on statistical selection and dataset grouping scheme) of a set of elements (genes/samples) where the expression data is used as inputs to produce a set of rules or reference weights as standards for the testing phase. The testing phase uses the standards created during "training" to assign a discriminator score to each element. Based on this score, each element is placed "into" or "out of" the class. A "leave-one-out" cross-validation is usually incorporated to validate the goodness of the model to avoid "over-fitting" it (i.e., only good for predicting the training dataset but not sufficiently generalized to work well on other new and unknown datasets). These procedures can be performed using *K*-nearest neighbor (KNN) classification or support vector machine (SVM). SVM is inherently binary (i.e., it only distinguishes between two classes whereas KNN, in principle, can separate any number of classes). Class prediction had been applied to zebrafish datasets for predicting toxicants (*12*).

(a) Training Phase

1. Format the selected dataset (usually differentially-expressed gene set) according to the requirement of the software. The samples must be labeled according to their respective classes.
2. Load data onto a software with SVM or KNN algorithm (MeV has both algorithms while GeneSpring™ has SVM and other class prediction algorithms). Use the default parameter settings and perform "leave-one-out" cross-validation.
3. Run the software to generate a classification matrix/table, a report and graphic view(s) of the analysis. The sensitivity, specificity, and precision of the model can be determined. The information will help the user to gauge the goodness of the prediction model, its strength and weakness. If it needs to be further tuned by changing the parameters, rerun the "leave-one-out" cross-validation (*see* **Note 12**).
4. Save the dataset together with the parameter settings that give the best prediction performance (high sensitivity and/or specificity) as the prediction model.

(b) Testing Phase

1. Load the test dataset that has not been used for the training of the prediction model.
2. Load the prediction model.
3. Run the software and it will generate a classification matrix/table, a report and graphic view(s) of the prediction output **(Fig. 4a)**.

Fig. 4. Data analysis outputs of toxicant-treated zebrafish for class prediction, gene set enrichment analysis, biologic function classification, inference of pathways, and networks (unpublished data). (**a**) Prediction models for estrogenic compounds using κ-nearest neighbors (kNN) and SVM correctly predicted most of the samples except for few cases indicated by the mismatched color cell between "Expected Results" and the "Predicted Results" columns. *Red cell* indicates presence or positive identification of a potent estrogenic compound whereas *green cell* indicates absence or negative identification of it. (**b**) A gene set enrichment analysis (GSEA) report of liver transcriptome from mercury-treated zebrafish showing detailed enrichment results of gene sets, and (**c**) graphic display of an enriched gene set associated with proteasome degradation pathway [*vertical bars* indicate distribution of proteasome-associated genes in the top ranking gene region (*red zone*)]. (**d**) Canonical pathways involved with various biologic processes, (**e**) a proteasome ubiquitination pathway (*partially shown*) and (**f**) a network identified as significantly associated with liver from mercury-treated zebrafish. Histograms correspond to –log(*p*-value) and line plots correspond to ratio of over-represented genes to total genes associated with the pathway. *Red* and *green* symbols represent up- and down-regulation of the genes, respectively. *Solid* and *dashed arrows* represent direct and indirect action on the molecules, respectively (*see Color Plates*).

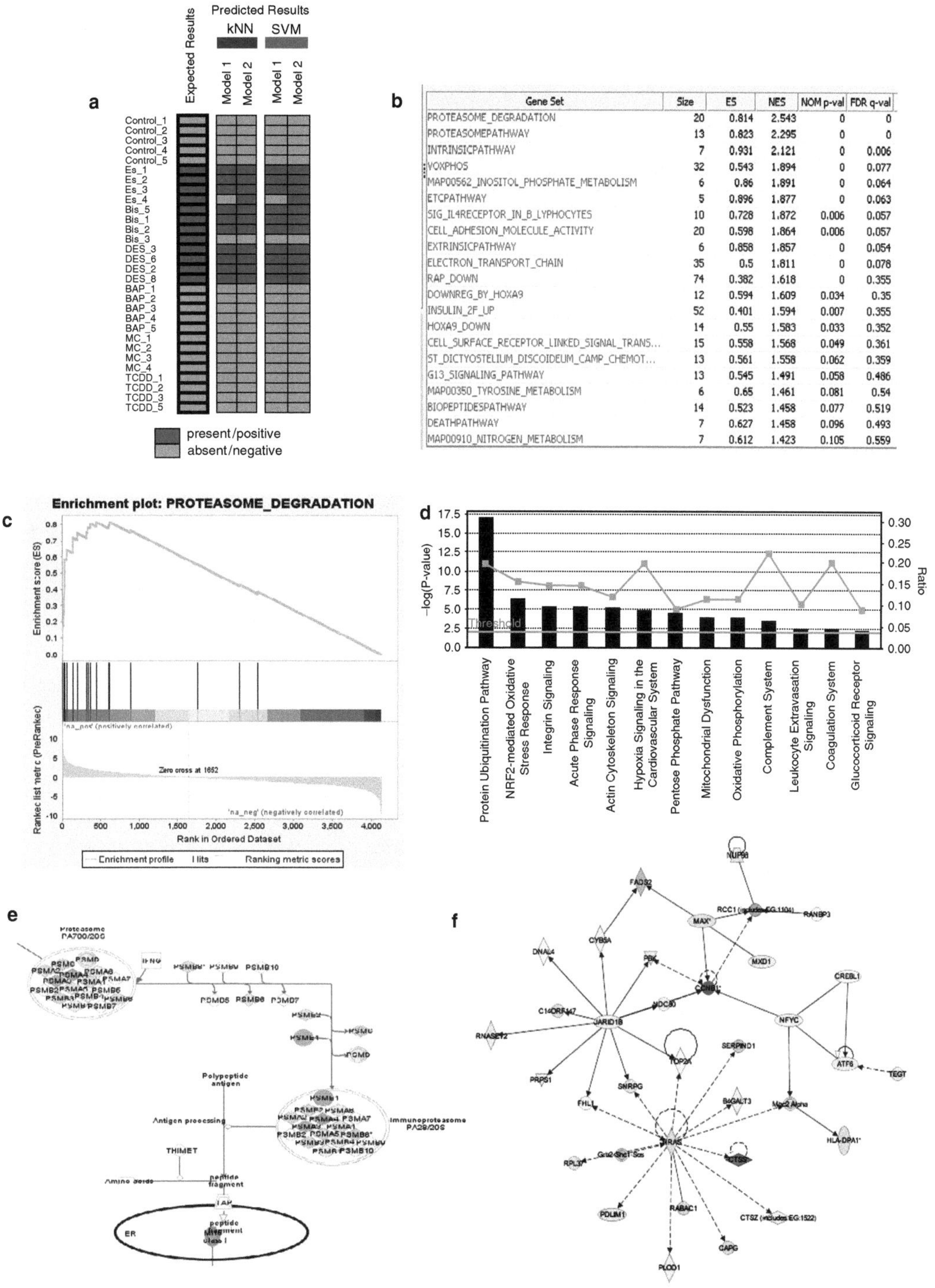

Gene Set	Size	ES	NES	NOM p-val	FDR q-val
PROTEASOME_DEGRADATION	20	0.814	2.543	0	0
PROTEASOMEPATHWAY	13	0.823	2.295	0	0
INTRINSICPATHWAY	7	0.931	2.121	0	0.006
VOXPHOS	32	0.543	1.894	0	0.077
MAP00562_INOSITOL_PHOSPHATE_METABOLISM	6	0.86	1.891	0	0.064
ETCPATHWAY	5	0.896	1.877	0	0.063
SIG_IL4RECEPTOR_IN_B_LYPHOCYTES	10	0.728	1.872	0.006	0.057
CELL_ADHESION_MOLECULE_ACTIVITY	20	0.598	1.864	0.006	0.057
EXTRINSICPATHWAY	6	0.858	1.857	0	0.054
ELECTRON_TRANSPORT_CHAIN	35	0.5	1.811	0	0.078
RAP_DOWN	74	0.382	1.618	0	0.355
DOWNREG_BY_HOXA9	12	0.594	1.609	0.034	0.35
INSULIN_2F_UP	52	0.401	1.594	0.007	0.355
HOXA9_DOWN	14	0.55	1.583	0.033	0.352
CELL_SURFACE_RECEPTOR_LINKED_SIGNAL_TRANS...	15	0.558	1.568	0.049	0.361
ST_DICTYOSTELIUM_DISCOIDEUM_CAMP_CHEMOT...	13	0.561	1.558	0.062	0.359
G13_SIGNALING_PATHWAY	13	0.545	1.491	0.058	0.486
MAP00350_TYROSINE_METABOLISM	6	0.65	1.461	0.081	0.54
BIOPEPTIDESPATHWAY	14	0.523	1.458	0.077	0.519
DEATHPATHWAY	7	0.627	1.458	0.096	0.493
MAP00910_NITROGEN_METABOLISM	7	0.612	1.423	0.105	0.559

3.4.3. Homology Mapping for Comparative Data Mining and Analysis

Because human/mouse databases are better annotated, and most available software does not support zebrafish data, it is useful to map zebrafish genes to human/mouse homologs for data mining, especially for procedures **Subheadings 3.4.4** and **3.4.5 (Fig. 1)**. Moreover, there is a strong motivation for comparing fish data with higher vertebrate data (human/mouse) to capture conservation and to apply the findings from zebrafish into human/mouse data (*9–12*). We have developed a simple procedure by using the National Center for Biotechnology Information (NCBI, USA; http://www.ncbi.nlm.nih.gov/) HomoloGene and UniGene databases for human/mouse homology mapping of the zebrafish genes. A PERL script was written to enable automated mapping of the GenBank Accession Number of zebrafish to their respective UniGene cluster which are then mapped to human/mouse UniGene cluster(s) that has been identified as homolog(s) by HomoloGene database. In some occasions, a zebrafish gene will be mapped to multiple homologs, usually belonging to the same protein group/family and therefore the researcher is required to select the putative homolog and interpret the data judiciously. This automated procedure is part of the Genome Institute of Singapore Zebrafish Microarray Annotation Database (http://giscompute.gis.a-star.edu.sg/~govind/zebrafish/version2/).

1. Load either the GenBank Accession Numbers or UniGene IDs for the list of genes of interest. Select for "Species" (Danio rerio), the latest UniGene Build and Input ID (GenBank or UniGene ID).
2. Submit the list of GenBank/UniGene IDs and choose the outputs (human or mouse). The web-tool will generate a corresponding list of human/mouse homologs with their respective UniGene ID(s) and description.
3. The data can be downloaded and saved in Excel format for mapping back to their expression values.

3.4.4. Gene Set Enrichment Analysis

The purpose of Gene Set Enrichment Analysis (GSEA) is to determine whether a predefined set of genes is significantly associated with the distinction between two biological states/groups (e.g., disease and healthy). This procedure uses statistics performed on the expression data to rank the genes (e.g., from highly significant to non-significant based on *p*-values between disease and healthy). It then compares a predefined gene set (e.g., genes belonging to a pathway or genes identified to be deregulated in a biologic/experimental condition) with the rank-ordered gene list to determine if the genes in the predefined set are randomly distributed throughout the rank-ordered gene list or correlated with the hig- ranking genes that are differentially-expressed between the two biological states. Thus, a collection of gene sets with biologic significance can be tested and those that are associated with the distinction of the two biologic states can be identified. GSEA is useful for revealing biologic insights when few genes pass the statistic significance criterion or when a unifying biological theme is difficult to identify (*13*). In addition, GSEA can be used for comparative analysis between two independently-derived gene expression datasets where direct integration of expression data is not appropriate (as a result of different platforms, biologic systems, or experimental set-up). For example, we have used GSEA to compare a zebrafish liver tumor enriched gene set with datasets from different human tumor types to identify genes that are significantly correlated with top ranking genes of human liver tumors (*10*). This algorithm is freely available as GSEA or is incorporated with minor variation in SAM and GeneSpring™.

1. Format the selected data files according to the requirement of the software. It is important that similar gene identifiers are used in all data files (*see* **Subheading 3.4.3**).
2. Load the formatted data files. Gene sets can also be downloaded from a database linked to the software (e.g., Molecular Signatures Database: http://www.broad.mit.edu/gsea/msigdb/index.jsp) and a collection of gene sets can be loaded and analyzed in a single run. Custom preranked gene lists can also be loaded. Specify the number of permutations (begin with 100 for a test run and increase to 1000 for the actual analysis), select for appropriate "phenotype label" (GSEA; Broad Institute) or "interpretation and pairing options" (Gene Spring™) and use default values for other additional parameters.
3. Run the software. GSEA ranks the genes in the expression dataset and analyzes the distribution of a pre-defined set of genes across the rank-ordered gene list.
4. View and interpret analysis. A GSEA report containing detailed enrichment results **(Fig. 4b)** and graphic display of the enriched gene set is generated **(Fig. 4c)**. An enrichment score (*ES*) computed for each gene set reflecting the degree of over-representation at the extremes (top or bottom) of the entire ranked list. A positive *ES* indicates gene set enrichment at the top of the ranked list; a negative *ES* indicates gene set enrichment at the bottom of the ranked list. *ES* is normalized to account for the size of the gene set, yielding a normalized enrichment score (*NES*) that can be used to compare analysis results across gene sets. A statistical significance (nominal p value) of the *ES* is computed for each gene set and is adjusted for false discovery rate (FDR) or q-value corresponding to each *NES*. An FDR of 5–25% is considered appropriate.

3.4.5. Biologic Function Classification and Inference of Pathways and Networks

The purpose of these procedures is to place the selected differentially-expressed gene set in the context of predefined biologic themes and pathways and to generate possible new networks based on known interactions between molecules. This will help in determining how the experimental conditions affect certain biological processes and pathways. It will also help in gaining novel biologic insights and developing new hypotheses. Most existing software, both freely available (e.g., GOTM, DAVID/EASE) and commercially available (e.g., GeneSpring™, Ingenuity®, MetaCore™, and Pathway Studio®), relies on similar statistical concepts to find over-representation of functional classes (e.g., biologic themes, gene ontology, pathways/networks) in a gene list derived from the expression data. They involve the calculation of statistic significance of nonrandom representation of genes belonging to a functional class in the differentially-expressed gene list and comparing it with a reference list (e.g., the entire list of gene probes on the array, the whole genome or all the genes/molecules in the entire database). In this manner, statistically significant functional classes with over-represented genes are identified to suggest affected biologic processes, to infer involvement of pathways and new networks, and help in prioritizing biologic areas for further study or validation. These procedures are highly knowledge-based dependent and therefore homology mapping to human and rodent databases is necessary when using zebrafish. These procedures had been performed on datasets from zebrafish *mind bomb* (*mib*) mutants (*11*) and toxicant-treated zebrafish (*12, 14*) to identify biologic processes and pathways associated with the experimental conditions.

1. Format the gene list of interest in a file according to the requirement of the software (usually in a tab delimited text or Excel format). Zebrafish genes are mapped to human homologs (*see* **Subheading 3.4.3**) and hence the gene identifiers can be human Unigene ID or gene symbol. Replicate expression values are usually consolidated to a mean/median value.

2. Upload the data file of interesting genes and the reference list (the entire arrayed probe set or select for options from the software database). Other parameters are set to default.

3. Run the software and it will perform a Fisher's exact test or hypergeometric test comparing the gene set belonging to individual functional categories with the gene set of interest (and the reference set) to determine if there is any over-representation of genes in any of the functional categories.

4. View and interpret analysis. The software generates tables or expandable trees for browsing the genes in enriched functional classes, bar charts at different annotation levels as well as graphs, pathways and networks for visualizing significant

functional classes (**Fig. 4d–f**). A functional class is considered significantly over-represented when the ratio of the observed number of genes of interest over the expected number of genes for the category surpasses a threshold value. A statistical significance is computed for each functional class and usually only the significant ($p < 0.01$ or 0.05) functional classes will be listed. Networks are generated from the available gene list of interest by maximizing the specific connectivity of the molecules, which is their interconnectedness with each other relative to all molecules they are connected to in the database. Each network is limited to a fixed number of molecules (e. g., 35 for Ingenuity® **Fig. 4f**) to keep it to a functional size and a network score is generated based on the hypergeometric distribution and is calculated with the right-tailed Fisher's exact test.

4. Notes

1. If the slide has high background, the minimum intensity can be increased to 1024 or more. If the signal intensity is weak (and the background intensity is also low), the minimum intensity can be lowered to 256. The key for two-color intensity adjustment is to minimize over-saturation (too many spots having intensity units >65,000 which will appear as bright white spots on the scanned image) without sacrificing good data (of low-expressing genes) into background.
2. The aim is to use the highest PMT setting with minimal saturated spots. Initial PMT settings should be high enough to see your features, but not so high that your image has a lot of saturated spots (about 0.1% of total number of spots is acceptable). Ideally, the same amount of red and green signal is acquired in each channel, resulting in a pixel ratio of approximately 1.0. Because post-acquisition normalization can correct for variations in relative signal intensities, a ratio between 0.8–1.2 is acceptable (*see* **Subheading 3.2**). Sometimes it is best to save multiple scans at different settings and sort it out during image analysis, although multiple scans can lead to dye bleaching (resulting in reduced signal).
3. The following information is required to support MIAME-compliant publications (http://www.mged.org/Workgroups/MIAME/miame_2.0.html): (a) the raw array data generated by the image analysis software for each of the hybridizations; (b) the final processed data for the set of arrays in the experiment which were used to draw the conclusion

in the study; (c) the essential experimental information for each sample (e.g., compound, time, concentration, physiological state, tissue source of sample); (d) the experimental design including sample data relationships (e.g., relationship between samples, arrays, and raw data files, groupings, and comparisons, technical and/or biological replicates); (e) the array design with sufficient annotation (e.g., gene identifiers, probe sequence information or commercial array catalog number); and (f) the experimental and data processing protocols for background correction, normalization, or data transformation used to obtain the final processed data.

4. In some cases, background correction can introduce more noise to low intensity spots and lead to the loss of data (from lower abundant genes). Therefore, it may be better not to subtract background (*1*).
5. Total intensity normalization assumes an equal amount of starting RNA material from samples, equal efficiencies of dye-labeling, and that the arrayed-probes are a random representation of genes in the organism. Hence, about same number of labeled molecules from the samples should hybridize to the array and the total fluorescent intensities summed over all arrayed probes should be the same for both channels. Some level of total intensity normalization has been performed during image acquisition.
6. An easy way to visualize intensity-dependent variation is to plot the calculated $\log_2(\mathrm{Cy5}/\mathrm{Cy3})$ against the $\log_2\sqrt{(\mathrm{Cy5}\times\mathrm{Cy3})}$ for individual arrayed probes on a scatter plot **(Fig. 2f)**. LOWESS normalization assumes that most of the genes are equivalently expressed (only relatively few are differentially expressed) and there is no systematic relationship between differential gene-expression and intensity or location of the spots. Therefore, in theory, most of the data-points should center around zero (*y*-axis) for $\log_2(\mathrm{Cy5}/\mathrm{Cy3})$ across all intensities (*x*-axis). However, this is usually not true as different patterns of deviation from zero $\log_2$(ratio) are observed across spots with different intensities. Thus, LOWESS normalization essentially subtracts the best-fit curve from the original $\log_2$(ratio) data and centralize the data points around zero $\log_2$(ratio) across all intensities **(Fig. 2g)**.
7. The type of experiment is mostly decided by the hypothesis driving the experiment but the number of replicates in each group is decided by technical feasibility and/or cost constraints. Even so, it is important to note that minimum number of replicates in each group depends on the number of groups to be analyzed and whether the variable is continuous or discrete/nominal. In general, a minimum of three biologic replicates in each group is mandatory and some

statistical tests such as ANOVA or time-course experiments require the number of replicates to be more than the number of individual groups/time-points.

8. In a microarray experiment, *p*-values are measures of statistic significance of the differential expression of genes, whereby lower values indicate greater confidence in the reality of differential expression. Hence, *p*-value of a gene indicates the probability of it being a false positive(i.e., a gene calculated statistically to be differentially expressed but in reality it is not). The incidence of false positives is proportional to the number of tests performed and the significance threshold (the usual significance threshold for *p*-values for a single hypothesis testing is 0.05 or 0.01). So in a single hypothesis testing situation (e.g., for a gene, a *p*-value of 0.05 indicates that there is a probability of 5% that the gene is a false-positive and if the test is repeated 10,000 times on the gene, it would be expected to show up as a false-positive 500 times). Similarly, in a microarray experiment, if a hypothesis test were to be repeated on 10,000 genes at a *p*-value of 0.05 would mean that 500 genes are expected to be false positives. The aim of multiple testing correction is to account for the occurrence of false positives resulting from multiple gene (hypotheses) testing. This correction is performed on the *p*-values by taking into account the number of tests (equivalent to the number of genes whose *p*-values are computed) in the microarray experiment. There are several ways to correct for multiple testing such as Bonferroni, Bonferroni step-down (Holm), Westfall and Young Permutation, and Benjamini and Hochberg False Discovery Rate (listed in the order of decreasing stringency) (*4*). We recommend using an approach with lesser stringency, such as Benjamini and Hochberg False Discovery Rate, to strike a balance between discovering statistically significant genes and limiting the occurrences of false positives, because an overly stringent approach may result in many false negatives.

9. Typical thresholds for FDR are 1, 5, and 10%, and for fold-change are twofold or 1.5-fold (up- or down-regulation). The decision on what statistical and fold-change thresholds to use as cutoffs depends on underlying biologic assumptions and/or the experimental aims. In a biologic system, it is assumed that majority of the genes do not change their expression levels. For signaling molecules or transcriptional regulators, the fold-change of these genes may not be very high (<2-fold) because they are likely to be tightly regulated. If the aim of the experiment were to identify several robust genes or biomarkers where extensive down-stream validation is to be performed, higher statistical significance (smaller

FDR threshold) and/or high fold-change will be the criteria for selection of a small number of genes. However, if the aim were to capture biological mechanisms or affected processes associated with a condition, a greater number of differentially-expressed genes (bigger FDR threshold) would provide a better picture (since secondary statistical tests will be performed on the selected genes). Likewise, instead of a cut-off, the list of genes can be rank-ordered according to their statistical significance and analyzed using GSEA (*see* **Subheading 3.4.4**).

10. Clustering analysis requires selection of distance metric, linkage, and type of clustering. A distance metric is a mathematical formula used to determine the similarity of the expression patterns for a pair of genes or samples/experiments. Distance metric choice depends on the notion of similarity. Pearson centered correlation is used for capturing trends, especially if the trend of the response is more important than their absolute values. Pearson uncentered correlation combines the similarity of trend as well as the average levels of expression profiles and is useful when reference sample is biologically related to the test samples (e.g., zero-hour sample is used as reference for all other samples in a time-point series). Spearman-rank and Kendall's tau are non-parametric versions of Pearson centered correlation (i.e., they ignore the actual gene expression values but instead use their rank order to determine similarity). An Euclidian distance metric is useful for clustering samples only if it accounts for the changes of absolute values from one sample to the other. An Euclidian metric is not meant for trend similarity measurement.

11. Linkage methods will determine how the distance between clusters are measured when constructing the hierarchical tree. The distance between clusters can be taken to be the distance between the closest members of the clusters (known as single linkage clustering), furthest members (complete linkage), the distance between the centers of the clusters (centroid linkage, which is the default setting in GeneSpring™), or the average distance of all members between clusters (average linkage). Single linkage clustering tends to form loose clusters whereas complete linkage clustering tends to form tighter clusters. Average and centroid linkage clustering is generally used for clustering because it is the less sensitive to noise as it takes into account of the distances/positions of all the members between clusters.

12. Sensitivity (%) [number of true positives/(number of true positives + number of false negatives)] is the proportion of true positives classified to be positive. High sensitivity indicates that the prediction model can identify positive samples

very well (low false negative rate). Specificity (%) [number of true negatives/(number of true negatives + number of false positives)] is the proportion of true negatives classified to be negative. High specificity indicates that the prediction model can classify negative samples very well (low false-positive rate). Precision or Positive Predictive Value (%) [number of true positives/(number of true positives + number of false positives)] is the proportion of the samples classified to be positive are actually true positives. A high precision indicates that the prediction model has a high probability of classifying a positive sample correctly (low false-discovery rate). The combination of these three parameters will allow the user to gauge on the strength and/or weakness of the prediction model (as a positive predictor or a negative predictor) and if further tuning is required.

Acknowledgments

Our zebrafish microarray research is supported by Biomedical Research Council (BMRC) and Environment and Water Industry (EWI) Development Council, Singapore.

References

1. Yang, Y.H., Buckley, M.J., and Speed, T.P. (2001). Analysis of cDNA microarray images. *Brief. Bioinform.* **2**, 341–349.
2. Brazma, A., Hingamp, P., Quackenbush, J., Sherlock, G., Spellman, P., Stoeckert, C., Aach, J., Ansorge, W., Ball, C.A., Causton, H.C., Gaasterland, T., Glenisson, P., Holstege, F.C., Kim, I.F., Markowitz, V., Matese, J.C., Parkinson, H., Robinson, A., Sarkans, U., Schulze-Kremer, S., Stewart, J., Taylor, R., Vilo, J., and Vingron, M. (2001). Minimum information about a microarray experiment (MIAME)-toward standards for microarray data. *Nat. Genet.* **29**, 365–371.
3. Quackenbush, J. (2002). Microarray data normalization and transformation. *Nat Genet.* **32** Suppl, 496–501.
4. Slonim, D.K. (2002). From patterns to pathways: gene expression data analysis comes of age. *Nat. Genet.* **32** Suppl, 502–508.
5. Storey, J.D. and Tibshirani, R. (2003). Statistical significance for genomewide studies. *Proc. Natl. Acad. Sci. U S A.* **100**, 9440–9445.
6. Santos, E.M., Paul, G.C., Van Look, K.J., Workman, V.L., Holt, W.V., van Aerle, R., Kille, P., and Tyler, C.R.(2007). Gonadal transcriptome responses and physiological consequences of exposure to oestrogen in breeding zebrafish (*Danio rerio*). *Aquat. Toxicol.* **83**, 134–142.
7. Yang, L., Kemadjou, J.R., Zinsmeister, C., Bauer, M., Legradi, J., Müller, F., Pankratz, M., Jäkel, J., and Strähle, U. (2007). Transcriptional profiling reveals barcode-like toxicogenomic responses in the zebrafish embryo. *Genome Biol.* **8**, R227.
8. Mathavan, S., Lee, S.G., Mak, A., Miller, L.D., Murthy, K.R., Govindarajan, K.R., Tong, Y., Wu, Y.L., Lam, S.H., Yang, H., Ruan, Y., Korzh, V., Gong, Z., Liu, E.T., and Lufkin, T. (2005). Transcriptome analysis of zebrafish embryogenesis using microarrays. *PLoS Genet.* **1**, 260–276.
9. Lam, S.H. and Gong, Z. (2006). Modeling liver cancer using zebrafish: a comparative oncogenomics approach. *Cell Cycle* **5**, 573–577.
10. Lam, S.H., Wu, Y.L., Vega, V.B., Miller, L.D., Spitsbergen, J., Tong, Y., Zhan, H., Govindarajan, K.R., Lee, S., Mathavan, S.,

Murthy, K.R., Buhler, D.R., Liu, E.T., and Gong, Z. (2006). Conservation of gene expression signatures between zebrafish and human liver tumors and tumor progression. *Nat. Biotechnol.* **24**, 73–75.

11. Hegde, A., Qiu, N.C., Qiu, X., Ho, S.H., Tay, K.Q., George, J., Ng, F.S., Govindarajan, K.R., Gong, Z., Mathavan, S., and Jiang, Y.J. (2008). Genomewide expression analysis in Zebrafish *mind bomb* alleles with pancreas defects of different severity identifies putative notch responsive genes. *PLoS ONE* **3**, e1479.
12. Lam, S.H., Mathavan, S., Tong, Y., Li, H., Murthy, K.R., Wu, Y.L., Vega, V.B., Liu, E.T., and Gong, Z. (2008). Zebrafish whole-adult-organism chemogenomics for large-scale predictive and discovery chemical biology. *PLoS Genet.***4**, e1000121.
13. Subramanian, A., Tamayo, P., Mootha, V.K., Mukherjee, S., Ebert, B.L., Gillette, M.A., Paulovich, A., Pomeroy, S.L., Golub, T.R., Lander, E.S., and Mesirov, J.P. (2005). Gene set enrichment analysis: a knowledge-based approach for interpreting genome-wide expression profiles. *Proc. Natl. Acad. Sci. U S A.* **102**, 15545–15550.
14. Lam, S.H., Winata, C.L., Tong, Y., Korzh, S., Lim, W.S., Korzh, V., Spitsbergen, J., Mathavan, S., Miller, L.D., Liu, E.T., and Gong, Z. (2006). Transcriptome kinetics of arsenic-induced adaptive response in zebrafish liver. *Physiol. Genomics* **27**, 351–361.

Chapter 14

Validating microRNA Target Transcripts Using Zebrafish Assays

Luke Pase and Graham J. Lieschke

Summary

Hundreds of tiny noncoding RNAs known as microRNAs (miRNAs) have been identified in the genomes of plants and animals. Studies are increasingly demonstrating that individual miRNAs are important in normal development and physiology. miRNAs are regulators of gene expression that bind target mRNAs and modulate their translation and turnover. The specificity of this regulation is achieved by partial sequence complementarity between the miRNA and its target mRNA. Understanding which mRNAs are targeted by each particular microRNA is critical to an understanding of the biologic role of any particular miRNA. Bioinformatic approaches can be used to predict mRNAs that may be miRNA targets, but each of these predictions requires experimental validation. We describe a method for a reporter assay based on a fluorescence intensity readout that uses transient techniques in zebrafish to easily deliver the reporter assay components. In addition, we describe a rigorously controlled strategy for determining the bona fide miRNA binding sites in the 3′UTR of mRNAs.

Key words: microRNAs, RNA, small interfering, Transient transgenesis, RNA duplex, Microscopy, fluorescence, Microinjections, Zebrafish.

1. Introduction

The development and differentiation of diverse cell types are achieved by the intricate regulation of an organism's genes. The coordination of gene expression is possible through the interplay between *trans*-acting elements (e.g., transcription factors) with *cis*-acting elements (e.g., promoter region, TATA box). A recently identified class of *trans*-acting elements regulating the transcriptome *(1–3)* is the large number of microRNAs (miRNAs). miRNAs bind to *cis*-acting elements within the 3′ untranslated region (UTR) of mRNA.

Graham J. Lieschke et al. (eds.), *Zebrafish,* Methods in Molecular Biology, vol. 546
DOI: 10.1007/978-1-60327-977-2_14

Currently, there are more than 800 different miRNAs discovered in a broad range of metazoans and many highly conserved across species *(1)*. miRNA expression is dynamically regulated, both temporally and spatially, throughout development *(4–6)*. Single-stranded, 19–25 nucleotide (nt), mature miRNAs are the product of a multistep maturation process. Newly synthesised miRNA precursor transcripts are cleaved by protein complexes containing the ribonucleases III Drosha or Dicer to produce a double-stranded miRNA–miRNA* duplex *(7)*. The mature miRNA strand, selected from the duplex by virtue of its lower thermal stability at its 5′ end, is incorporated into the miRNA RNA-induced silencing complex (miRISC) *(7)*.

The specificity of miRNA-mediated repression is achieved by sequence complementarity between the miRNA and mRNA 3′UTR *(8–10)*. In animals, although the binding is weak, complementarity (allowing for G:U mismatches) with the miRNA seed sequence (residues 2–7) is sufficient, for promoting miRNA-mediated repression *(10–12)*. The efficiency of repression is improved by expanding the mRNA seed sequence complementarity to seven nucleotides (residues 1–7 or 2–8), or even better to eight nucleotides (residues 1–8), and/or having expanded mRNA complementarity also embracing the 3′ portion of the miRNA *(9, 11, 12)*. Furthermore, multiple miRNA binding sites within a single mRNA result in an additive effect on repression, and if these are closely spaced (8 to ~40 nt), they often act in synergy *(11)*. Although these are key principles, other factors are also important, such as: secondary mRNA structure impairing access to the binding site; the position of the binding site along the 3′UTR; and the A/U composition surrounding the binding sites.

Target prediction algorithms incorporating a set of miRNA binding site parameters are excellent starting points for generating a list of potential mRNAs targeted by an individual miRNA, or of which miRNAs might target a particular mRNA of interest. Although these algorithms are continually being refined as more experimentally-proven interactions are demonstrated, there is still a high level of discrepancy between predicted and actual interactions. The false positive rate of the prediction algorithms is high, and false negatives also occur (e.g., gene silencing is observed despite gaps and mismatches in the seed sequence *[9, 13, 14]*). Therefore it is important to support a predicted miRNA–mRNA interaction as functionally capable by experimentation. One experimental approach is to demonstrate the capacity of the miRNA–mRNA interaction to function in a reporter assay.

This chapter provides a method for a highly controlled, fluorescent reporter assay for determining that a mRNA sequence contains a bona fide miRNA binding site, and for verifying the actual binding site. The assay is based on our experience *(15)* and is a refinement of the in vivo zebrafish assays initially reported *(16, 17)*. The delivery of a reporter mRNA encoding a fluorescent protein (e.g., green fluorescent protein, GFP) fused to a 3′UTR of interest and a synthetic miRNA duplex (that mimics the Dicer-processed miRNA–miRNA* duplex) tests for an interaction between the two that modulates fluorescence intensity. The reagents are simply delivered by microinjection into zebrafish embryos (**Fig. 1**). With the first needle, the GFP reporter mRNA is delivered to all embryos, minimizing inconsistencies in delivery. Then a random 50% of these embryos is reinjected from a second needle to deliver the miRNA duplex along with the red fluorescent tracer, rhodamine. The next day, the GFP fluorescence intensity between embryos in the two groups is compared to determine if an interaction between the miRNA and the GFP reporter mRNA has occurred. If there is miRNA mediated repression, one expects a destabilization of the reporter mRNA and/or inhibition of translation *(12)*, resulting in reduced abundance of GFP protein and subsequently

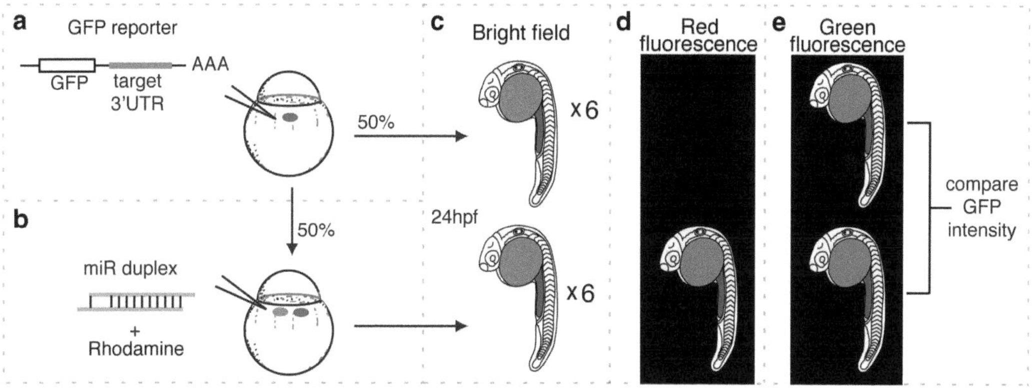

Fig. 1. Schema depicting the reporter assay methodology evaluating the capability of miRNA mediated repression of a selected 3′ UTR. (**a**) In vitro synthesized mRNA encoding GFP fused to the 3′UTR of interest is microinjected into the embryonic streaming of a single cell zygote. Approximately 50% of injected embryos are removed and allowed to develop. These embryos are the reference for maximum GFP fluorescent intensity for the injection session. (**b**) The remaining 50% of embryos are microinjected a second time to deliver a synthetic RNA duplex that mimics the miRNA of interest (miR duplex). Prior to injection, the miR-duplex is mixed with rhodamine to allow subsequent verification of miRNA delivery. (**c**) Once the embryos have developed for approximately 24 h, six embryos from each group are randomly selected under bright field and arrayed for assessment and documentation: six embryos from the GFP reporter injected alone group (*upper row*) and six embryos injected with the GFP reported and then with miR-duplex + rhodamine (*lower row*). (**d**) To confirm the delivery of miR-duplex to the selected embryos of the *bottom row*, embryos are assessed for red fluorescence. Only embryos that fluoresce red because of the presence of rhodamine contain the miR duplex. (**e**) To determine if there is miRNA mediated repression on the GFP reporter mRNA, embryos are assessed for green fluorescence. The GFP intensity between the *top row* and *bottom* of embryos is compared.

Table 1
Full Set of GFP Reporter Constructs for Evaluating a miRNA/mRNA 3′UTR Interaction with Delineation of the mRNA Target Binding Site

Reporter mRNA	Description	Schematic diagram
GFP reporter backbone	Capped mRNA encoding the GFP open reading frame flanked by vector sequence and polyadenylated through the SV40 late polyadenylation site. Subsequent reporter mRNAs contain the same backbone	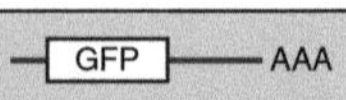
GFP:3′UTR	mRNA encoding GFP fused with a 3′UTR containing predicted miR-X binding sites	
GFP:3′UTR mutated	mRNA encoding GFP fused with a 3′UTR of which the seed sequence of the predicted miRNA binding site/s has been mutated	
GFP:3x binding site	mRNA encoding GFP fused with a predicted miRNA binding site in triplicate. Adjacent binding sites are separated by two nucleotides	
GFP:3x perfect site	mRNA encoding GFP fused with three sequential canonical miR-X binding sites. This reporter mRNA is more sensitive to the presence of miR-X	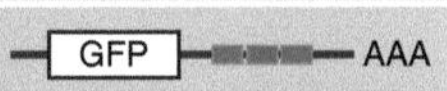

1 *Note*: In the schematic diagrams, the *grey box* indicates the fused test 3′**UTR** containing a putative miR-X target sequence (*white box*). *X* indicates a mutated site, and *darker grey boxes* indicate canonical binding sites (i.e., sites with 100% complementarity to miR-X)

weaker green fluorescence. The documentation of fluorescent intensity is sensitive to many technical factors, including image acquisition time, magnification, photo-bleaching due to prolonged exposure and single embryo variation. To achieve consistency in the documentation of the comparison between the control group (GFP reporter only) and experimental group (GFP reporter and then miRNA duplex injected), we recommend that several randomly chosen embryos of each group are arranged in two rows and documented together (*15*). We outline the following methods for undertaking this reporter assay: preparation of good injection needles, preparation of embryos for injection, microinjection of mRNA and miRNA duplex, and controlled documentation of the result. In addition, we provide a table (**Table 1**) and flowchart (**Fig. 2**) to assist in designing an appropriately controlled set of experiments to delineate a bona fide miRNA binding site. Throughout this protocol "miR-X" refers to the miRNA of interest that is the subject of study.

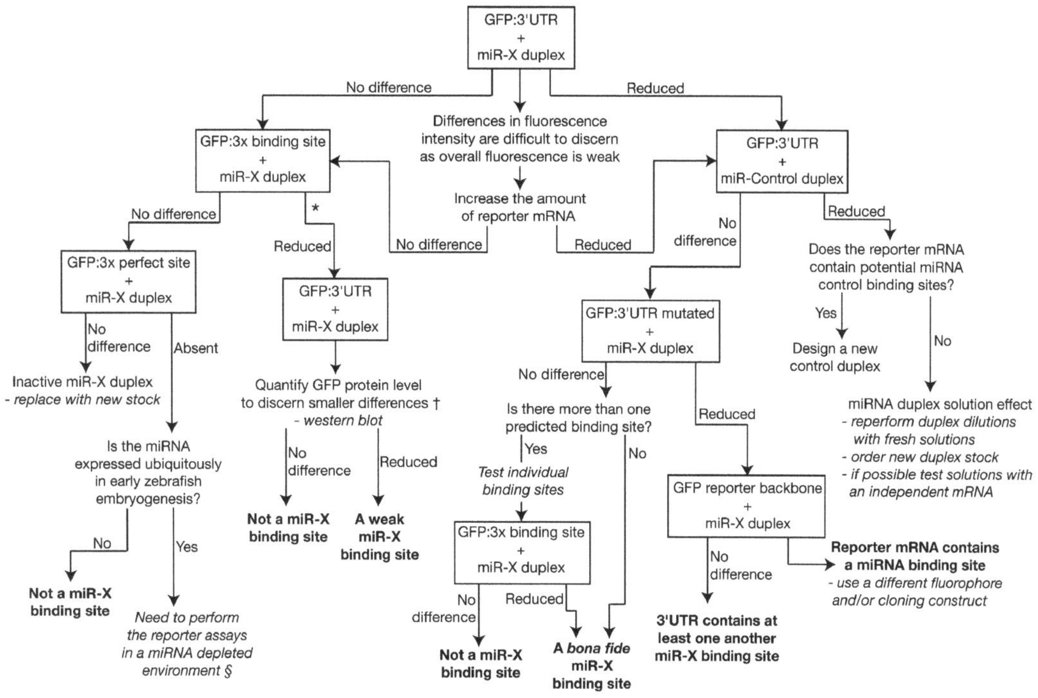

Fig. 2. Decision algorithm for evaluating a miRNA/mRNA interaction and delineating a bona fide functionally-capable miRNA binding site. Reporter assay components to be tested are *boxed*. *Arrows* indicate decision pathways, usually based on a comparison of fluorescence intensity between experimental and control groups in the reporter assay: "Reduced", GFP fluorescence levels are lower in the experimental group (GFP reporter + miR-X duplex) compared with control group (GFP reporter alone); "No difference", GFP fluorescence levels in the control and experimental groups are the same; "Absent", there is no or very little GFP fluorescence in the experimental group. Other decision questions and interpretations are unboxed. Comments are *italicized*. Definitive interpretations are **bolded**. *The multiple and closely spaced sites of the GFP: 3X binding site reporter mRNA provides a more sensitive capability test of a binding site/miRNA interaction. Interaction with the binding site needs to be confirmed in the context of its native 3′UTR. †Experiment requires an appropriate control (e.g., miR-Control duplex). §An embryo depleted of all miRNAs can be achieved by creating maternal-zygotic *dicer* mutant (*16*) or a specific miRNA knockdown with a miRNA morpholino oligonucleotide antagonist (*13, 19, 23*).

2. Materials

2.1. Zebrafish

1. Zebrafish wild-type lines (e.g. AB, WIK, and Tuebingen).

2.2. Instrumentation and Equipment

1. Needle puller: P-87 Flaming/Brown Micropipette Puller (Sutter Instruments; Navato, California) or equivalent.
2. Microinjection apparatus: Pico Injector PLI-100 (Harvard Apparatus; Holliston, MA) or equivalent.
3. Micromanipulator e.g. Standard Manual Control Micromanipulator, BS4-60-0570, (Harvard Apparatus) or equivalent.
4. Micromanipulator holder (GJ-8 magnetic stand, Narishige Scientific Instruments; Setagaya-ku, Tokyo) or equivalent.

5. Stereomicroscope (Leica – Wild Heerbrugg MDG-17; Heerbrugg, Switzerland) or equivalent.
6. Fluorescence stereomicroscope: Olympus SZX16; Mercury lamp device: Olympus U-RFL-T; Reflecting light fluorescence illuminator: Olympus SZX2-RFA16; Fluorescence filters: High performance GFP – SZX2-FGFPHQ, RFP2 – SZX2-FRFP2 (Shinjuku-ku, Tokyo). Or equivalent combination.
7. Camera: Olympus DP71 (Shinjuku-ku, Tokyo) or equivalent.
8. Image acquisition software (e.g., Olympus DP Controller) and analysis software (e.g., Olympus DP Manager).
9. Borosilicate filamented injection needles: 1.0 mm OD, 0.78 mm ID, 150 mm L (Harvard Apparatus, GC100TF-15).
10. 150-mm Petri dish or plastic box with plasticine/modelling clay.
11. 100-mm Petri dish.
12. Fine dissecting forceps: A.Dumont & Fils (no. 5).
13. Micropipet: P2, Gilson.
14. Injection tray mold (*see* **Note 1**).
15. Plastic transfer pipets (Samco; San Fernando, CA).
16. Sterile 1.5-mL microcentrifuge tubes.

2.3. Reagents, Media, and Solutions

1. In vitro transcribed, capped, polyadenlyated mRNA encoding GFP fused to the 3′YTR of the gene of interest, prepared using a kit according to the manufacturer's methods e.g. Ambion SP6 mMESSAGE MACHINE kit (Applied Biosystems; Austin, TX). Purify product with Quick Spin Columns (G-50 Sephadex Columns, Roche, Indianapolis, IN; cat. no. 1274015). Quantify mRNA and store aliquots at –80°C.
2. Synthetic miRNA duplex (siRNA Duplex Oligo 50 μM, Sigma-Proligo; St. Louis, MO) (*see* **Note 2**). Store in aliquots at –80°C. As well as a duplex to the miRNA of interest (miR-X), a control duplex should be procured (miR-Control) (*see* **Note 3**).
3. 0.5 mg/mL Phenol red solution (Sigma; cat. no. P0290).
4. Rhodamine-Dextran Stock: Rhodamine B isothiocyanate – Dextran (Sigma; cat. no. R8881) 50 mg/mL, prepared in 0.2 M KCl. Remove particles using Spin-X centrifuge tube filter, 0.22 μm (Costar, Corning, NY; cat. no. 8160).
5. E3 embryo medium (without methylene blue): 5 mM NaCl, 0.17 mM KCl, 0.33 mM $MgSO_4$, 0.33 mM $CaCl_2$, pH 7.5.
6. Benzocaine stock solution: Dissolve benzocaine powder (Sigma; cat. no. E1501) in 100% ethanol at 0.1 g/mL. Use at 1/1000–1/500 dilution to anesthetize embryos. Wrap in aluminium foil and store in the dark at room temperature.

7. PTU Stock Solution (100X): 0.3 g 1-phenyl-2-thiourea per 100 mL MilliQ water. Use at 1/100 dilution in E3 egg medium (i.e., working concentration of 0.003%).

3. Methods

The assay requires prior preparation, and is set up on one day and scored and documented on the next.

3.1. Prior to Injection Day

3.1.1. Panel of Reporter mRNAs

To thoroughly investigate the capability of a miRNA to mediate repression of a predicted target transcript mRNA, and to define the 3′UTR sequences through which the interaction is occurring, a set of reporter mRNAs needs to be tested. **Table 1** provides a comprehensive list of the reporter mRNAs required to fully evaluate a 3′UTR for miRNA interaction, using assays based on GFP fluorescence as the readout. The capped mRNA for microinjection is synthesized in vitro from DNA plasmid constructs using standard techniques.

3.1.2. Preparation of Microinjection Needles

1. Load borosilicate glass filamented capillary into a micropipette puller.
2. Pull capillaries using settings that produce a needle with a long, ridged, tapered tip (**Fig. 3**, and *see* **Note 4**). Multiple

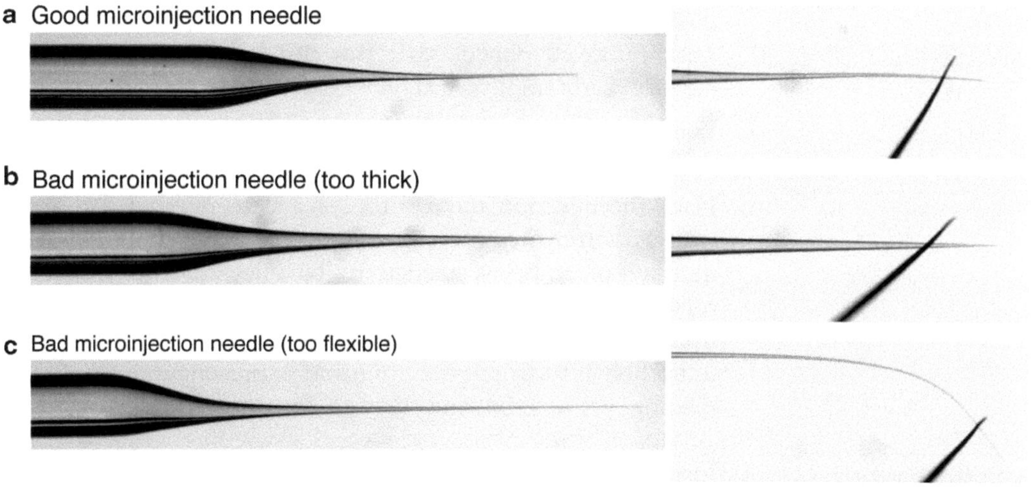

Fig. 3. Features of a good microinjection needle. *Left panels*, showing the taper and tip thickness of a good (**a**) and two different bad (**b** and **c**) microinjection needles. *Right panels* illustrate the flexibility in the tips of good and bad needles, demonstrated by a short strand of hair pressing against the tip. (**a**) A good microinjecting needle has a long thin tip to minimize damage to the embryo yet has minimal flexibility, so allowing easy piercing of the chorion and oocyte. (**b**) This needle will have a larger opening once broken because of the long thick taper. Although the tip is very rigid, the large opening will cause more injection trauma to the oocyte. (**c**) This needle has a very long thin tip that will cause little damage to the oocyte but bends easily making it very difficult to pierce the chorion.

needles will be used during each experiment, therefore it is necessary to pull many needles.

3. Store needles in a Petri dish on a row of plasticine or modeling clay to keeping the tips raised and prevent the needles from moving, hence minimizing breakage.

3.1.3. Preparation of Microinjection Tray

1. Prepare a 3% agarose solution in E3 embryo medium and dissolve by heating.
2. On a level bench, place a standard injection mold (*see* **Note 1**) on the bottom of a Petri dish (teeth up) and poor in the melted agarose to submerge the mold by approximately 3–4 mm.
3. Allow the agarose to set and then, using a scalpel, cut around the border of the injection mold.
4. Remove the agarose injection tray and store at 4°C submersed in E3 medium or water.

3.2. Day 1: Injection Day

3.2.1. Preparation of Injectates and Embryos

1. Prewarm microinjection tray to 28.5°C.
2. In a sterile microcentrifuge tube, dilute the reporter mRNA to 50 ng/μL (*see* **Note 5**) with nuclease-free water with or without phenol red (*see* **Note 6**) in a volume ≥2 μL. Vortex to mix and keep on ice.
3. In a different sterile microcentrifuge tube, dilute miR-X duplex to 10 μM with rhodamine-dextran stock in a volume ≥2 μL (*see* **Note 7**). Vortex to mix and keep on ice. Repeat if more than one duplex is to be tested.
4. Collect embryos from matings set in breeder boxes (*see* **Note 8**) in E3 medium (*see* **Note 9**) 15 min after the dividers were removed. Check that the clutch of embryos is of a uniform early single cell stage.
5. Load the first injection needle with the injectate containing the reporter mRNA.
6. Place the injection tray on top of a Petri dish lid and array within its trenches the early single cell stage embryos. The number of embryos needed to be injected with a specific reporter mRNA depends on the number of duplexes interaction to be tested (*see* **Note 10**). Aspirate excess E3 medium until there is sufficient water tension to prevent embryos freely moving, yet not allowing them to dry out.

3.2.2. Microinjection of Embryos with Reporter mRNA and miRNA Duplex

1. Insert the loaded microinjection needle into the needle holder of the injection apparatus, making sure an airtight seal is formed. We position the needle at an approximate 45° angle to the bench.
2. Under a stereomicroscope, break the very tip of the injection needle with forceps (*see* **Note 11**).
3. Adjust the chair height, microscope distance and eye pieces to a comfortable position.

4. Place the Petri dish lid containing the injection tray with arranged embryos on the microscope base.
5. Using controls on the micromanipulator, adjust the needle depth so its tip is just touching the top/right hand side of the chorion (top/left hand side if the needle is manipulated by the left hand) in the centre of the field of view.
6. Adjust the focus to the top surface of the embryo under approximately ×15 magnification.
7. Place one hand on the micromanipulator control device that only moves the needle in a stabbing motion. Place the other hand on the base of the Petri dish.
8. Twist the micromanipulator control to force the needle through the chorion and yolk cell (*see* **Note 12**).
9. At the same time, manipulate the Petri dish to position the needle tip within the embryonic streaming of the yolk (the ideal position is just under the embryonic cell).
10. Press the inject button to deliver a bolus ≤100 μm in diameter (*see* **Note 13** regarding bolus size). To increase efficiency, it is best to use the foot pedal to control this action.
11. Retract the needle using the micromanipulator. The embryo should be held in place by water tension.
12. Move Petri dish to position the next embryo in line with the needle and repeat the process.
13. Remove injected embryos from the injection tray by flooding with E3 embryo medium and collect in a Petri dish.
14. Load miR-X duplex into a microinjection needle.
15. Randomly select ≥30 injected embryos and array them into the injection tray trenches (*see* **Note 14**).
16. Repeat the procedure above to deliver an approximately uniform bolus size of miR-X duplex solution to all embryos.
17. Remove injected embryos from the injection tray with E3 embryo medium and collect into a new Petri dish.
18. Repeat for any other duplex to be tested with this reporter mRNA.
19. Incubate embryos at 28.5°C in E3 media (*see* **Note 15**) until they reach 22–28 h post fertilization (hpf) (*see* **Note 16**).

3.3. Day 2: Assessing the Outcome and Documenting the Assay Result

3.3.1. A Controlled Representation of a miR-X/GFP: 3′ UTR Interaction

1. Check that the microscope's fluorescence light source is distributed evenly across the field of view. This is important to allow accurate photographic documentation of the reporter assay. To test this, place a piece of white paper on the microscope base and look at the distribution of the fluorescent light. The light should be evenly distributed across the paper under the field of view (*see* **Note 17**).

2. Dechorionate the embryos and collect in clean PTU-supplemented E3 embryo medium.
3. Randomly select six embryos from the reporter mRNA alone group and six embryos from the reporter mRNA + miR-X duplex group into a clean Petri dish containing E3 embryo medium with benzocaine (*see* **Note 18**).
4. With a fluorescent stereomicroscope, evaluate the effectiveness of miR-X duplex delivery by determining the presence of the rhodamine tracer in the body of injected embryos. Discard embryos that contain strong rhodamine fluorescence in the yolk, as they are likely to have poor uptake of the miR-X duplex, and replace with further randomly-selected embryos.
5. Arrange the embryos into two rows, the top row comprising embryos injected with only the reporter mRNA, the bottom row comprising embryos injected with both reporter mRNA and miR-duplex traced with rhodamine (*see* **Note 19**).
6. Select a magnification that permits the capture of all twelve embryos and adjust the focus. Acquire three images: bright field, red fluorescence and green fluorescence (*see* **Note 20**).
7. Visual comparison of the GFP fluorescence level (green fluorescence image) between the two groups indicates if there is miRNA-mediated repression on the reporter mRNA (**Fig. 4**).
8. Assess the remaining embryos to confirm that the acquired image is representative of the overall experimental result (*see* **Note 21**).
9. Interpret the data with reference to **Fig. 2**.

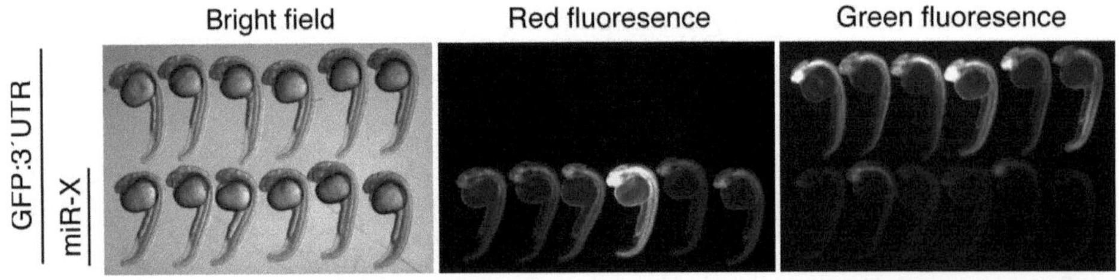

Fig. 4. Experimental example of miRNA mediated repression. Comparison of 22 hfp embryos injected with only GFP:3′UTR reporter mRNA (*top row*) with embryos receiving GFP:3′UTR and miR-X (*bottom row*). Brightfield photograph (*left panel*) demonstrates normal embryo development. miR-X delivery is traced by rhodamine and confirmed by positive red fluorescence in bottom row (*middle panel*). Green fluorescence brightness is reduced in embryos injected with miR-X, showing miRNA mediated repression of the GFP-encoding mRNA *right panel*; the lower row embryos has duller fluorescence. The array of representative embryos of both groups was photographed together in a single image to ensure valid comparison of relative green fluorescence intensity between the two groups.

4. Notes

1. The standard injection mold may be made from a plastic or machined from a brass plate. Full specifications of the standard injection mold can be found in The Zebrafish Book (http://zfin.org/zf_info/zfbook/chapt5/5.1.html).
2. Our experience (*15*) and that of others (*16*) is with synthetic siRNA designed with two 3′ DNA base overhangs and a mismatch with the second 5′ nucleotide of the miRNA. This mismatch promotes the correct miRNA strand to be incorporated into the miRISC complex. Others have successfully used full RNA duplexes with overhangs and mismatches with the first two 5′ nucleotides of the miRNA (*17, 18*), RNA duplexes that are identical to the endogenous miRNA/miRNA* (*19*), synthetic RNA oligonucltide representing the miRNA and a RNA with two 3′ DNA bases for the miRNA* (*20*) or just single strand RNA mimicking the miRNA (*13*).
3. Various control miRNA duplexes have been used: an irrelevant miRNA not expected to interact; one with ≥3 nt mutated within the seed sequence; one with a scrambled nt sequence.
4. Needles that have a thin but not very flexible tip are ideal for injections into the yolk of a recently fertilized embryo as it undergoes embryonic streaming (**Fig. 3**). Depending on the individual needle puller machine and the age of the filament, the settings that result in optimal needles will vary. As a guide, we commonly use the following settings on our machine: $P = 300$; Heat = 675; Pull = 200; Vel = 30; Time = 250.
5. We routinely use 50 μg/μL of reporter mRNA with reproducible results, however the use of concentrations between 50 and 250 μg/μL is reported (*19*).
6. If it is found difficult to monitor the delivery of uniformly-sized boluses of the clear reporter mRNA solution, Phenol Red Solution may be used as a diluent. Phenol Red is a non-fluorescent dye that can assist in accurately estimating the uniformity of bolus size during injections. Phenol Red will not interfere with the estimation of the subsequent miR-duplex injection containing rhodamine as phenol red diffuses quickly throughout the oocyte.
7. 10 μM miR-X duplex is commonly used, but up to 30 μM has been delivered (*21*).
8. Although it is common to use mass matings for microinjections, we recommend the use of breeder boxes. Breeder boxes (or equivalent) allow separation of two females from two males by a divider on the day before the injection day. This approach has the advantage that the spawning and mating

times can be controlled by removing the divider, resulting in more uniformly-staged, recently-fertilized oocytes.

9. Methylene blue is best avoided as it can complicate the fluorescence readout of the reporter assay.
10. Although six embryos are typically arranged for analysis, we recommend that for each experimental group ≥30 embryos to be injected. This generates a sufficient number of embryos for a confident assessment of the reporter assay outcome. For example, an interaction test of miR-X and miR-control with the reporter construct GFP:3′UTR requires ≥90 embryos to be injected with the reporter construct; ≥30 embryos only receive the reporter construct (i.e., control group), ≥30 embryos receive both the reporter construct and miR-X duplex, and ≥30 embryos receive both the reporter construct and miR-Control duplex.
11. We find that gently pressing the flat surface of the forceps handle perpendicular to the needle tip allows a small break of the needle tip.
12. At times, a gentle tap of the micromanipulator can assist the piercing of stubborn chorions.
13. In this reporter assay, only the embryos microinjected with the reporter mRNA mix in one injection session are compared. Therefore, consistency in bolus size for one injection session is more importance than delivering the same bolus size on different days. To maintain a consistent bolus size we recommend that all embryos that need the reporter mRNA mix to be injected sequentially with only one needle. We find that a correctly broken needle (*see* **Note 11**) should produce boluses of approximately ≥100 μm diameter at the following microinjector settings: 13 psi for approximately100 ms with a 1 psi forward pressure. It is recommended not to use an injection time lower than 3 ms as this results in increased variation between injections. Refer to Soroldoni et al. (*see* Chapter "Simple and Efficient Transgenesis with Meganuclease Constructs in Zebrafish") for a detailed method for calibrating an injection bolus.
14. Embryos randomly selected for miR-X duplex delivery will be older, however the experiment should only proceed if they are no later than the 2-cell stage.
15. To maintain embryos transparency, we recommend incubating embryos in E3 containing PTU-supplemented E3 embryo medium to block pigmentation. PTU should be added to the embryos after the first 6–8 h of development to prevent the induction of developmental abnormalities. At this stage, necrotic and unfertilised embryos should be removed before overnight incubation at 28.5°C.
16. Although miRNA mediated repression of a reporter GFP:3′ UTR is evident during epiboly and throughout convergence

and extension, it is difficult to uniformly document this in a group of embryos. This results primarily from the uneven distribution of cells around the yolk, together with the random axis on which they lie, which results in a variable density of fluorescent cells across the image. Embryos >22 hpf are preferred for scoring and documentation, as their protruding tail enables embryos to be uniformly positioned, providing a uniform density of cells from which fluorescence intensity can be estimated.

17. Refer to manufacturer's instruction manual to correct a poor microscope set-up resulting in uneven fluorescent light distribution.
18. It is common for injected groups to contain abnormal embryos. This may be a result of many factors, including mechanical injury during the injection process or a direct effect of the injectate. GFP mRNA and the subsequent translated GFP protein does not affect zebrafish development at the concentrations recommended. Therefore abnormal embryos in the group injected with reporter mRNA alone are likely to be a result of technical factors. These embryos may be excluded from the analysis. Experienced microinjectors can achieve abnormality rates due to technical factors of <10%. On the other hand, the ectopic expression of miR duplexes may potentially have profound effects on development. These may potentially affect the validity of the reporter assay, although not necessarily so (*17, 22*). miRNA-related abnormalities should be consistent across the embryos; care should be taken that the comparison of fluorescence intensity between the two groups is not compromised.
19. Arranging embryos into two groups is solely to simplify the labeling of acquired images (e.g., as in Fig. **4**).
20. A shorter exposure time is preferred to reduce overexposed/ saturated acquisitions.
21. We recommend that the assay be repeated on three independent occasions to verify reproducibility.

Acknowledgments

We thank the following for advice, valuable discussions and/or encouragement: W. Alexander, J. Heath, J. Layton, N. Nicola, P. Waterhouse. The work on which our experience is based was supported by NIH (R01 HL079545) and the NHMRC (461222, 461208). LP was supported by an Australian Postgraduate Award and a CSIRO Enhancement Stipend.

References

1. Bartel, D. P. (2004). MicroRNAs: genomics, biogenesis, mechanism, and function, *Cell* **116**, 281–297.
2. Nilsen, T. W. (2007) Mechanisms of microRNA-mediated gene regulation in animal cells, *Trends Genet* **23**, 243–249.
3. Stefani, G., and Slack, F. J. (2008) Small noncoding RNAs in animal development, *Nat Rev Mol Cell Biol* **9**, 219–230.
4. Landgraf, P., Rusu, M., Sheridan, R., Sewer, A., Iovino, N., Aravin, A., Pfeffer, S., Rice, A., Kamphorst, A. O., Landthaler, M., Lin, C., Socci, N. D., Hermida, L., Fulci, V., Chiaretti, S., Foa, R., Schliwka, J., Fuchs, U., Novosel, A., Muller, R. U., Schermer, B., Bissels, U., Inman, J., Phan, Q., Chien, M., Weir, D. B., Choksi, R., De Vita, G., Frezzetti, D., Trompeter, H. I., Hornung, V., Teng, G., Hartmann, G., Palkovits, M., Di Lauro, R., Wernet, P., Macino, G., Rogler, C. E., Nagle, J. W., Ju, J., Papavasiliou, F. N., Benzing, T., Lichter, P., Tam, W., Brownstein, M. J., Bosio, A., Borkhardt, A., Russo, J. J., Sander, C., Zavolan, M., and Tuschl, T. (2007). A mammalian microRNA expression atlas based on small RNA library sequencing, *Cell* **129**, 1401–1414.
5. Wienholds, E., Kloosterman, W. P., Miska, E., Alvarez-Saavedra, E., Berezikov, E., de Bruijn, E., Horvitz, H. R., Kauppinen, S., and Plasterk, R. H. (2005). MicroRNA expression in zebrafish embryonic development, *Science* **309**, 310–311.
6. Chen, P. Y., Manninga, H., Slanchev, K., Chien, M., Russo, J. J., Ju, J., Sheridan, R., John, B., Marks, D. S., Gaidatzis, D., Sander, C., Zavolan, M., and Tuschl, T. (2005). The developmental miRNA profiles of zebrafish as determined by small RNA cloning, *Genes Dev* **19**, 1288–1293.
7. Rana, T. M. (2007). Illuminating the silence: understanding the structure and function of small RNAs, *Nat Rev Mol Cell Biol* **8**, 23–36.
8. Lewis, B. P., Shih, I. H., Jones-Rhoades, M. W., Bartel, D. P., and Burge, C. B. (2003). Prediction of mammalian microRNA targets, *Cell* **115**, 787–798.
9. Brennecke, J., Stark, A., Russell, R. B., and Cohen, S. M. (2005). Principles of microRNA-target recognition, *PLoS Biol* 3, e85.
10. Lewis, B. P., Burge, C. B., and Bartel, D. P. (2005). Conserved seed pairing, often flanked by adenosines, indicates that thousands of human genes are microRNA targets, *Cell* **120**, 15–20.
11. Grimson, A., Farh, K. K., Johnston, W. K., Garrett-Engele, P., Lim, L. P., and Bartel, D. P. (2007). MicroRNA targeting specificity in mammals: determinants beyond seed pairing, *Mol Cell* **27**, 91–105.
12. Giraldez, A. J., Mishima, Y., Rihel, J., Grocock, R. J., Van Dongen, S., Inoue, K., Enright, A. J., and Schier, A. F. (2006). Zebrafish MiR-430 promotes deadenylation and clearance of maternal mRNAs, *Science* **312**, 75–79.
13. Flynt, A. S., Li, N., Thatcher, E. J., Solnica-Krezel, L., and Patton, J. G. (2007). Zebrafish miR-214 modulates Hedgehog signaling to specify muscle cell fate, *Nat Genet* **39**, 259–263.
14. Li, N., Flynt, A. S., Kim, H. R., Solnica-Krezel, L., and Patton, J. G. (2008). Dispatched homolog 2 is targeted by miR-214 through a combination of three weak microRNA recognition sites, *Nucleic Acids Res* **36**, 4277–4285.
15. Pase, L., Layton, J. E., Kloosterman, W. P., Carradice, D., Waterhouse, P. M., Lieschke, G. J. miR-451 regulates zebrafish erythroid maturation in vivo via its target gata2. Blood. 2008 Oct 10. [Epub ahead of print] DOI 10.1182/blood-2008-05-155812.
16. Giraldez, A. J., Cinalli, R. M., Glasner, M. E., Enright, A. J., Thomson, J. M., Baskerville, S., Hammond, S. M., Bartel, D. P., and Schier, A. F. (2005). MicroRNAs regulate brain morphogenesis in zebrafish, *Science* **308**, 833–838.
17. Kloosterman, W. P., Wienholds, E., Ketting, R. F., and Plasterk, R. H. (2004) Substrate requirements for let-7 function in the developing zebrafish embryo, *Nucleic Acids Res* **32**, 6284–6291.
18. Woltering, J. M., and Durston, A. J. (2008). MiR-10 represses HoxB1a and HoxB3a in zebrafish, *PLoS ONE* **3**, e1396.
19. Eberhart, J. K., He, X., Swartz, M. E., Yan, Y. L., Song, H., Boling, T. C., Kunerth, A. K., Walker, M. B., Kimmel, C. B., and Postlethwait, J. H. (2008). MicroRNA Mirn140 modulates Pdgf signaling during palatogenesis, *Nat Genet* **40**, 290–298.
20. Leucht, C., Stigloher, C., Wizenmann, A., Klafke, R., Folchert, A., and Bally-Cuif, L. (2008). MicroRNA-9 directs late organizer activity of the midbrain-hindbrain boundary, *Nat Neurosci* **11**, 641–648.
21. Mishima, Y., Giraldez, A. J., Takeda, Y., Fujiwara, T., Sakamoto, H., Schier, A. F., and Inoue, K. (2006). Differential regulation of germline mRNAs in soma and germ cells by zebrafish miR-430, *Curr Biol* **16**, 2135–2142.
22. Zhao, X. F., Fjose, A., Larsen, N., Helvik, J. V., and Drivenes, O. (2008). Treatment with small interfering RNA affects the microRNA pathway and causes unspecific defects in zebrafish embryos, *Febs J* **275**, 2177–2184.
23. Kloosterman, W. P., Lagendijk, A. K., Ketting, R. F., Moulton, J. D., and Plasterk, R. H. (2007). Targeted inhibition of miRNA maturation with morpholinos reveals a role for miR-375 in pancreatic islet development, *PLoS Biol* **5**, e203.

Part V

Imaging

Chapter 15

Multiple Embryo Time-Lapse Imaging of Zebrafish Development

Leah Herrgen, Christian Schröter, Lola Bajard, and Andrew C. Oates

Summary

Understanding the dynamics of developmental and cellular processes requires documentation of their changes with appropriate temporal and spatial resolution. Furthermore, simultaneous recording from a population of embryos under identical conditions allows statistical estimates of precision and variability to be made. This chapter describes a protocol for time-lapse microscopy of multiple embryos in parallel developing under tightly controlled conditions. This method is currently best suited to follow tissue-scale morphogenetic movements with temporal resolution in the minute range, for hours or even days. Applications of the method include the comparison of the dynamics of a process of interest between groups of wild-type embryos and their mutant siblings or between embryos treated with different chemical compounds. Temperature control allows for the investigation of the temperature dependence of a process of interest.

Key words: Zebrafish, Embryogenesis, Brightfield imaging, Time-lapse, Morphogenetic movements, Population statistics, Quantification, Temperature control.

1. Introduction

Time-lapse imaging of developing embryos has become a widely used tool in cell biological and developmental studies. Zebrafish embryos are well suited for this approach because their transparency and external development allow for easy visualization of a wide range of biological processes without adversely perturbing the embryo. As detailed elsewhere in this volume, techniques for observing single cells as well as cell populations in single embryos have reached a high level of sophistication. However, to enable the quantitative comparison of potentially noisy or variable processes

Graham J. Lieschke et al. (eds.), *Zebrafish*, Methods in Molecular Biology, vol. 546
DOI: 10.1007/978-1-60327-977-2_15

within and between populations of embryos, multiple embryos need to be observed simultaneously. We here report a protocol for time-lapse imaging of multiple entire embryos throughout development. This setup is particularly useful for analysis of global morphogenetic processes on a population level.

For multiple embryo time-lapse imaging, embryos are held in defined positions in agarose molds and a motorized stage enables the recording of single images or *z*-stacks of individual embryos at defined time intervals. Temperature in the dish can be monitored throughout the entire experiment using a thermocouple device. Movies are generated from images of individual embryos, and used for further processing and analysis. The high number of embryos observed in one experiment allows statistical treatment of the data. The temporal resolution of the system, determined by the duration of an imaging cycle, is in the range of minutes, and embryos can be continually monitored for 30 h or longer (**Fig. 1**). Brightfield imaging was the method of choice in most of our experiments. However, we anticipate that the power of this technique may increase considerably if combinations of different illumination settings are used, like multiple color fluorescence imaging, or combinations of brightfield and fluorescence imaging.

We have used this method to carry out an extensive characterization of the dynamics of zebrafish somitogenesis *(1)*, and we venture that this technique will be so versatile as to enable observation of a wide range of developmental processes. We anticipate

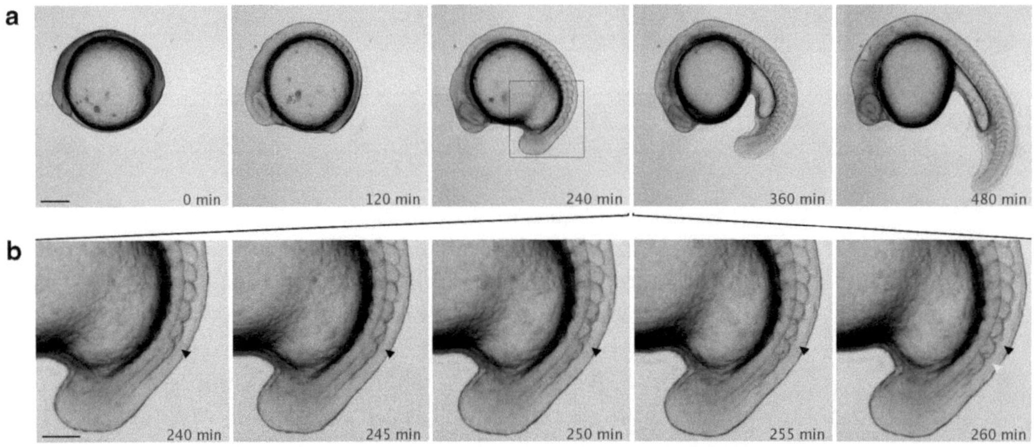

Fig. 1. Selected set of images from a time-lapse movie of a laterally mounted embryo. (**a**) Imaging of somitogenesis from bud stage until the 25-somite stage. The embryo shown is *nic*^{b107} homozygous *(2)*, and therefore genetically paralyzed, which makes it suited for imaging of late developmental stages over extended periods of time. *Scale bar* represents 200 μm. (**b**) Temporal resolution is in the range of minutes, which enables investigation of the dynamics of somite formation. A *black arrowhead* highlights the most recently formed somite boundary at 240 min. After approximately 20 min, a new somite boundary (*white arrowhead*) has formed. *Scalebar* represents 100 μm.

applications such as observation of the dynamics of eye morphogenesis, brain ventricle formation, epiboly, or other such general morphogenetic changes in early development.

2. Materials

2.1. Aluminum or Lucite Cones and Silicone Molds

1. Aluminum or Lucite plates (40 mm × 40 mm × 2 mm).
2. Drilling device from in-house workshop.
3. Silicone elastomer Sylgard 184 (Dow Corning).
4. Adhesive tape.
5. Medium-sized Petri dishes (55 mm × 15 mm).
6. 2% agarose in E3 medium.
7. E3 medium: 5 mM NaCl, 0.17 mM KCl, 0.33 mM $CaCl_2$, 0.33 mM $MgSO_4$, and 10^{-5}% methylene blue (*3*).

2.2. Imaging

1. Manipulation tools for orienting embryos. We use eyelash hair on a pencil, blunted glass probes, slightly blunt watchmaker's forceps, or similar non-damaging probes.
2. Sharpened watchmaker's forceps (size 5) for removing chorions.
3. Large plastic Petri dishes (94 mm × 16 mm) coated with 2% agarose/E3 for dechorionation.
4. Microscope system with motorized stage. We use a Zeiss Axioskop 200M with a Zeiss MCU 28 motorized stage. Lenses, filter sets, phase rings and illumination settings depend on the desired imaging technique.
5. Digital camera. We use a Photometrics HQ Coolsnap Camera.
6. Software package for driving the motorized stage and for image acquisition. We use MetaMorph software (version 6.2r4).
7. Binocular dissection scope for manipulating and orienting embryos. We use an Olympus SZ40 with magnification ×0.67 – 4.0.

2.3. Temperature Control and Monitoring

1. Temperature control devices: air conditioning and heater. We heat the room with an electric radiator (AKO-ISMET, type T909 TSIII) and achieve temperature regulation with a built-in air conditioning system (Silent; Axair).
2. Device to ensure even mixing of air in imaging room. We use a rotary ventilator or fan.

3. Thermocouple. We use a K-type thermocouple (Voltcraft Plus K202).
4. Data logger and software for temperature recording (Voltcraft Plus K202, ThermoLogger).

2.4. Optional Materials for Specific Adaptations

1. 96-well deep well plate (nerbe plus) for further treatment of time-lapsed embryos.
2. Separating plastic plates for subdividing one Petri dish into up to four different chambers (53 mm × 10 mm × 2 mm), for observation of embryos treated with different chemicals.
3. Ethyl-*m*-aminobenzoate methanesulphonate (Tricaine; Sigma-Aldrich). Prepare as 4 mg/mL stock solution in 1% NaH_2PO_4 buffer, pH 7.4. Store at −20°C and avoid exposure to light. Anesthetic to assist observation of late developmental stages.
4. Heterozygous *nic*^{b107} carrier fish (*2*); an alternative to using anesthetic.
5. Danieau's buffer: 58 mM NaCl, 0.7 mM KCl, 0.4 mM $MgSO_4$, 0.6 mM $Ca(NO_3)_2$, 5 mM HEPES, pH 7.1–7.3 (*2*).

3. Methods

Before the start of the first time-lapse experiment, a number of devices that are not commercially available need to be prepared, such as silicone cones for production of agarose molds for embryo mounting (*see* **Subheading 3.1**) and separating plates for observation of differentially treated embryos (*see* **Subheading 3.4, step 2**). Furthermore, the microscope system (*see* **Subheading 3.2**) and temperature control devices, if required (*see* **Subheading 3.3**), need to be set up and checked for constant functioning before the start of imaging.

Because of the strong temperature dependence of most, if not all, developmental processes in the zebrafish embryo, care must be taken when comparing and interpreting imaging results from environments with unknown or variable temperatures. Nevertheless, the core imaging protocol can be run without temperature control (*see* **Subheading 3.2**). Controlling and recording the temperature also is covered below (*see* **Subheading 3.3**).

3.1. Production of Aluminum or Lucite Cones and Silicone Molds

1. Drill depressions into Lucite or aluminium plate. Size and shape of the depressions have to be adjusted to the desired developmental stage and orientation of the embryo during imaging (*see* **Note 1** and **Fig. 2 a, a′, b, b′**).

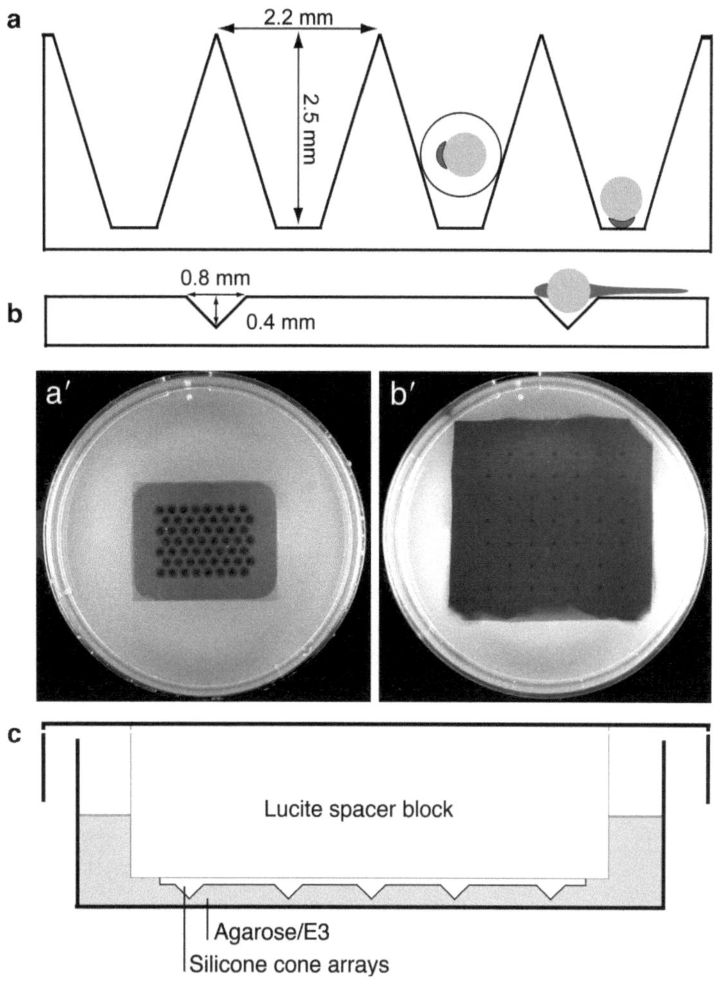

Fig. 2. Production of custom-milled agarose molds for imaging of zebrafish embryos. (**a**) Schematic drawing and dimensions of agarose molds suited for imaging embryos up to the 18-somite stage, as viewed from the side. (**a′**) Petri dish with ready-to-use array of agarose molds as depicted in (**a**), as viewed from above. (**b**) Schematic drawing and dimensions of agarose molds suited for imaging embryos from bud stage onward, as viewed from the side. (**b′**) Petri dish with ready-to-use array of agarose molds as depicted in (**b**), as viewed from above. (**c**) A silicone plate is fixed on a spacer block of Lucite glued into the lid of a Petri dish. Applying the lid to the agarose-containing Petri dish then creates molds of the desired size and spacing.

2. Mix the two components of the silicone elastomer kit with a spatula or similar device according to the manufacturer's instructions. Remove resulting air bubbles by centrifugation for 2 min at 860*g*.
3. Surround the depression plate with adhesive tape and cast a 5-mm layer of the silicone mixture onto the depression plate. Remove air bubbles by centrifugation for 1 min at 210*g* on plate buckets.

4. Let the silicone mixture solidify (2 h at 70°C or overnight at room temperature). The mixture will remain fluid for more than 12 h when stored at 4°C.
5. After the mixture has solidified, carefully remove the silicone plate with the cones from the depression plate. Remove excess silicone. The depression plate can be used repeatedly for production of silicone cone plates.

3.2. Imaging

1. Cast 6–8 mL of 2% agarose/E3 into a medium-sized Petri dish and place the silicone cone plate into the liquid agarose. Fixing of the silicone plate onto a spacer block of Lucite glued into the lid of the Petri dish is useful to minimize the thickness of the agarose layer between cone depressions and the bottom of the dish **Fig.** (**2c**).
2. Remove silicone cones after the agarose has set (*see* **Note 2**).
3. If required, dechorionate embryos with sharpened watchmaker's forceps in an agarose-coated Petri dish to avoid perturbation of the embryo resulting from adhesion to the plastic (*see* **Note 3** or **Subheading 3.6**).
4. Place one embryo in each conical well of the imaging dish when embryos have reached the desired developmental stage. Under a stereomicroscope set up right next to the time-lapse microscope, orient embryos as desired shortly before the beginning of recording (*see* **Note 4**), using slightly blunt forceps or some other gentle embryo manipulator.
5. Mount the Petri dish containing the oriented embryos onto the time-lapse microscope, and adjust illumination settings.
6. Use MetaMorph's MultiDimensional Acquisition tool to specify the variables of the time-lapse; these include number of time points, time interval between time points and, optionally, number and spacing of *z* planes (*see* **Notes 5** and **6**).
7. Select the positions to be monitored during the time-lapse using the manual control of the motorized stage, and save them via MetaMorph software.
8. Start the recording. It may be worthwhile to occasionally check the progress of the time-lapse.
9. Use the "Open Special/Build Stack" commands in MetaMorph to generate TIFF-stacks from the recorded images, and the "Make Movie" command to produce QuickTime movies (*see* **Note 7**).
10. The resulting stacks and movies as well as individual images are then used for further analysis, either by visual inspection or by automated image processing, depending on the experimental requirements.

3.3. Temperature Control

1. To obtain a stable temperature for the experiment, switch on heater and air conditioning system at least 2 h before starting the time-lapse. This allows benches, microscopes, and walls to reach equilibrium (*see* **Note 8**). Place empty imaging dish onto the microscope stage and switch on microscope, camera, and motorized stage (*see* **Note 9**).
2. Insert the temperature sensor of the thermocouple device into the agarose through a small hole in the lid after the dish containing the embryos has been mounted on the microscope (*see* **Note 10**).
3. Ensure air circulation around the body of the microscope (e.g., by placing a ventilator or fan in front of the microscope [*see* **Note 11**]).
4. Start temperature recording simultaneously with the time-lapse. Many commercially available thermocouples come with data logger software capable of monitoring temperature at specified intervals throughout the duration of the time-lapse. Save temperature data in a separate text file after the end of the time-lapse.

3.4. Further Treatment of Time-Lapsed Embryos

It may be important to identify the genotype or other distinguishing characteristic of the embryos for later correlation with the time-lapse movies.

1. Carefully remove the Petri dish from the stage after the time-lapse movies have been recorded.
2. Place embryos individually into a multi-well plate and let them develop until they have reached the desired stage.
3. Depending on the experimental requirements, these embryos can then be subject to higher magnification DIC brightfield or fluorescence microscopy. Alternatively, they can be fixed for immunostaining or *in situ* hybridization, or used for DNA extraction and subsequent polymerase chain reaction (PCR) genotyping *(3)*.

3.5. Differential Treatment with Chemicals in One Dish

If the experimental aim is the comparison of chemically treated and untreated embryos, two or four individual chambers can be created in one Petri dish with separating plates **(Fig. 3)**.

1. For creating four chambers, insert two plastic plates with small square incisions in the middle of one of the long sides **(Fig. 3a)**, which enables interdigitation of the two plates **(Fig. 3b)**.
2. For creating two chambers, grease three sides of a size-matched plastic plate with silicone grease, and position the plate in the Petri dish **(Fig. 3c)**.

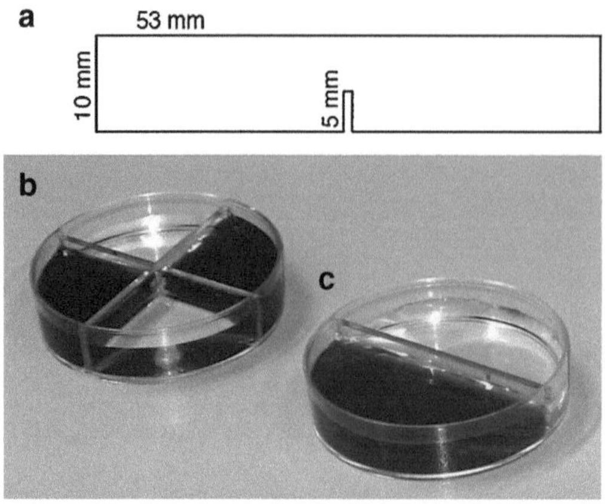

Fig. 3. Creation of multiple chambers in one Petri dish by insertion of separating plates. (**a**) Schematic drawing and dimensions of separating plastic plate with small square incision. (**b**) Four chambers are created by two interdigitating plastic plates. (**c**) Two chambers are created by one plastic plate. Different colored liquids may help to detect leaks between the chambers.

3. Check whether chambers are leak proof by filling differently colored liquids into the chambers and test for mixing after overnight incubation.
4. Use 4 mL of agarose/E3 for one of two chambers, and 2 mL for one of four chambers, and cover the agarose with E3.
5. Add the chemical to the desired experimental concentration. Calculate the amount of chemical to be added according to the total volume of agarose/E3 and E3.
6. Let the concentrations of chemicals in agarose/E3 and E3 equilibrate for at least 1 h before time-lapsing.

3.6. Documenting Late Developmental Stages

At approximately the 18-somite stage, zebrafish embryos start moving. At the same time, tail elongation results in bending of the axis as long as the embryos are in the chorion. Furthermore, shortly before the end of the segmentation process, embryos start to produce melanophores, which may preclude observation of some processes occurring beyond that stage (*4*). To accommodate these morphological changes in the imaging setup, integrate the following modifications into the core protocol (*see* **Subheadings 3.2** and **3.3**).

1. To accommodate the pronounced axial elongation following mid-somitogenesis stages, the distance between the silicone cones must be large enough (*see* **Note 12**). Use 0.5-mm deep conical depressions to fit the yolk of the embryo and guide tail elongation over the surface of the agarose (**Fig. 2b**).

2. The length of the zebrafish embryo at late-somitogenesis stages and beyond may be problematic if the entire embryo is to be documented. Choose a very low magnification objective, a digital camera with a large chip or an adapter to obtain a large enough field of view.
3. Cast agarose depression dishes as described above, but dissolve the agarose in 0.3X Danieau's buffer instead of E3. Cover the agarose with Danieau's and determine the volume of agarose and medium by weighing. Add Tricaine from the stock to a final concentration of approximately 0.016% (*see* **Note 13**) and let the anesthetic diffuse into the agarose for at least 30 min. Instead of chemical anesthetics, homozygous *nic*^{b107} embryos can be used, which are genetically paralyzed *(2)* (*see* **Note 14**).
4. Dechorionate embryos at any time before starting the recording using sharpened watchmaker forceps in an agarose-coated Petri dish. Transfer embryos to the depression dish and orient them laterally. If embryos are mounted at bud stage, the tail will elongate approximately in the direction of the tail bud. Orient all embryos with the bud pointing towards the same direction and set the region of imaging such that the tail will then elongate across the imaging area **(Fig. 1)**.

4. Notes

1. We found that depressions 2.2-mm wide and 2.5-mm deep **(Fig. 2a, a′)** are suited for imaging embryos both with and without chorion, either dorsally or laterally, respectively, up to the 18-somite stage *(1, 5)*. Conical depressions 0.8-mm wide and 0.4-mm deep **(Fig. 2b, b′)** are suited to image embryos laterally from bud stage onward. Finding the optimal depression design for other imaging purposes may require some experimentation. Number and arrangement of cones can be chosen as required.
2. For use on the same day, cover agarose molds with E3 medium immediately to prevent desiccation. A number of agarose molds can be prepared in advance and stored at 4°C for up to 2 wk.
3. Whether embryos need to be dechorionated or not depends on the scope of the experiment. Dechorionation may be advisable if embryos are to be mounted in an orientation that they do not normally adopt inside their chorions, such as dorsal imaging of somite stage embryos, which tend to lie on their sides inside the chorion. Furthermore, for all embryos to be imaged after the 16-somite stage, dechorionation is

necessary because tail outgrowth would otherwise result in bending of the embryo axis.

4. Allow sufficient time for dechorionation of embryos to start recording at the desired developmental stage. When mounting the embryos, take into account the morphogenetic changes that might alter the orientation of the embryo in the course of the time-lapse. Make sure the stereomicroscope and the time-lapse microscope are right next to each other to minimize transport distance between microscopes.
5. The number of images to be taken and saved in each round of imaging depends on the number of embryos and the number of *z* planes recorded, and this is the prime limiting factor of temporal resolution. In our setup, the time required for saving the individual images seemed to contribute the largest part to the time required to monitor one embryo. When imaging forty embryos and five *z* planes with 50 μm intervals per embryo, the minimal interval between time points is approximately 5 min. If the system cannot complete an imaging round in the time interval specified in the MetaMorph software, there will be no error message, but the system will go through repeated rounds of imaging without interruption. This results in disagreement of the supposed and real time interval, the latter being larger by an unknown amount. Therefore, it is important to match the time resolution of the experiment to the other parameters determining the number of images per round before the beginning of recording. Higher temporal resolution may be achieved by increasing the speed of the motorized stage, by reducing the number of embryos and the number of *z* planes per stack and by reducing file size of individual images.
6. The speed of the motorized stage must be adjusted to avoid dislodging the embryos form their oriented positions. We found a speed of approximately 4 mm/s appropriate for the cones described here.
7. Manual generation of stacks or movies in MetaMorph can be time-consuming if a large number of embryos is to be analyzed, and if selection of different *z* planes has to be performed. To automate this procedure, we used the freely available ImageJ software to produce focused and time-stamped stacks and movies *(1)*. The ImageJ Stack Focuser plugin (http://rsb.info.nih.gov/ij/plugins/stack-focuser.html) selects and combines the in-focus areas from different *z* planes of one time point. All focused images from one position are combined into a time-stack, time-stamped and saved both as a TIFF-stack and QuickTime movie. We wrote a multithreaded Java plugin, combining Stack Focuser, Time Stamper, and Movie Writer plugins to simultaneously

Table 1
Reporter Zebrafish Lines with Marked Myelomonocytic Compartments

Reporter line	Gene promoter	Myelomonocytic cells marked	References
Tg(fli1a:EGFP)y1	fli1a	Macrophages	*(18, 31)*
Tg(−5.3spi1:EGFP)gl21	spi1/pu.1	Early myeloid precursors	*(32)*
Tg(−9.0spi1:EGFP)zdf11	spi1/pu.1	Early myeloid precursors	*(33)*
Tg(mpx:GFP)i113/i114	mpx/mpo	Neutrophils	*(8)*
Tg(mpx:GFP)uwm1	mpx/mpo	Neutrophils	*(9)*
Tg(lyz:EGFP)nz117	lyz/lysC	Neutrophils/macrophages	*(19)*
Tg(lyz:DsRED2)nz50	lyz/lysC	Neutrophils/macrophages	*(19)*
Tg(gata1:DsRed)sd2	gata1	Erythromyeloid progenitors	*(17, 34)*
Tg(lmo2:EGFP)	lmo2	Erythromyeloid progenitors	*(17, 35)*
CLGY463(YFP)	myc enhancer	Neutrophils	*(13)*

2. Materials

2.1. Generating Ventral Fin Wounds for Inflammatory Response and Cell Tracking

1. E3 medium. 5 mM NaCl, 0.17 mM KCl, 0.33 mM $CaCl_2$, 0.33 mM $MgSO_4$.
2. Tricaine (Sigma-Aldrich). 0.168 mg/mL in E3, to anesthetize embryos and larvae.
3. 1-phenyl-2-thiourea (PTU, Sigma-Aldrich). Use at 0.003% in E3 to inhibit pigmentation in embryos older than 24 h post fertilization (hpf).
4. Fine forceps (Dumont no. 55).
5. Pronase (Roche). Use at 1 mg/mL in E3 to dechorionate embryos as previously described *(25)*.

2.2. Microinjection of CMNCBZ-Caged Carboxy-Q-Rhodamine

1. E3 medium. 5 mM NaCl, 0.17 mM KCl, 0.33 mM $CaCl_2$, 0.33 mM $MgSO_4$.
2. Tricaine (Sigma-Aldrich). 0.168 mg/mL in E3, to aaesthetize embryos and larvae.
3. PTU (Sigma-Aldrich). Use at 0.003% in E3 to inhibit pigmentation in embryos older than 24 hpf
4. CMNCBZ-caged carboxy-Q-rhodamine (cat. no. D34678) is supplied as a lyophilized reagent (Molecular Probes, Invitrogen). We have successfully used this reagent in lineage tracing experiments up to several months following initial resuspension and

storage at –20°C (5 mg/mL in sterile water supplemented with 100 mM KCl). All resuspensions should be protected from light and stored at –20°C.

5. FemtoJet pressure injection system (Eppendorf).
6. Microinjection needles (Warner Instruments Inc.).

2.3. Microinjection of pHrodo/Alexa Fluor 488/594 BioParticles

1. E3 medium. 5 mM NaCl, 0.17 mM KCl, 0.33 mM $CaCl_2$, 0.33 mM $MgSO_4$.
2. Tricaine (Sigma-Aldrich). Use at 0.168 mg/mL in E3 to anesthetize embryos and larvae.
3. PTU (Sigma-Aldrich). Use at 0.003% in E3 to inhibit pigmentation in embryos older than 24 hpf.
4. pHrodo *Escherichia coli* (E. coli) BioParticles Conjugate (Molecular Probes, Invitrogen, cat. no. P35361). Manufacturer recommends storage of lyophilized product at –20°C (should be stable for at least 6 mo). We have used pHrodo to live image phagocytosis up to 5 mo following initial resuspension and storage at –20°C (1 mg/mL in phosphate buffered saline [PBS] supplemented with 20 mM HEPES, pH 7.4). All pHrodo-containing solutions should be protected from light and stored at –20°C.
5. Alexa Fluor-488 (cat. no.E-13231) and –594 (cat. no.E-23370) *E. coli* BioParticles are supplied as lyophilized reagents (Molecular Probes, Invitrogen). When stored at –20°C, lyophilized product is stable for at least 1 yr. When resuspended according to manufacturer's instructions, product is stable for up to several weeks at 4°C. We have used Alexa Fluor *E. coli* BioParticles to live image phagocytosis up to 4 mo following initial resuspension and storage at –20°C (5 mg/mL in PBS supplemented with 2 mM sodium azide). All resuspensions should be protected from light and stored at –20°C.
6. 2% Methyl cellulose in E3.
7. FemtoJet pressure injection system (Eppendorf).
8. Microinjection needles (Warner Instruments Inc.).

2.4. Mounting Embryos/Larvae for Live Confocal Imaging

1. E3 medium. 5 mM NaCl, 0.17 mM KCl, 0.33 mM $CaCl_2$, 0.33 mM $MgSO_4$.
2. Tricaine (Sigma-Aldrich). Use at 0.168 mg/mL in E3 to anesthetize embryos and larvae.
3. PTU (Sigma-Aldrich). Use at 0.003% in E3 to inhibit pigmentation in embryos older than 24 hpf.
4. Low melting temperature agarose (Invitrogen).
5. 35 mm plastic Petri dish (Falcon).
6. 145 × 20 mm Plastic Petri dish (Greiner Bio-One, GmbH).

7. Sterile scalpel blade.
8. Eyelash manipulator.

3. Methods

3.1. Generating Ventral Fin Wounds for Inflammatory Response and Cell Tracking

The fin is well suited for time-lapse imaging of cell migration as it is only a few cell layers thick. This facilitates the rapid collection of *z*-series images owing to the small number of *z* sections required to completely image through the entire tissue. Between 12–15 *z* sections of no more than 3 μm are sufficient to capture all infiltrating leukocytes and track their migration throughout their contribution to the inflammatory response. The ventral fin is appropriate for these studies because of the adjacent population of leukocytes in the posterior blood island (PBI). The ventral fin wounding assay we use is an adaptation of that previously described for a neutrophil-specific reporter line *(9)*. A small wound inflicted at the ventral-most surface is sufficient to reproducibly mobilize cells to participate in the ensuing inflammatory event. Using this technique, within the Tg(lyz:EGFP/DsRED2) backgrounds, it is possible to identify a number of distinct leukocyte behaviors *(19)*. Dissecting the genetic determinants that drive these behaviors by coupling this live imaging with the inherent genetic tractability of the zebrafish system positions the zebrafish as a model with potential to significantly enhance the understanding of inflammation.

1. Raise transgenic reporter lines to appropriate age in E3 medium supplemented with PTU to inhibit pigmentation, if not bred into a pigment mutant background.
2. Dechorionate larvae, either manually using forceps or via pronase digestion. Take care to avoid damaging the larval tissue; such damage can result in inflammation and subsequent leukocyte recruitment, potentially diluting the inflammatory response to the induced ventral fin wound.
3. Anesthetize larvae in E3 medium supplemented with tricaine (0.168 mg/mL) before generating a wound using clean fine forceps (Dumont no. 55, tip dimensions of 0.05 × 0.02 mm) by gently applying pressure to the ventral-most surface of the ventral fin (**Fig. 1**). We induce wounds within a 2-somitewide domain, 5–6 somite lengths from the cloaca (**Fig. 1a**). Using this technique it is possible to generate a wound that is reproducible both in size and inflammatory response, as measured by numbers of marked infiltrating leukocytes within the Tg(lyz:EGFP) line (*see* **Note 1**).

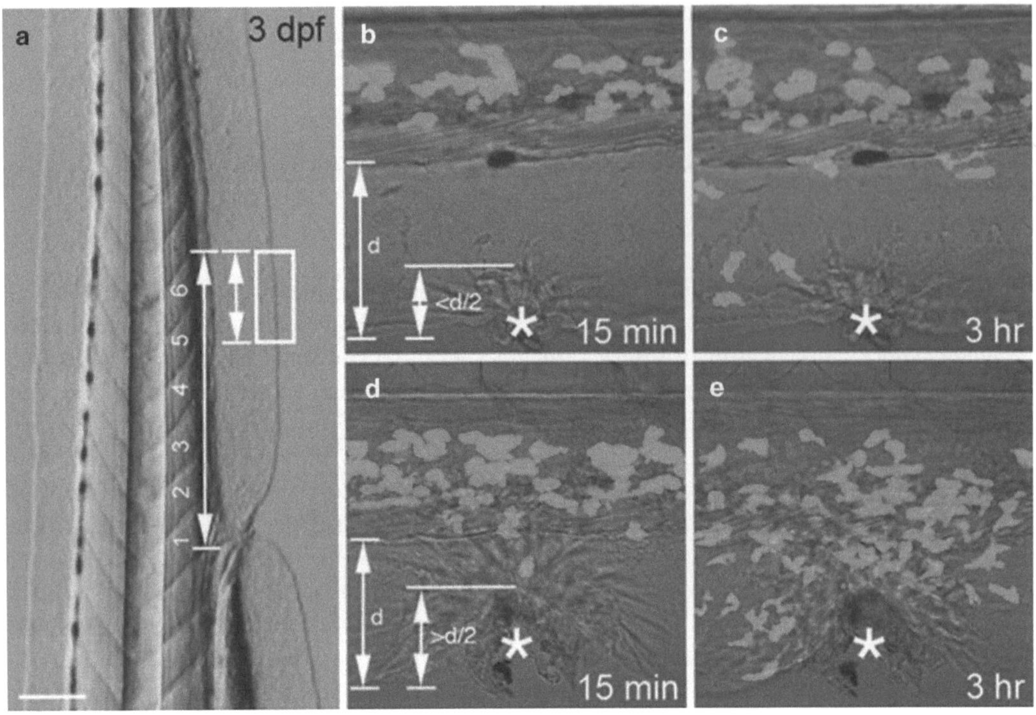

Fig. 1. Magnitude of inflammatory response correlates with size of induced wound. (**a**) Tail of three dpf larva (lateral view, anterior towards bottom) demonstrating the region in which ventral fin wounds are generated (*white box*). Somites are numbered posteriorly from cloaca. (**b–e**) Confocal imaging of EGFP-marked cells (merged with bright field) in ventral fin region of three dpf Tg(lyz:EGFP) larvae (lateral view, anterior to left) 15 min (**b** and **d**) and 3 h (**c** and **e**) following wounding. *denotes wound. Measurement *d* represents the length from ventral most fin surface to the caudal vein. The size of the wounds are expressed as a fraction of *d*, either < *d*/2 (**b** and **c**) or > *d*/2 (**d** and **e**). Scale bar: 100 μm in **a** (*see Color Plates*).

4. Immediately mount wounded larvae for live confocal imaging (*see* **Subheading 3.4** and **Note 2**).
5. The typical scanning settings we use for imaging ventral fin wounds in 3–6 dpf Tg(lyz:EGFP) and Tg(lyz:DsRED2/fli1a:EGFP) larvae are as follows:
   - ×60 Water immersion lens.
   - Sequential detection.
   - Multi-line Argon (488 nm) and green Helium Neon (543 nm) lasers set between 1 and 3%.
   - Scan speed of 8.0 μs/pixel.
   - Image format of 320 × 320.
   - To reduce noise, the Kalman filter was used with averaging set to four cycles.
   - 12–15 *z* sections (depending on age of larvae), no greater than 3 μm apart (*see* **Note 3**).
   - 1.0–1.5 min between time points (*see* **Note 4**).

6. Incrementally increase the laser intensity during imaging to compensate for photobleaching during long time-lapse runs (*see* **Note 5**).
7. Because larval tissues around the wound contract during wound healing, the region of interest (ROI) occasionally drifts (typically along the *x*- and *z*-axes). To compensate for shift along the *x*- and *y*-axes, small manual adjustments to the microscope stage may be necessary to keep the ROI stationary. When the ROI drifts along the *z*-axis, momentarily pause imaging and set new upper and lower *z* limits (*see* **Note 6**).

3.1.1. Image Analysis and 4D Reconstruction

1. Generate Z projections for each time point using ImageJ ×1.38 *(26)* (http:///www.rsb.info.nih.gov/ij/) then align individual time frames using the "Align Slice" plugin to compensate for any slight movements of the ROI during the course of the time-lapse.
2. Individual migratory paths of the marked leukocytic compartment can be determined using the ImageJ Manual Tracking plugin (http://www.rsb.info.nih.gov/ij/plugins/index.html) **(Fig. 2a)**.
3. Using the Imaris 5.7.2 (Bitplane) 3-D and 4-D image analysis suite, a 3-D volume rendered representation of the collected data can be generated **(Fig. 2b)**. To reveal more surface detail an Isosurface can also be applied to the data **(Fig. 2c–e)**. Rotation of such data sets exposes positional information in the *z*-axis that would otherwise go undetected using more traditional 2-D time-lapse analysis. For example, in **Fig. 2a**, a 2-D projection reveals two cells (denoted by arrow and arrowhead) that appear to be touching. Following 3-D reconstruction using the same data set, a separation in the *z*-axis between these two cells is clearly evident **(Fig. 2d, e)**. As mentioned earlier, immune cell compartments demonstrate intimate cell-to-cell contacts that are fundamentally important in their managing of the inflammatory response. The ability to reliably identify genuine cell-to-cell contacts between infiltrating immune cell compartments during inflammation is important to fully appreciating how these cells participate in the infection/wound healing event.

3.2. Microinjection of CMNCBZ-Caged Carboxy-Q-Rhodamine, Uncaging and Confocal Imaging for Live Cell Tracking and Lineage Tracing

A number of recent studies using the zebrafish have utilized photoactivatable caged fluorescent dyes to trace specific cell populations and explore their differentiation potential during early development *(11, 17, 27, 28)*. Novel insights into early definitive hematopoiesis have been obtained from such studies, in which putative blood stem cells have been traced from their region of origin to various hematopoietic niches where their potential to contribute to definitive blood lineages has been assessed.

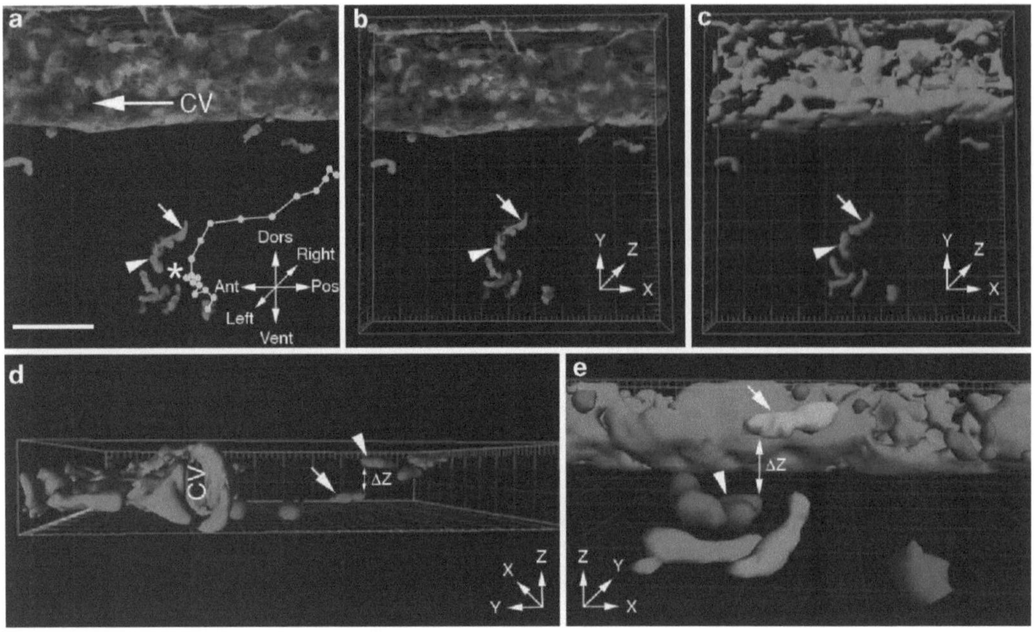

Fig. 2. Multi-dimensional imaging of leukocytes during inflammation. (**a**) Projection of a Z series collected during a time-lapse analysis within a wounded ventral fin of a six dpf Tg(lyz:DsRED2/fli1a:EGFP) larva. Large *white arrow marks* the location and direction of circulation within the caudal vein (CV). * marks wound. *Yellow lines* mark migratory path of an infiltrating leukocyte. *Small white arrowhead* and *arrow* mark DsRED2-expressing leukocytes that appear to be in contact with each other in the Z projection view. (**b** and **c**) Volume and Isosurface rendered views, respectively, of the same data set shown in **a**. (**d**) Panel (**c**) rotated 90° around the *x*- and *y*-axes to reveal distribution of infiltrating leukocytes along the *z*-axis. (**e**) Panel (**d**) rotated 90° around the *Z*-axis to reveal that the two DsRED2-expressing leukocytes (marked with *small white arrow* and *arrowhead* in all views) are separated in the *z*-axis (ΔZ). Scale bar: 50 μm in **a** (*see Color Plates*).

This technology also has utility in tracking specific immune cells before, during, and after their contribution to an inflammatory response. Live imaging of these events has the potential to enhance the understanding of the resolution of inflammation, a process that must be exquisitely controlled so as to ensure that immune cells disperse only after any risk for infection has been adequately managed *(29)*. Understanding how and why neutrophils and macrophages "decide" to disperse or die and the determinants that control these fate decisions has tremendous potential for therapeutic applications.

1. For photolysis experiments, we use a 10,000 molecular weight (MW) dextran conjugate of CMNCBZ-caged carboxy-Q-rhodamine. We are able to successfully uncage this dye using a 405 nm diode laser attached to an Olympus FV1000 confocal laser scanning microscope, despite the CMNCBZ caging agent possessing an absorption maximum of 310 nm. Once uncaged, this dye exhibits orange fluorescence with excitation/emission maxima of 545/575 nm.

2. Prepare the CMNCBZ-caged carboxy-Q-rhodamine for microinjection as follows:
   (a) Resuspend the caged dye to a concentration of 5 mg/mL in sterile water supplemented with 100 mM KCl.
   (b) Use a 1:5 dilution (1 mg/mL) for microinjection.
   (c) Protect solutions from light.
3. Inject 1 nL of the injection mixture into one-cell stage Tg(lyz:EGFP) embryos (*see* **Note 7**). To protect the dye from light, inject using minimal illumination. Allow injected embryos to recover at 28.5°C in E3 medium supplemented with PTU and protected from light until required for either lineage tracing experiments or leukocyte tracking (using the ventral fin wounding assay).
4. For leukocyte tracking experiments, generate a ventral fin wound (*see* **Subheading 3.1**) using minimal illumination and mount the larvae for uncaging and imaging (*see* **Subheading 3.4**).
5. Alternatively, for lineage tracing experiments, raise injected embryos to the desired age and similarly mount (*see* **Subheading 3.4**), using minimal illumination.
6. For leukocyte tracking experiments, monitor the wound site every 2–3 min to detect the first-arriving EGFP-marked leukocytes **(Fig. 3a)**. Target individual leukocytes for uncaging using the ×60 water immersion objective and additional ×5–7 magnification (*see* **Note 8**). The SIM scanner enables illumination of any designated area with the 405 nm excitation laser (*see* **Note 9**). The size of the area to be uncaged can be set manually using an on-screen ROI drawing function **(Fig. 3b)**. Once the correct region is set to suit the size/shape of the targeted cell, it is exposed to the 405 nm laser (*see* **Note 10**). Typical settings are 25–30% laser intensity/voltage, 1 s illumination and 2–3 s resting time. Using these parameters, we detect rhodamine fluorescence following 2–3 illuminations **(Fig. 3c)**. If the intensity/voltage of the 405 nm laser is too high, rapid photobleaching of the EGFP signal will occur. This photobleaching is a manageable side-effect of the uncaging process due to high EGFP expression driven by the lyz promoter **(Fig. 3d)**. The number of illuminations required will vary from cell to cell and depend upon the amount of caged dye within the cell and the EGFP fluorescence intensity of the marked cell (cells strongly expressing EGFP, if required, can be exposed more frequently to the excitation laser as they can tolerate more photobleaching).
7. For lineage tracing experiments, uncage cells in the same fashion then remove from the 1% agarose mounts and leave to

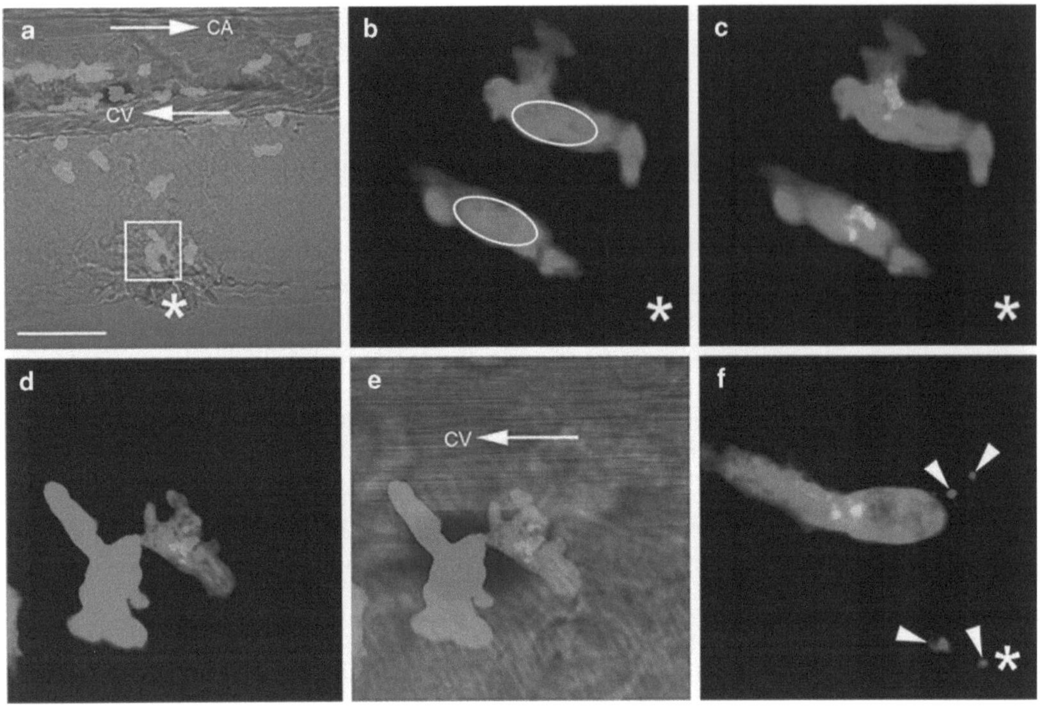

Fig. 3. Tracking/lineage tracing uncaged leukocytes in Tg(lyz:EGFP) larvae. (**a–f**) Confocal images (merged with bright field in **a** and **e**) before (**a** and **b**) and after (**c–f**) uncaging of CMNCBZ-caged carboxy-Q-rhodamine within three dpf Tg(lyz:EGFP) larvae following wounding (all lateral views, anterior to left). * marks the wound region. *Arrows* highlight location and direction of circulation within the caudal artery (CA) and caudal vein (CV). (**b** and **c**) Higher magnification of boxed region in A before (**b**) and immediately following (**c**) uncaging. White boundaries in B mark the region of interest to be targeted with the 405 nm (uncaging) laser. (**d** and **e**) An uncaged leukocyte that has migrated back towards the CV demonstrating photobleaching of the EGFP signal relative to a leukocyte that has not been exposed to the 405 nm laser. (**f**) Rhodamine dye (*arrowheads*) uncaged outside of the targeted EGFP-expressing leukocyte. Scale bar: 50 μm in **a** (*see Color Plates*).

recover in E3 medium supplemented with PTU at 28.5°C while being protected from light. At the desired stage mount larvae and determine distribution of uncaged cells.

3.3 Microinjection of pHrodo/Alexa Fluor 488/594 BioParticles and Live Confocal Imaging of Phagocytosis

To live image the phagocytic capacity of early myelomonocytes and their potential to clear the early embryo/larvae of infection we employ fluorescently labeled heat- or chemically-killed *E. coli* (K-12 strain) conjugated with either Alexa Fluor 488 or Alexa Fluor 594 dyes (when imaged within Tg(lyz:DsRED2) or Tg(lyz:EGFP) lines, respectively). These reagents, with natural antigenicities, permit direct observation of phagocytosis as well as the tendency of these dead bacteria to adhere extensively to the extracellular surface of the leukocytes when administered via the circulation.

We also employ another Molecular Probes BioParticle called pHrodo. This product was originally designed as an indicator of phagocytosis for cell culture applications. Like the Alexa Fluor-labeled particles described above, these are heat- or chemically-killed *E. coli*. However, their conjugated dye emits red fluorescence only when exposed to an acidic environment, such as that experienced within an intracellular phagocytic compartment. We have adapted this reagent to mark the phagosomal vesicles within zebrafish leukocytes following delivery into the early circulatory system. In addition to live imaging applications, due to the stability of the Alexa dyes and that of pHrodo, applications could extend to isolating pure populations of activated fluorescently-labeled leukocytes also marked with phagocytosed fluorescent bacteria for further analysis, such as expression profiling.

1. Prepare the pHrodo *E. coli* BioParticles for microinjection:
   - (a) Resuspend a single vial containing 2 mg of lyophilized pHrodo reagent in PBS (1 mg/mL) supplemented with 20 mM HEPES, pH 7.4, to minimize background fluorescence from non-phagocytosed particles.
   - (b) Use a 1:5 dilution (200 μg/mL in PBS/HEPES solution) for microinjection.
   - (c) Protect solutions from light.
2. Prepare the Alexa Fluor 488 and 594 *E. coli* BioParticles for microinjection:
   - (a) Resuspend the Alexa Fluor BioParticles in PBS (5 mg/mL) supplemented with 2 mM sodium azide. The manufacturer (Invitrogen) recommends the addition of sodium azide for prolonged storage (several weeks at 4°C).
   - (b) Vortex (3 × 15 s at maximum power) to generate a homogenous suspension and disrupt aggregated bacteria.
   - (c) Use a 1:20 dilution (250 μg/mL in PBS) for microinjection.
   - (d) Protect solutions from light.
3. Microinject the Alexa Fluor and pHrodo BioParticles into the circulation (via the sinus venosa) of anesthetized Tg(lyz:EGFP/DsRED2) embryos and larvae immobilized in 2% methyl cellulose in E3. Inject approximately 40–50 nL through 5–10 separate injection pulses. For microinjections we use an Eppendorf FemtoJet microinjection system. Perform microinjections under minimal illumination to preserve the fluorescence signal of the BioParticles. To monitor injection success, immediately following microinjection, analyze the embryos/larvae under the appropriate filter set using a fluorescent stereomicroscope.
4. Immediately mount for confocal imaging (*see* **Subheading 3.4**).
5. The excitation/emission maxima for the different BioParticles are:
   - (a) Alexa Fluor 488, 495 nm/519 nm.

(b) Alexa Fluor 594, 590 nm/617 nm.

(c) pHrodo, 560 nm/585 nm.

6. EGFP- and DsRED2-expressing leukocytes can be observed phagocytosing the fluorescently labeled *E. coli* from as early as 20 min post-injection (**Fig. 4a–h**). When imaging Tg(lyz:EGFP) embryos/larvae injected with pHrodo, phagosomal compartments within marked leukocytes begin to emit

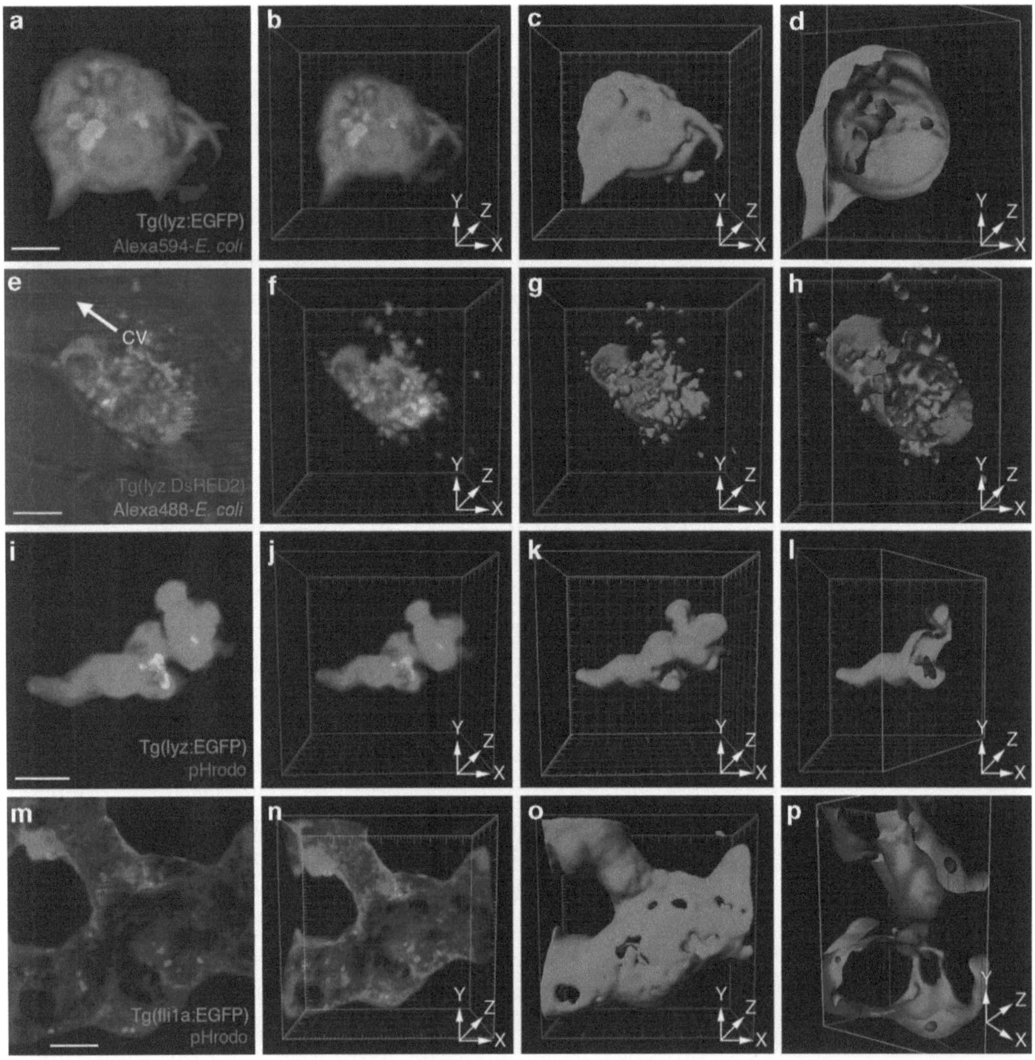

Fig. 4. Live multi-dimensional imaging of phagocytosis by leukocytes. *z* projection (**a**, **e** (merged with bright field), **i** and **m**), volume rendered (**b**, **f**, **j** and **n**) and Isosurface (**c/d**, **g/h**, **k/l** and **o/p**) views of an EGFP-expressing leukocyte on the yolk containing Alexa Fluor 594 *E. coli* (**a–d**), a DsRED2-expressing leukocyte in the PBI surrounded by, and containing, Alexa Fluor 488 *E. coli* (**e–h**), an EGFP-expressing leukocyte within the PBI containing red-fluorescent pHrodo (**i–l**) and EGFP-expressing endothelial cells of the caudal vascular plexus containing red fluorescent pHrodo (**m–p**). *Arrow* in **e** marks the direction of circulation in the caudal vein (CV). Views in **d**, **h**, **l** and **p** represent cutting planes through the marked cells revealing the phagocytosed fluorescent contents. Scale bars: 3 μm in **a**; 5 μm in **e** and **i**; 10 μm in **m** (*see Color Plates*).

intense red fluorescence from approximately 30 min post injection **(Fig. 4i–l)** (*see* **Notes 11** and **12**).

7. To slow the often rapid migration demonstrated by fluorescently-marked leukocytes within Tg(lyz:EGFP/DsRED2) larvae and permit longer scanning periods (either to enable higher resolution images or when collecting multiple *z* series time-lapse data), place mounted embryos/larvae on ice to slow their metabolism.

3.4. Mounting Embryos/Larvae for Live Confocal Imaging

1. Raise embryos/larvae in E3 supplemented with PTU until the desired age, then anesthetize with tricaine.
2. Mount in 1% low melting temperature agarose (in E3 medium supplemented with PTU and tricaine) within a 35 mm plastic petri dish **(Fig. 5a)**.
3. First, create an agarose bed approximately 5 mm deep within the 35-mm Petri dish. Once set, excavate a small trench from the surface of the agarose (using a small scalpel blade) to support the embryo/larvae in the orientation desired for imaging. Place the anesthetized specimen into the trench immediately followed by enough agarose/tricaine/PTU solution to provide a 1 mm-deep layer **(Fig. 5a)**. While this layer is setting manipulate the embryo (using an eyelash manipulator) within the trench to obtain the desired orientation.

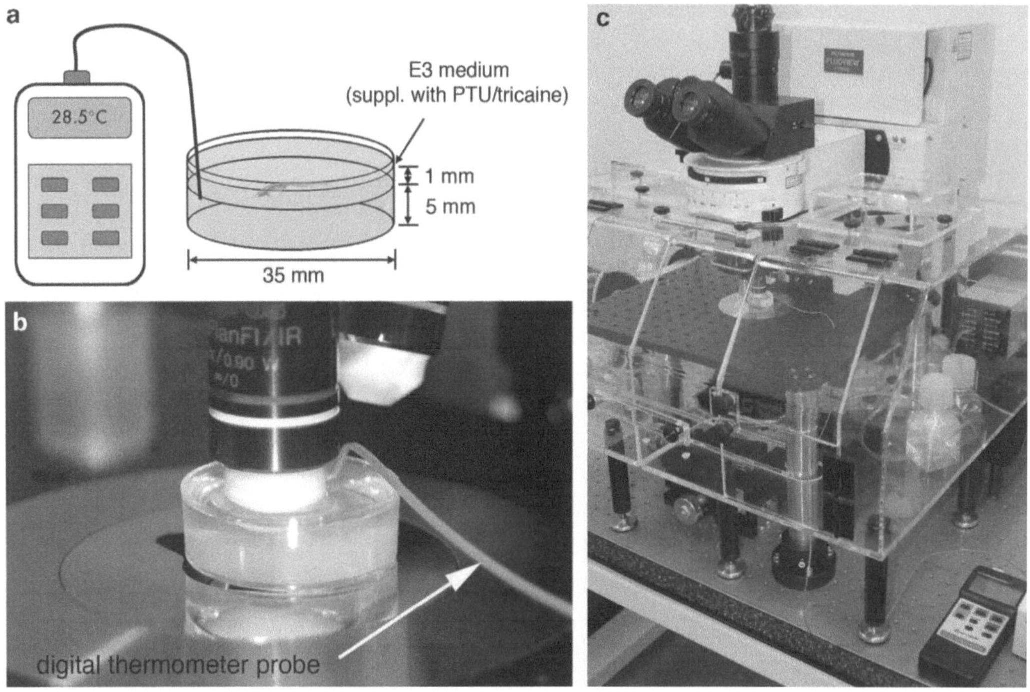

Fig. 5. Live confocal imaging setup. (**a**) Schematic diagram of mounted embryo. (**b** and **c**) Views of mounted embryo within incubation chamber of an Olympus FluoView FV1000 confocal microscope (*see Color Plates*).

4. When the agarose has polymerized, apply a small volume of E3 (supplemented with PTU and tricaine) to the surface of the agarose-mounted specimen to prevent the embryo/larvae from drying out (*see* **Note 13**).
5. Next, insert a digital thermometer probe into the agarose to monitor the temperature of the mounted specimen during imaging **(Fig. 5a–c)**.

4. Notes

1. When generating the ventral fin wounds care must to taken to generate a wound that does not extend dorsally beyond one-half the width of the fin. Wounds that are too large and extend too far towards the caudal vein result in a severe inflammatory response, where large numbers of responding cells make it impossible to identify and subsequently track individual leukocyte migratory paths (compare **Fig. 1b, c** with **Fig. 1d, e)**. Furthermore, in large wounds, contraction of the epithelial cells during subsequent wound healing can physically obstruct circulation in the immediately dorsal segment of the caudal vein.
2. All live confocal imaging described in this chapter was performed on an Olympus FV1000 confocal laser scanning microscope equipped with long working distance ×20 and ×60 water immersion lenses and an incubation chamber, the temperature in which was adjusted to maintain the larvae at a constant temperature of 28.5°C (as measured using a digital thermometer probing the embedding agarose). The optimal settings were empirically determined and depended upon the strength of the fluorescent signal in the marked cells, the length of the intended imaging run, the ROI size, and the image resolution required. In general, the laser voltage and degree of resolution were such that the specimen spent minimal time exposed to the least intense illumination that still generated images with an acceptable signal-to-noise ratio. All settings reported here were optimized for the Olympus FV1000 system and will need to be adapted to other confocal setups.
3. Typically *z* sections collected are no greater than 3 μm apart to ensure that all labeled migrating leukocytes are detected throughout the inflammatory response.
4. The time between each *z* series was routinely 1 min (never greater than 1.5 min) so that migratory paths of individual marked leukocytes could be maintained. When the time is

too great, differentiating between individual cell trajectories becomes increasingly difficult given that leukocytes can migrate at up to 15 μm/min and dramatically alter their migratory path.

5. We routinely collect time-lapse *z* series for up to 5 h, after which the larvae begin to deteriorate. This likely results from their sensitivity to prolonged tricaine exposure. Imaging within a paralyzed mutant such as the *nic1* background may overcome this obstacle (*30*).
6. Manually adjusting the microscope stage along the *x*- and *y*-axes (as well as setting new *z*-section limits) is rarely required and depends upon how the induced wound heals.
7. Microinjection of the CMNCBZ-caged carboxy-Q-rhodamine, in our hands, is toxic at amounts greater than 1 ng/embryo.
8. As a result of the often rapid migration of infiltrating leukocytes, multiple illuminations of specific cells often requires moving the microscope stage to maintain the ROI/EGFP-marked leukocyte within the uncaging region.
9. Rapid real-time monitoring of the success of uncaging is permitted through the SIM (SIMultaneous) scanning function of the Olympus FV1000 system. This system incorporates two separate laser scanners, one dedicated to stimulation, the other to simultaneous imaging.
10. When the set illumination region exceeded the area of the marked leukocyte or the region was not maintained directly over the cell during repeated stimulations, uncaged dye was detected outside of the marked cell (**Fig. 3f**). A very small amount of uncaged dye outside of the marked leukocyte was unavoidable due to inevitable exposure of tissue immediately above and below the cell with the excitation laser. However, in such events, this uncaged dye did not interfere with subsequent cell tracking.
11. A small amount of very dim background fluorescence following pHrodo microinjection may be detectable prior to uncaging. However, this dim fluorescence is readily distinguishable from the intense red fluorescence emitted from the pHrodo BioParticles following uptake into the cells.
12. Delivery of pHrodo through the circulation also results in intense staining of vesicles within endothelial cells of the caudal vascular plexus. The endothelial identity of these marked cells has been confirmed by injecting this reagent into the Tg(fli1a:EGFP) background (**Fig. 4m–p**).
13. When imaging is conducted for periods greater than 4–5 h (up to 12 h) agarose is carefully removed from the anterior and

posterior regions of the larva to facilitate specimen growth while a small bridge is maintained over the yolk/trunk domain to ensure immobilization. In addition, the 35 mm petri dish is submerged within a 145 × 20 mm dish to create a larger reservoir for the E3.

Acknowledgments

The authors would like to thank Jacqui Ross and other members of the Biological Imaging Research Unit (The University of Auckland) for imaging assistance; Alhad Mahagaonkar for management of the zebrafish facility; and Annie Chien and Lisa Pullin for expert technical assistance and Makoto Kamei for imaging advice. This work was supported by a grant from the Foundation for Research Science and Technology.

References

1. Martin, P., and Leibovich, S. J. (2005). Inflammatory cells during wound repair: the good, the bad and the ugly. *Trends Cell Biol* **15**, 599–607.
2. Anghelina, M., Moldovan, L., Zabuawala, T., Ostrowski, M. C., and Moldovan, N. I. (2006). A subpopulation of peritoneal macrophages form capillarylike lumens and branching patterns in vitro. *J Cell Mol Med* **10**, 708–715.
3. Hume, D. A. (2006). The mononuclear phagocyte system. *Curr Opin Immunol* **18**, 49–53.
4. Germain, R. N., Miller, M. J., Dustin, M. L., and Nussenzweig, M. C. (2006). Dynamic imaging of the immune system: progress, pitfalls and promise. *Nat Rev Immunol* **6**, 497–507.
5. Germain, R. N., Castellino, F., Chieppa, M., Egen, J. G., Huang, A. Y., Koo, L. Y., and Qi, H. (2005). An extended vision for dynamic high-resolution intravital immune imaging. *Semin Immunol* **17**, 431–441.
6. Scheinecker, C. (2005). Application of in vivo microscopy: evaluating the immune response in living animals. *Arthritis Res Ther* **7**, 246–252.
7. Herbomel, P., Thisse, B., and Thisse, C. (1999). Ontogeny and behaviour of early macrophages in the zebrafish embryo. *Development* **126**, 3735–3745.
8. Renshaw, S. A., Loynes, C. A., Trushell, D. M., Elworthy, S., Ingham, P. W., and Whyte, M. K. (2006). A transgenic zebrafish model of neutrophilic inflammation. *Blood* **108**, 3976–3978.
9. Mathias, J. R., Perrin, B. J., Liu, T. X., Kanki, J., Look, A. T., and Huttenlocher, A. (2006). Resolution of inflammation by retrograde chemotaxis of neutrophils in transgenic zebrafish. *J Leukoc Biol* **80**, 1281–1288.
10. Mathias, J. R., Dodd, M. E., Walters, K. B., Rhodes, J., Kanki, J. P., Look, A. T., and Huttenlocher, A. (2007). Live imaging of chronic inflammation caused by mutation of zebrafish Hai1. *J Cell Sci* **120**, 3372–3383.
11. Le Guyader, D., Redd, M. J., Colucci-Guyon, E., Murayama, E., Kissa, K., Briolat, V., Mordelet, E., Zapata, A., Shinomiya, H., and Herbomel, P. (2008). Live imaging of chronic inflammation caused by mutation of zebrafish Hai1. *Blood* **111**, 132–141.
12. Clay, H., Davis, J. M., Beery, D., Huttenlocher, A., Lyons, S. E., and Ramakrishnan, L. (2007). Dichotomous role of the macrophage in early Mycobacterium marinum infection of the zebrafish. *Cell Host Microbe* **2**, 29–39.
13. Meijer, A. H., van der Sar, A. M., Cunha, C., Lamers, G. E., Laplante, M. A., Kikuta, H., Bitter, W., Becker, T. S., and Spaink, H. P. (2008). Identification and real-time imaging

of a myc-expressing neutrophil population involved in inflammation and mycobacterial granuloma formation in zebrafish. *Dev Comp Immunol* **32**, 36–49.

14. Davis, J. M., Clay, H., Lewis, J. L., Ghori, N., Herbomel, P., and Ramakrishnan, L. (2002). Identification and real-time imaging of a myc-expressing neutrophil population involved in inflammation and mycobacterial granuloma formation in zebrafish. *Immunity* **17**, 693–702.
15. van der Sar, A. M., Musters, R. J., van Eeden, F. J., Appelmelk, B. J., Vandenbroucke-Grauls, C. M., and Bitter, W. (2003). Zebrafish embryos as a model host for the real time analysis of Salmonella typhimurium infections. *Cell Microbiol* **5**, 601–611.
16. van der Sar, A. M., Stockhammer, O. W., van der Laan, C., Spaink, H. P., Bitter, W., and Meijer, A. H. (2006). MyD88 innate immune function in a zebrafish embryo infection model. *Infect Immun* **74**, 2436–2441.
17. Bertrand, J. Y., Kim, A. D., Violette, E. P., Stachura, D. L., Cisson, J. L., and Traver, D. (2007). Definitive hematopoiesis initiates through a committed erythromyeloid progenitor in the zebrafish embryo. *Development* **134**, 4147–4156.
18. Redd, M. J., Kelly, G., Dunn, G., Way, M., and Martin, P. (2006). Imaging macrophage chemotaxis in vivo: studies of microtubule function in zebrafish wound inflammation. *Cell Motil Cytoskeleton* **63**, 415–422.
19. Hall, C., Flores, M. V., Storm, T., Crosier, K., and Crosier, P. (2007). The zebrafish lysozyme C promoter drives myeloid-specific expression in transgenic fish. *BMC Dev Biol* **7**, 42.
20. Davidson, A. J., and Zon, L. I. (2004). The 'definitive' (and 'primitive') guide to zebrafish hematopoiesis. *Oncogene* **23**, 7233–7246.
21. Zapata, A., Diez, B., Cejalvo, T., Gutierrez-de Frias, C., and Cortes, A. (2006). Ontogeny of the immune system of fish. *Fish Shellfish Immunol* **20**, 126–136.
22. Trede, N. S., Langenau, D. M., Traver, D., Look, A. T., and Zon, L. I. (2004). The use of zebrafish to understand immunity. *Immunity* **20**, 367–379.
23. Willett, C. E., Cortes, A., Zuasti, A., and Zapata, A. G. (1999). Early hematopoiesis and developing lymphoid organs in the zebrafish. *Dev Dyn* **214**, 323–336.
24. Lam, S. H., Chua, H. L., Gong, Z., Lam, T. J., and Sin, Y. M. (2004). Development and maturation of the immune system in zebrafish, Danio rerio: a gene expression profiling, in situ hybridization and immunological study. *Dev Comp Immunol* **28**, 9–28.
25. Westerfield, M. (2000). The Zebrafish Book. University of Oregon Press, Eugene, 4th ed.
26. Abramoff, M. D., Magelhaes, P. J., and Ram, S. J. (2004). Image Processing with. ImageJ. *Biophoton Int* **11**, 36–42.
27. Kissa, K., Murayama, E., Zapata, A., Cortes, A., Perret, E., Machu, C., and Herbomel, P. (2007). Live imaging of emerging hematopoietic stem cells and early thymus colonization. *Blood* **111**, 1147–1156
28. Murayama, E., Kissa, K., Zapata, A., Mordelet, E., Briolat, V., Lin, H. F., Handin, R. I., and Herbomel, P. (2006). Tracing hematopoietic precursor migration to successive hematopoietic organs during zebrafish development. *Immunity* **25**, 963–975.
29. Winsauer, G., and de Martin, R. (2007). Resolution of inflammation: intracellular feedback loops in the endothelium. *Thromb Haemost* **97**, 364–369.
30. Sepich, D. S., Wegner, J., O'Shea, S., and Westerfield, M. (1998). An altered intron inhibits synthesis of the acetylcholine receptor alpha-subunit in the paralyzed zebrafish mutant nic1. *Genetics* **148**, 361–372.
31. Lawson, N. D., and Weinstein, B. M. (2002). In vivo imaging of embryonic vascular development using transgenic zebrafish. *Dev Biol* **248**, 307–318.
32. Ward, A. C., McPhee, D. O., Condron, M. M., Varma, S., Cody, S. H., Onnebo, S. M., Paw, B. H., Zon, L. I., and Lieschke, G. J. (2003). The zebrafish spi1 promoter drives myeloid-specific expression in stable transgenic fish. *Blood* **102**, 3238–3240.
33. Hsu, K., Traver, D., Kutok, J. L., Hagen, A., Liu, T. X., Paw, B. H., Rhodes, J., Berman, J. N., Zon, L. I., Kanki, J. P., and Look, A. T. (2004). The pu.1 promoter drives myeloid gene expression in zebrafish. *Blood* 104, 1291–1297.
34. Traver, D., Paw, B. H., Poss, K. D., Penberthy, W. T., Lin, S., and Zon, L. I. (2003). Transplantation and in vivo imaging of multilineage engraftment in zebrafish bloodless mutants. *Nat Immunol* **4**, 1238–1246.
35. Zhu, H., Traver, D., Davidson, A. J., Dibiase, A., Thisse, C., Thisse, B., Nimer, S., and Zon, L. I. (2005). Regulation of the lmo2 promoter during hematopoietic and vascular development in zebrafish. *Dev Biol* **281**, 256–269.

Chapter 17

Imaging Zebrafish Embryos by Two-Photon Excitation Time-Lapse Microscopy

Lara Carvalho and Carl-Philipp Heisenberg

Summary

The zebrafish is a favorite model organism to study tissue morphogenesis during development at a subcellular level. This largely results from the fact that zebrafish embryos are transparent and thus accessible to various imaging techniques, such as confocal and two-photon excitation (2PE) microscopy. In particular, 2PE microscopy has been shown to be useful for imaging deep cell layers within the embryo and following tissue morphogenesis over long periods. This chapter describes how to use 2PE microscopy to study morphogenetic movements during early zebrafish embryonic development, providing a general blueprint for its use in zebrafish.

Key words: Zebrafish, Two-photon excitation microscopy, Time-lapse, Cell migration, YSL nuclei, Gastrulation.

1. Introduction

During embryonic development, coordinated morphogenetic movements shape and organize internal tissues and organs. In the last years, the zebrafish embryo has proven to be an ideal vertebrate model organism to study tissue morphogenesis during development. Zebrafish offer key advantages over other vertebrate model organisms as they are transparent and easy to image and development takes places quickly and externally.

Recent advances in optical microscopy, combined with in vivo fluorescence labeling techniques, have catalyzed a rapidly expanding field of imaging studies in living tissues. Confocal

Graham J. Lieschke et al. (eds.), *Zebrafish,* Methods in Molecular Biology, vol. 546
DOI: 10.1007/978-1-60327-977-2_17

microscopy has been successfully used to study cell movements in live zebrafish embryos, but some major limitations still exist: (a) cell and tissue layers located deep within the embryo are difficult to image due to the decreased resolution resulting from the scattering of light through the tissue; and (b) long term time-lapses are not possible because of photobleaching of the fluorophores and consequent phototoxicity. The discovery of two-photon excitation (2PE) microscopy *(1)* and its application in living tissues has been helpful to image cell movements and other cellular processes in the developing zebrafish *(2–7)*, particularly in the deep cell layers. 2PE microscopy has many advantages compared with other techniques such as confocal microscopy (for reviews see *(8–10)*). Most importantly, the excitation wavelengths used in 2PE (deep red and near infrared) penetrate the tissue better than the visible wavelengths used in confocal microscopy *(11)*. Moreover, 2PE excitation is mostly limited to a small focal volume (it can be as small as 0.1 μm^3 *(12)*), which reduces overall photobleaching and phototoxicity, a crucial requirement for long-term imaging. There are, however, a couple of significant drawbacks with 2PE—photobleaching can be high in the focal plane, thus reducing applicability to thin samples; and the most commonly used lasers perform relatively poorly at wavelengths optimal for excitation of red fluorescent proteins (RFP) (>1,000 nm) *(12)*, thus limiting the use of fluorophores.

This chapter describes a protocol to monitor tissue morphogenesis during zebrafish gastrulation using 2PE time-lapse microscopy. The same techniques can also be easily adapted to other embryonic stages and processes. In zebrafish, several approaches are available to fluorescently label cells. Commonly, in vitro prepared mRNA is injected into the early embryo to transiently express genetically engineered fluorescent proteins such as green fluorescent protein (GFP) and its derivatives. Alternatively, stable transgenic lines expressing GFP under the control of a specific promoter are used. This technique is particularly useful as it allows imaging of distinct cell populations in the embryo. In our protocol, we take advantage of a transgenic line expressing a membrane-bound form of GFP under the control of a ubiquitous promoter *(13)*. We explain in detail how to prepare zebrafish gastrulae for 2PE microscopy, and also lay out the basic principles of operating a 2PE setup. How to use 2PE microscopy to visualize nuclear movements within the yolk syncytium (yolk syncytial layer, YSL) during gastrulation *(14)* will be described in detail. These YSL nuclei are located very deep in the embryo, therefore 2PE is an ideal method to follow their movements with high resolution over long periods.

2. Materials

2.1. Embryos

1. Wild-type embryos. TL (Tupfel Long fin) background.
2. Membrane-bound GFP transgenic embryos. Transgenic line Tg(*β-actin:hras-egfp*), constructed and described before *(13)*.

2.2. Media

1. E3 medium. 5 mM NaCl, 0.17 mM KCl, 0.33 mM $CaCl_2$, 0.33 mM $MgSO_4$, methylene blue 0.2% (*see* **Note 1**). Make a 60X stock solution and add the methylene blue when preparing the 1X solution.
2. 1X Danieau's buffer. 58 mM NaCl, 0.7 mM KCl, 0.4 mM $MgSO_4$, 0.6 mM $Ca(NO_3)_2$, 5 mM HEPES, pH 7.2. Prepare each component as a 100X stock solution, except for NaCl, which should be 50X concentrated. Add 5 mL of a mixture of 10,000 units penicillin/mL and 10,000 μg/mL of streptomycin (Invitrogen) to 1,000 mL of solution when preparing a 1X buffer (*see* **Notes 2** and **3**).
3. 0.3X Danieau's buffer. Dilute 300 mL of 1X Danieau's buffer in 700 mL of water to make 1000 mL of buffer (*see* **Note 3**).

2.3. Dechorionating and Mounting Embryos

1. Stereomicroscope equipped with transmitted light.
2. 35 mm, 60 mm, and 94 mm plastic Petri dishes (Greiner).
3. Two sharp forceps (Dumont no. 5).
4. Plastic Pasteur pipets.
5. 150-mm Glass Pasteur pipets.
6. Electrophoresis grade agarose (Invitrogen).
7. Pronase (Roche Diagnostics). Dilute the pronase powder in E3 medium to make a final concentration of 2 mg/mL. Keep 2 mL aliquots at –20°C.
8. Low melting point (LMP) agarose (Invitrogen).
9. Heating block at 70°C.
10. Silicon grease (Beckman).
11. Glass rings (MNK-145-030 K, Fisher Scientific).
12. Coverslips 22 mm × 22 mm, 0.13–0.16 mm thickness (Menzel; cat. no. BB022022A1).
13. Microscope slides 76 mm × 22 mm (Menzel).

2.4. Two-Photon Excitation Time-Lapse Imaging

1. Two-photon excitation setup: beam conditioning unit and scanning head (Bio-Rad Radiance 2100MP); Pump laser with 10 W of power (Verdi); Mode-lock unit (Coherent Mira 900).

2. Nikon Eclipse TE300 inverted microscope.
3. Plan Fluor ×20 water-immersion objective (NA 0.75).
4. Super Fluor ×40 oil-immersion objective (NA 1.3)
5. Plan Apo ×60 water-immersion objective (NA 1.2).
6. Objective heater and objective collar (Bioptechs).
7. Computer with LaserSharp 2000 software (Zeiss).

2.5. Labeling of YSL Nuclei

1. Plastic mold with grooves (**Fig. 4a**).
2. Glass capillaries (GC100F-15, Harvard).
3. Magnet holder (MB-B, Kanetec).
4. Microinjector and Pico-Pump with foot pedal (PV820, World Precision Instruments).
5. Micromanipulator (MN-151, Narishige).
6. Needle puller (Flaming/Brown P87, Sutter).
7. Watchmaker forceps.
8. Histone H1 Alexa Fluor® 488 conjugate (Invitrogen). Dilute the powder (1 mg) in 100 μL of water to obtain a final concentration of 10 mg/mL. Make 5 μL aliquots and store at -20°C in the dark. The aliquots can be frozen and thawed several times.
9. Gel loader tips of 0.5–20 μL (Eppendorf; cat. no. 0030001.222,).
10. Objective micrometer with 0.01 mm divisions (Fine Science Tools).
11. Stereomicroscope equipped with transmitted and ultraviolet (UV) light.

3. Methods

3.1. Imaging Cell Movements During Gastrulation

3.1.1. Obtaining Embryos

To image the movement and morphology of mesendodermal (hypoblast) and ectodermal (epiblast) precursor cells during gastrulation, transgenic embryos expressing membrane-bound GFP – Tg:(*β-actin:hras-egfp*) – can be used (*see* **Note 4**) *(13)*. Because the GFP is fused to a sequence that targets it to plasma membranes, all cell membranes in these transgenic embryos have high GFP levels (**Fig. 1a**).

1. Collect embryos after the fish have laid and keep them in 94-well Petri dishes with E3 medium, at 28.5°C.
2. After 1–2 h, sort out unfertilized eggs and embryos that look unhealthy using a plastic pipet and keep around 80 embryos per dish.

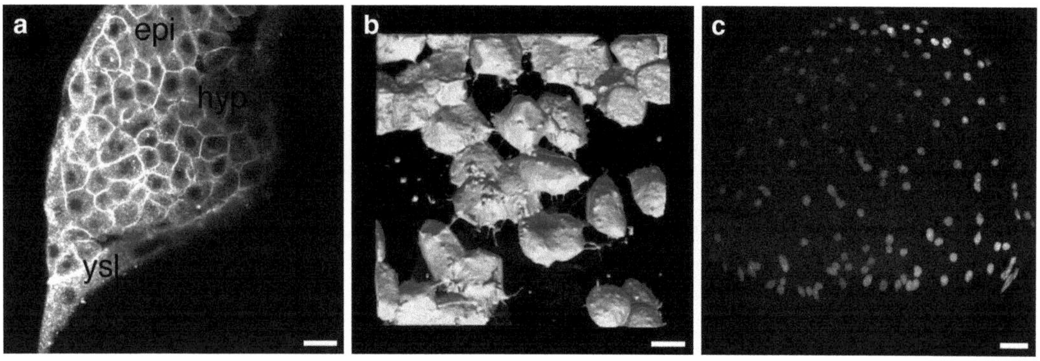

Fig. 1. Examples of images obtained using 2PE microscopy. (**a**) Optical section (single *z* slice) through the dorsal margin of an embryo at the onset of gastrulation. Mesendoderm (hyp, hypoblast) and ectoderm (epi, epiblast) progenitor cells express membrane-bound GFP under control of the β-actin promoter. (**b**) 3-D rendering of ectoderm progenitor cells labeled with a mixture of cytoplasmic GFP and membrane-bound GFP. (**c**) Histone-labeled nuclei within the YSL during gastrulation (maximum *z* projection of a stack of 36 slices). (**a**, **b**) were obtained using the Plan Apo ×60 objective; (**c**) was obtained using the Plan Fluor ×20 objective. (**a**) is a lateral view; (**b**,**c**) are dorsal views (*see* **Fig. 2c**). The animal pole is always on top. *epi* epiblast; *hyp* hypoblast; *ysl* yolk syncytial layer. Scale bars: (**a**, **b**) 10 μm, (**c**) 50 μm.

3.1.2. Dechorionating Embryos

In order to image embryos during gastrulation stages, it is necessary to remove their chorions before mounting. To dechorionate the embryos, either a manual or enzymatic technique can be used. Manual dechorionation is in general less invasive to the embryos and is thus preferable; however, use of the proteolytic enzyme pronase also gives good results when done carefully and is advantageous when large numbers of dechorionated embryos are needed. Both procedures are described in detail below.

Handling Dechorionated Embryos

Dechorionated embryos are very sensitive and stick to plastic, thus they must be handled with care and only manipulated with blunt forceps or "hair loops." Importantly, they must be kept in 2% agarose-coated Petri dishes (to prevent sticking) and be transferred with fire-polished Pasteur pipets. In this section, we describe how to prepare these different tools.

1. To obtain hair loops, use a hair or a tungsten wire (diameter 0.125 mm) and glue it to the tip of a glass Pasteur pipet or glass capillary (**Fig. 2a**).
2. To obtain fire-polished Pasteur pipets, melt the tip of a glass Pasteur pipet for a few seconds using a flame from a Bunsen burner.
3. To prepare agarose-coated Petri dishes, dilute agarose in E3 medium to a final concentration of 2% agarose *(m/v)*. Boil the solution in a microwave oven and swirl it to remove air bubbles. Pour the solution into plastic Petri dishes in order to form a thin layer of agarose and let it solidify for some minutes.
4. The dishes can be stored at 4°C for a few days in a dry environment (e.g., inside a plastic bag with a paper towel that will absorb water condensation).

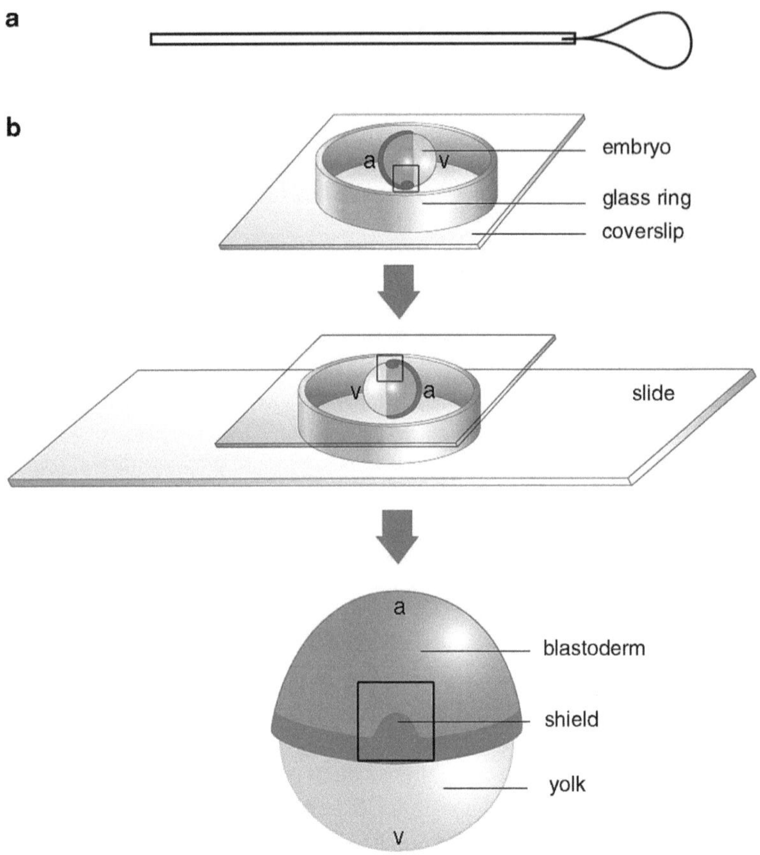

Fig. 2. Mounting embryos in agarose. (**a**) Schematic representation of hair loop, a tool used to mount embryos. (**b**) Mounting chamber and embryo orientation used to image embryos on an inverted microscope. The mounting chamber consists of a coverslip attached to a glass ring in which the embryo is placed in agarose medium. The region of the embryo that is going to be imaged (here the dorsal side or "shield") should face the coverslip (*square region*). *a* animal pole; *v* vegetal pole.

Manual Dechorionation of Embryos

1. Use a plastic Pasteur pipet to transfer up to 20 embryos to a 2% agarose-coated 35 mm Petri dish containing E3 medium.
2. Chorions can be easily removed with two sharp forceps. Gently make small holes in the chorion until the embryo falls out.

Enzymatic Dechorionation of Embryos

1. Transfer up to 80 embryos to a 60 mm Petri dish containing E3 medium.
2. Remove most of the medium and add 2 mL of pronase 2 mg/mL (in E3 medium). Keep it at room temperature for 10 min (*see* **Note 5**).
3. Transfer the embryos into an agarose-coated 94 mm Petri dish containing 0.3X Danieau's buffer by gently pouring the embryos into the dish or using a fire-polished glass Pasteur pipette. During this step, most of the embryos will fall out of their chorions.

4. Rinse the embryos thoroughly, at least five times, with 0.3X Danieaus's buffer in order to wash out all pronase enzyme and broken chorions (*see* **Note 6**).

3.1.3. Mounting Embryos in Agarose

In order to perform time-lapse imaging of embryos, it is necessary to mount them in a medium that is sufficiently rigid to immobilize the embryos, but at the same time allows them to develop normally. The mounting device used for inverted microscopes, such as the one in our 2PE setup, consists of a self-assembled imaging chamber in which the embryo is mounted in agarose-containing medium **(Fig. 2b)**.

1. To prepare 1% LMP agarose for mounting, dilute the agarose in E3 medium or 0.3X Danieau's buffer to obtain a final concentration of 1% (*m/v*) and dissolve it in a microwave oven. Make 0.5–1 mL aliquots in Eppendorf tubes and store at 4°C.
2. Before starting to mount the embryos, a few agarose aliquots should be melted at 70°C for at least 30 min.
3. Start to mount the embryos as soon as the dorsal side becomes visible under the stereomicroscope (shield stage, 6 h post-fertilization).
4. Glue a glass ring to a coverslip using silicon grease to make a mounting chamber **(Fig. 2b)**.
5. Place three drops of previously melted 1% LMP agarose on the coverslip with a glass Pasteur pipet (*see* **Notes** 7 and **8**).
6. Transfer one dechorionated embryo with a fire-polished Pasteur pipet from the dish to the tube containing the melted agarose and then transfer the embryo to the coverslip (*see* **Notes 9** and **10**).
7. Carefully orient the embryo with a hair loop or blunt forceps so that the region desired for imaging is facing the coverslip (*see* **Note 11**). For example, in order to observe movement of axial hypoblast cells, the embryos should be mounted on their dorsal side **(Fig. 2b)**.
8. Wait a few minutes until the agarose is solidified and cover it with E3 medium or 0.3X Danieau's buffer (*see* **Note 12**).
9. Use silicon grease to seal a microscope slide to the glass ring, taking care to avoid the formation of air bubbles (*see* **Notes 13** and **14**).

3.1.4 .Two-Photon Excitation Time-Lapse Microscopy

In order to image cell movements, we use two-photon excitation microscopy (2PE) **(Fig. 3)**. In many respects, a 2PE setup is similar to a conventional laser scanning confocal microscope. The main differences are the excitation laser and the detection pathway. Our 2PE setup is a Bio-Rad Radiance 2100 MP. It consists of a mode-locked Titanium:Sapphire (Ti:Sapphire) laser originating from a pump laser with a power of 10 W. The laser beam is scanned onto the sample by galvanometer mirrors and

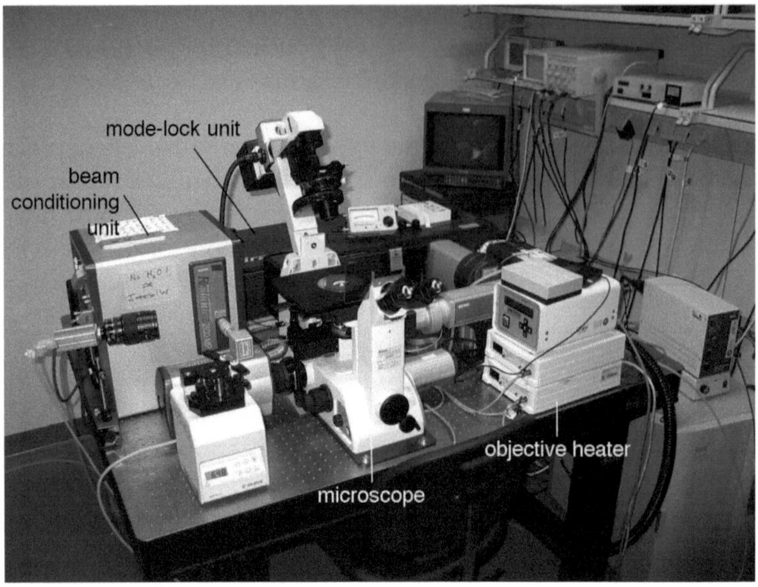

Fig. 3. 2PE microscope setup.

the detection system consists of non-descanned photomultiplier tubes. The software used to acquire the images is LaserSharp 2000. In this section, we describe the general procedure to setup a time-lapse on the 2PE microscope.

1. Turn on the laser at least 1 h before imaging.
2. Tune the laser to 890–900 nm wavelength (*see* **Note 15**), mode-lock it and adjust the output power to the maximum possible (in our setup this can be up to 500 mW).
3. Turn on the objective heater and set it to 28–29°C, at least 10 min before imaging (*see* **Note 16**).
4. Turn on all other devices needed (e.g., microscope, computer).
5. Place an agarose-mounted embryo on the microscope stage and focus on the desired region (*see* **Note 17**). To follow the movements and morphology of cells at high resolution, both the Plan Apo ×60 water-immersion objective (NA 1.2) and the Super Fluor ×40 oil-immersion objective (NA 1.3) give good results. The Plan Fluor ×20 water-immersion objective (NA 0.75) can also be used in order to obtain an overview of the movements occurring in the whole embryo.
6. Cover all light sources before starting to scan (*see* **Note 18**).
7. Start scanning the embryo using the LaserSharp 2000 software (*see* **Notes 19** and **20**).
8. Decrease laser power as low as possible to minimize photobleaching.

9. Select the first and last *z* slices of the stack and the space between each *z* slice. When using the ×40 and ×60 objectives, the space between *z* slices is typically 0.2–2 μm; when using the ×20 objective, the space between *z* slices is typically 1–5 μm.
10. Choose the speed and resolution of scanning, taking into account that the higher the image quality, the slower the scan. In our setup, we found that a speed of 166 lines per scan and a resolution of 512 × 512 pixels are usually suitable.
11. Adjust gain and offset, taking care not to saturate the image.
12. Select the time parameter. When using the ×40 and ×60 objectives, the period between two consecutive time points should not be longer than 2 min in order to be able to follow cell trajectories; when using the ×20 objective, the period between time points can be longer but should be adjusted according to the specific cellular aspect analyzed.
13. Start the time-lapse.
14. The time-lapse files are saved in PIC format by the LaserSharp 2000 software and can be imported and analyzed with Volocity (Improvision) or ImageJ (NIH) software.

3.2. Imaging YSL Nuclei Movement During Gastrulation

3.2.1. Labeling YSL Nuclei

To specifically label the nuclei in the YSL, a fluorescent-tagged version of the H1 histone protein can be injected in the YSL between high and sphere stage of development (3.3–4 h postfertilization at 28.5°C). At this stage, the cytoplasmic connections between the blastoderm and the yolk have closed *(15–17)*, thus the protein does not spread to the blastoderm. This method has been used before by D'Amico and Cooper *(14)* and causes no secondary effects on embryonic development (*see* **Note 21**). Below we describe the details of injecting this protein into the YSL.

Preparation of Injection Chambers

During the injection procedure, special agarose chambers containing vertical furrows are used to hold the embryos in the correct position. These chambers are prepared using a plastic mold that introduces a series of grooves in an agarose-coated dish where the embryos are then aligned (**Fig. 4a**).

1. Dissolve agarose in E3 medium to obtain a final concentration of 2% *(m/v)*, boil the solution in a microwave oven and swirl it to remove bubbles.
2. Pour 20 mL of the liquified agarose into the cover of a 94 mm Petri dish (*see* **Note 22**).
3. Float a mold in the agarose, taking care to avoid trapping air bubbles below the mold, and allow the agarose to solidify for a few minutes.

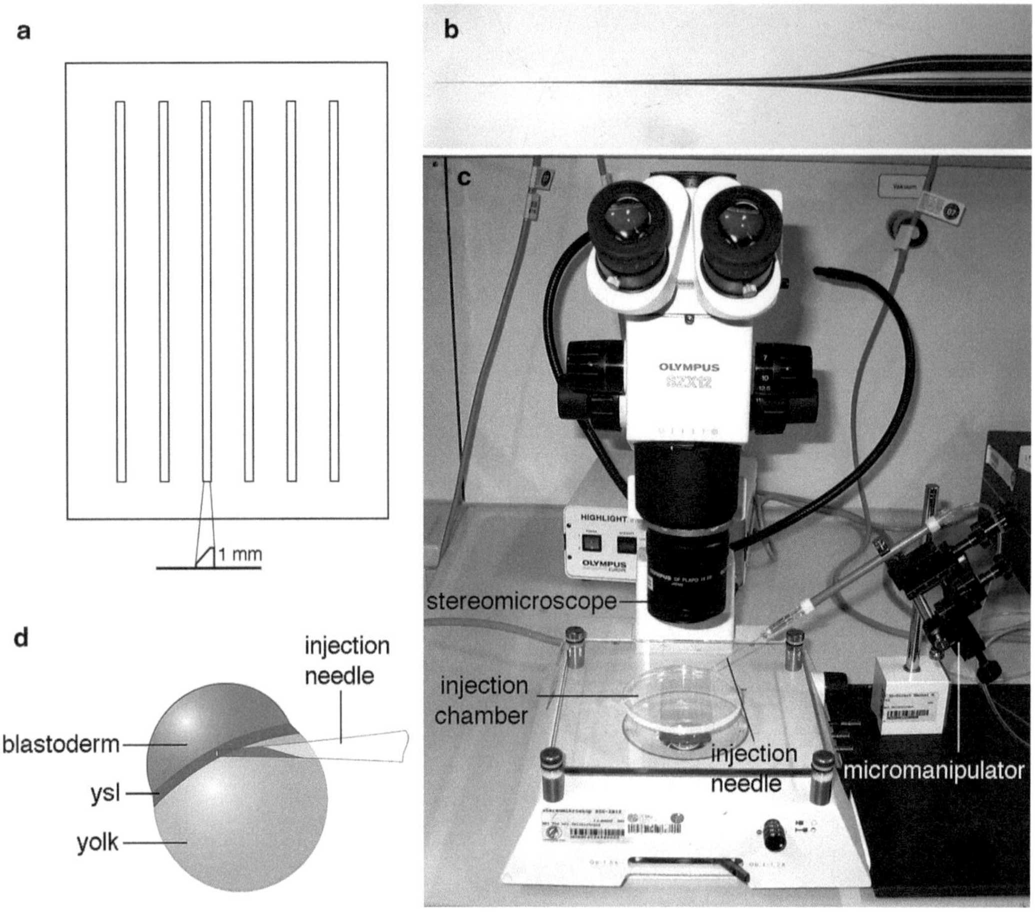

Fig. 4. YSL microinjection setup. (**a**) Schematic representation of the mold used to make agarose injection chambers. The dimensions of each groove are approximately 1 mm×1 mm. (**b**) Microinjection needle obtained after pulling. (**c**) Injection setup composed of a stereomicroscope, a micromanipulator connected to both a microinjector (not shown) and an injection needle. An injection plate is shown on the microscope stage. (**d**) Schematic representation of the orientation used to inject into the YSL of dechorionated embryos. *ysl* yolk syncytial layer.

Injection Procedure

1. Prepare the histone solution by diluting 0.5 μL of histone stock solution (10 mg/mL) in 4.5 μL of water to a final histone concentration of 1 mg/mL. Spin the solution in a mini-centrifuge for 5 min in order to separate out any precipitates that formed.
2. Prepare injection needles by pulling glass capillaries on a needle puller that creates a predefined tip size (*see* **Notes 23** and **24**) (**Fig. 4b**).
3. Load a needle with 2 μL of histone solution using a pipet tip designed to load sequencing gels (*see* **Note 25**).
4. Break the tip of the needle with forceps to obtain an opening of approximately 0.1 mm (*see* **Note 26**).

5. Place the needle on a micropipet holder mounted on a micromanipulator. The micropipet holder should be connected to a pneumatic microinjector with Teflon tubing (**Fig. 4c**).
6. Adjust the size of the spherical drop of solution by regulating the pulse duration of the pressure on the microinjector (*see* **Note 27**). Calibrate the diameter of the drop to 100 μm by expelling the solution close to an objective micrometer. This diameter should correspond to a volume of 0.5 nL (0.5 ng of histone solution).
7. Fill a previously prepared agarose injection chamber with 1X Danieau's buffer containing penicillin and streptomycin (*see* **Note 3**) and carefully transfer the dechorionated embryos into the furrows using a fire-polished Pasteur pipet.
8. To label YSL nuclei, inject one drop of histone in the YSL. The needle containing the histone should be inserted through the yolk into the center of the YSL when the embryo is between high and sphere stage (3.3–4 h post-fertilization, **Fig. 4d**).
9. After injecting all the embryos, transfer them to a new agarose-coated 94 mm Petri dish containing 1X Danieau's buffer plus penicillin and streptomycin and keep them at 28.5°C until the desired stage.
10. Check the embryos after 1–2 h under a stereomicroscope equipped with UV light. Discard embryos that have weak or no fluorescence in the YSL nuclei, or that are dead (*see* **Note 28**).

3.2.2. Mounting Embryos in Agarose

To image YSL nuclei movement, the embryos should be mounted in agarose as described in **Subheading 3.1.3**.

3.2.3. Two-Photon Excitation Time-Lapse Imaging

To image long-range movements of histone-labeled YSL nuclei by 2PE, the procedure described in **Subheading 3.1.4** should be followed. We used the Plan Fluor ×20 water-immersion objective (NA 0.75), because it covers the whole embryo and has good resolution. When choosing the space between *z* slices and consecutive time points, 5 μm and 2.5 min are usually suitable to follow the nuclei movement within the YSL. An example of YSL nuclei imaged using 2PE is shown in **Fig. 1c**.

4. Notes

1. Methylene blue prevents fungal and bacterial infections.
2. The addition of antibiotics minimizes bacterial infections, which happen more often in dechorionated and/or injected embryos. During and after injection, we found that embryos

incubated in 1X Danieau's buffer survive better than those in 0.3X Danieau's buffer.

3. 0.3X Danieau's buffer should be used when handling pronase dechorionated embryos because it increases survival rates in comparison to E3 medium. 1X Danieau's buffer should be used when injecting dechorionated embryos.
4. As an alternative to the transgenic zebrafish line, similar results can be obtained by injecting a mixture of mRNA encoding for cytoplasmic GFP and membrane-bound GFP at the one-cell stage of development. This early injection leads to ubiquitous GFP expression from blastula stages onwards **(Fig. 1b)**. This can be achieved by injecting the yolk of one-cell stage embryos with one 0.5 nL drop of a mixture of 0.24 μg/μL of *gap43-gfp* (membrane-bound GFP) mRNA (0.12 ng) and 0.08 μg/μL of cytoplasmic *gfp* mRNA (0.04 ng) diluted in RNAse-free water. This procedure has been described in detail before *(2)*. To obtain mRNA, DNA from a pCS2+ expression vector containing the GFP inserts (*gap43-gfp* from *(18)*; *gfp*, gift of S. Wilson, UCL) should be linearized by digestion with *Not*I restriction enzyme for 3 h at 37°C and purified using a polymerase chain reaction (PCR) purification kit (Qiagen) according to the manufacturer's instructions. The DNA should be eluted in RNAse-free water and the concentration determined using standard procedures. The linearized plasmid DNA can then serve as a template to synthesize mRNA in vitro using SP6 polymerase (mMessage mMachine kit, Ambion) following the manufacturer's instructions.
5. This incubation time is adequate for embryos between high and sphere stage (3.3–4 h post-fertilization), but should be shorter if treating younger embryos, or longer if treating older embryos. Incubation time can be critical for embryo survival, especially before gastrulation stages: if they stay too long in the pronase solution, the enzyme might start degrading proteins in the embryo. In contrast, if this period is too short, the embryos will squeeze through holes in the chorions and eventually die. An easy way to check whether the embryos are ready to fall out from their chorions is to pipette some embryos up and down with the glass Pasteur pipet while in pronase solution; if the embryos easily fall out of the chorion, the pronase should be washed out.
6. During the washing procedure, all embryos should fall out of their chorions. If not, it is possible that the incubation time in pronase was too short and has to be adjusted. Pipetting the embryos up and down with a glass pipet might also help some embryos to fall out of their chorions.
7. Before placing the drops of liquid agarose on the coverslip, wait until it cools down to around 30–40°C (this can be

checked by touching the agarose tube), otherwise the agarose will be too hot and the embryo will be damaged.

8. The mounting procedure should be performed as quickly possible, otherwise the agarose solidifies before you can properly orient the embryo.
9. This step is necessary because the medium that is pipeted together with the embryo can dilute the agarose that is placed on the coverslip if added directly.
10. Care should be taken that the embryo does not touch the walls of the agarose-containing tube, as dechorionated embryos stick to plastic.
11. The embryo should be placed as close as possible to the coverslip, otherwise the distance between the embryo and the objective will be too high.
12. While the agarose is solidifying, do not move the coverslip, otherwise the embryo might reorient. To check whether the agarose is solidified, touch its surface carefully with the hair loop or a forcep. The time that the agarose takes to solidify depends on the room temperature: if the room is too cold, the agarose will solidify too fast for proper embryo orientation, whereas if too warm, it will not solidify completely. In our experience, the ideal temperature is around 24°C.
13. Air bubbles might interfere with imaging.
14. Care should be taken not to cover the chamber with too much medium, otherwise attachment of the glass ring to the glass slide will not be tight enough.
15. In our experience, this is the wavelength that gives the best fluorescence intensity and resolution when imaging GFP proteins.
16. Make sure that the collar that connects the objective to the objective heater is attached to the objective; if not, the collar will be damaged. If an objective heater is not available, make sure the temperature of the room is between 24°C and 31°C and stable, or else the dynamics of cell migration might vary in different time-lapses.
17. It occasionally happens that you cannot focus deep enough or that the slide moves. This can be due to an embryo mounted too far away from the coverslip, preventing focusing of the objective. Alternatively, the attachment of the glass ring to the coverslip and/or to the microscope slide might not be tight enough.
18. Because the 2PE does not have a pinhole, in contrast to confocal microscopes, all stray light in the room will reach the detectors.
19. If no image appears on the screen when you start scanning, even though the sample has bright fluorescence, there are several points to check (a) the laser power dropped down or

the laser mode-lock is lost; (b) the laser beam might not be centered; (c) movement of the slide is causing the embryo to go out of focus (*see* **Note 17**); and (d) the connection to the detector is not enabled. If the screen appears white or has white stripes when you start scanning, it means there is some source of light in the room or in the microscope.

20. When scanning the embryo to obtain an overview of the sample or to select the first and last *z* slices of the stack, a fast scanning speed should be used in order to minimize bleaching.
21. The cell-impermeant nuclear dye Sytox Green (Molecular Probes) has also been used for the same purpose *(14)*, but we have experienced better results using fluorescently-tagged histone.
22. Petri dish covers provide better access for the injection needle and micromanipulator.
23. The parameters should be adjusted according to the type of needle puller, type of capillaries, and specific application. Under our conditions, we found that a pull strength of 70 units, a pulling velocity of 90 units, and a time delay of 120 units are the best suitable parameters to obtain good injection needles. The temperature should be adjusted every time the filament is changed (automatically determined by a "ramp test", as described by the manufacturer's instructions).
24. One glass capillary can be used to make two needles.
25. The glass capillaries used contain a filament inside that allows the solution to be pulled to the tip of the needle.
26. The size of the opening is important: when it is too big the embryo will be damaged, when it is too small it will be difficult to obtain a sufficiently big drop.
27. The drop size should be stable for a few milliseconds. This can be adjusted by tuning the hold and/or eject pressure of the microinjector.
28. During the injection procedure, some embryos might not incorporate enough histone protein in the YSL nuclei, and sometimes fluorescence is observed in the blastoderm or as a droplet inside the yolk. This usually results from problems in the injection procedure, and these embryos should be discarded.

Acknowledgments

We are grateful to P. Stockinger and F. Ulrich for providing us with images obtained by 2PE and J. Topczewski for providing us with the Tg(*β-actin:hras-egfp*) fish line. We also thank J. Peychl for advice and assistance with 2PE microscopy and G. Junghanns, E. Lehmann, and J. Hückmann for help with the fish care. We thank F. Friedrich for help with figures and L. Rohde for critical reading of the manuscript. This work was supported by grants from the Emmy Noether Program of the Deutschen Forschungsgemeinschaft, the European Union, and the Max Planck Society to CPH.

References

1. Denk, W., Strickler, J. H., and Webb, W. W. (1990). Two-photon laser scanning fluorescence microscopy. *Science* **248**, 73–6.
2. Montero, J. A., Carvalho, L., Wilsch-Bräuninger, M., Kilian, B., Mustafa, C. and Heisenberg, C. P. (2005). Shield formation at the onset of zebrafish gastrulation. *Development* **132**, 1187–98.
3. Ulrich, F., Concha, M. L. Heid, P. J., Voss, E., Witzel, S., Roehl, H., Tada, M., Wilson, S. W., Adams, R. J., Soll , D. R., and Heisenberg, C. P. (2003). Slb/Wnt11 controls hypoblast cell migration and morphogenesis at the onset of zebrafish gastrulation. *Development* **130**, 5375–84.
4. Blaser, H., Reichman-Fried, M., Castanon, I., Dumstrei, K., Marlow, F. L., Kawakami, K., Solnica-Krezel, L., Heisenberg, C. P., and Raz, E. (2006). Migration of zebrafish primordial germ cells: a role for myosin contraction and cytoplasmic flow. *Dev Cell* **11**, 613–27.
5. Das, T., Payer, B. Cayouette, M., and Harris, W. A.(2003). In vivo time-lapse imaging of cell divisions during neurogenesis in the developing zebrafish retina. *Neuron* **37**, 597–609.
6. Meyer, M. P., and Smith, S. J. (2006). Evidence from in vivo imaging that synaptogenesis guides the growth and branching of axonal arbors by two distinct mechanisms. *J Neurosci* **26**, 3604–14.
7. Yaniv, K., Isogai, S., Castranova, D., Dye, L., Hitomi, J., andWeinstein, B. M. (2006). Live imaging of lymphatic development in the zebrafish. *Nat Med* **12**, 711–6.
8. Diaspro, A., Chirico, G., and Collini, M. (2005). Two-photon fluorescence excitation and related techniques in biological microscopy. *Q Rev Biophys* 38, **97**–166.
9. Helmchen, F., and Denk, W. (2005). Deep tissue two-photon microscopy. *Nat Methods* **2**, 932–40.
10. Svoboda, K., and Yasuda R.(2006). Principles of two-photon excitation microscopy and its applications to neuroscience. *Neuron* **50**, 823–39.
11. Oheim, M., Beaurepaire, E., Chaigneau, E., Mertz, J., and Charpak, S. (2001). Two-photon microscopy in brain tissue: parameters influencing the imaging depth. *J Neurosci Methods* **111**, 29–37.
12. Zipfel, W. R., Williams, R. M., and Webb, W. W. (2003). Nonlinear magic: multiphoton microscopy in the biosciences. *Nat Biotechnol* **21**, 1369–77.
13. Cooper, M. S., Szeto, D. P., Sommers-Herivel, G., Topczewski, J., Solnica-Krezel, L., Kang, H. C., Johnson, I., and Kimelman, D. (2005). Visualizing morphogenesis in transgenic zebrafish embryos using BODIPY TR methyl ester dye as a vital counterstain for GFP. *Dev Dyn* **232**, 359–68.
14. D'Amico, L. A., and Cooper, M. S. (2001). Morphogenetic domains in the yolk syncytial layer of axiating zebrafish embryos. *Dev Dyn* **222**, 611–24.
15. Kimmel, C. B., and Law, R. D. (1985). Cell lineage of zebrafish blastomeres. I. Cleavage pattern and cytoplasmic bridges between cells. *Dev Biol* **108**, 78–85.
16. Kimmel, C. B., and Law, R. D. (1985). Cell lineage of zebrafish blastomeres. II. Formation of the yolk syncytial layer. *Dev Biol* **108**, 86–93.
17. Kimmel, C. B., Ballard, W. W., Kimmel, S. R., Ullmann, B., and Schilling, T. F. (1995). Stages of embryonic development of the zebrafish. *Dev Dyn* **203**, 253–310.
18. Okada, A., Lansford, R., Weimann, J. M., Fraser, S. E., and McConnell, S. K. (1999). Imaging cells in the developing nervous system with retrovirus expressing modified green fluorescent protein. *Exp Neurol* **156**, 394–406.

Chapter 18

Morphological Analysis of the Zebrafish Digestive System

Andrew J. Trotter, Adam C. Parslow, and Joan K. Heath

Summary

The zebrafish provides an ideal model for the study of vertebrate organogenesis, including the formation of the digestive tract and its associated organs. Despite optical transparency of embryos, the internal position of the developing digestive system and its close juxtaposition with the yolk initially made morphological analysis relatively challenging, particularly during the first 3 d of development. However, methodologies have been successfully developed to address these problems and comprehensive morphologic analysis of the developing digestive system has now been achieved using a combination of light and fluorescence microscope approaches—including confocal analysis—to visualize wholemount and histological preparations of zebrafish embryos. Furthermore, the expanding number of antibodies that cross-react with zebrafish proteins and the generation of tissue-specific transgenic green fluorescent protein reporter lines that mark specific cell and tissue compartments have greatly enhanced our ability to successfully image the developing zebrafish digestive system.

Key words: Zebrafish, Intestine, Epithelium, Goblet cells, Enteroendocrine cells, Histology, Immunohistochemistry, Transgenic lines.

1. Introduction

The development of vertebrate organs is a complex process, requiring the differentiation and integration of multiple cell types into stereotypic arrangements that are essential for organ function. The zebrafish provides an outstanding model system in which to study vertebrate organogenesis due to several favorable characteristics, most notably the external development of optically transparent embryos. In this chapter we aim to highlight the considerable opportunities afforded by the zebrafish model to delineate the morphological events that give rise to the organs

Graham J. Lieschke et al. (eds.), *Zebrafish,* Methods in Molecular Biology, vol. 546
DOI: 10.1007/978-1-60327-977-2_18

of the digestive system, with particular emphasis on the development of the intestine.

In zebrafish, the primitive gut tube which gives rise to the liver, pancreas, gall bladder and intestine is first discernible by histology at around 26 h post-fertilization (hpf). At this point, endoderm cells, first recognized at 4.5 hpf by *cas* expression *(1)*, have converged at the midline of the developing organism to form an endodermal rod that lies directly dorsal to the yolk. A series of well-characterized morphological events over the following 4 days of development give rise to a fully functional digestive system allowing exotrophic nutrition to ensue *(2–6)*.

Despite the advantages usually conferred by the optical transparency of zebrafish embryos, the visualization of the developing zebrafish digestive system was initially problematic. Impeding issues included the relatively deep internal position of the developing digestive tract and its intimate juxtaposition with the yolk and its extension. We, and others, have used a combination of approaches to circumvent these problems and highly satisfactory analyses have been achieved using minor variations of established techniques—such as histology, histochemistry, and immunohistochemistry—that have previously been extensively applied to the study of other model organisms. These zebrafish-specific adaptations, while subtle, have proved critical to achieving required outcomes and are likely to be useful for studying other organ systems in zebrafish. In particular, we found that standard histology followed by hematoxylin and eosin (H + E) staining provides an extremely powerful tool to examine zebrafish intestinal development *(3)*. While often outsourced, histologic techniques can be established with little specialized equipment (apart from a microtome and embedding console) and, once mastered, rarely prove problematic. Alcian blue-periodic acid Schiff staining is particularly useful to visualize the differentiation of intestinal epithelial cells. The application of histologic techniques with immunohistochemistry and assays of cell behavior, such as bromodeoxyuridine incorporation analysis to monitor cell proliferation *(3)*, can be used to build a more comprehensive picture of zebrafish intestinal development. As well as analysis on thin (5–10 μm) sections, immunohistochemistry can be performed on thick (150–200 μm) sections of zebrafish embryos prepared on a vibrating microtome, or on wholemount preparations. These approaches require a relatively gentle fixation process that can prove very useful when working with problematic epitopes and can accommodate a very rapid turnaround time from fixation to imaging. Immunohistochemistry of thick sections also circumvents reagent penetration issues by allowing direct access to the tissues of interest. Wholemount analysis is extremely straightforward and affords assessment in a three dimensional (3-D) cellular context.

The image quality produced from live specimens has improved greatly through recent advances in brightfield, fluorescence/confocal, and differential interference contrast (DIC)/Nomarski microscopy. Coupled with the creation of transgenic reporter lines such as the *Tg(gutGFP)*^{s854} and *Tg(nkx2.2a-EGFP)* lines, fluorescence microscopy affords exquisite imaging opportunities of digestive organ formation. Not only do these lines permit the clear visualization of the liver, pancreas, and intestinal primordia in wild-type, mutant, and morphant embryos but they can also be used to quantify whether specific cell types, for example enteroendocrine cells, are differentially affected between wild-type and mutant embryos. The *Tg(nkx2.2a-EGFP)* line, which we have used extensively, allows for monitoring of the differentiation of endocrine cells in both the developing pancreas and intestine *(3, 7)* and a number of lines based on a variety of tissue-specific promoters, such as the liver (*lfabp*) and intestinal fatty acid binding protein (*ifabp*) promoters *(8–10)*, *elastase A(11)*, *ptf1a(12)* and *pdx1(13)* have been developed and exploited in elegant studies *(12, 14, 15)*. This chapter provides detailed descriptions of the methods routinely used to analyze the development of the digestive system in our laboratory. A table (**Table 1**) summarizing some of the additional reagents and tools available to analyze digestive organ formation is also provided.

Table 1
Selected Targets and Reagents Useful for Marking Compartments Within the Zebrafish Digestive System

Cell/tissue type	Antibody/histochemical stain	Transgenic line
Intestinal epithelium Goblet cells (mucins)	Alcian Blue (acidic mucins) PAS (neutral mucins) Wheatgerm agglutinin-Alexa 488 (Molecular Probes)	No
Intestinal epithelium Enteroendocrine cells	Mouse anti-Glucagon (Sigma) Rabbit polyclonal antibody against Somatostatin	*Tg(nkx2.2a-EGFP) (3)*
Intestinal epithelium Secretory cells (goblet cells and enteroendocrine cells)	Mouse monoclonal antibody, 2F11 *(16)*. Also marks the hepatopancreatic ductal system and gallbladder *(14)*	*Tg(nkx2.2a-EGFP) (3)* (enteroendocrine cells only)

(continued)

Table 1 (continued)

Cell/tissue type	Antibody/histochemical stain	Transgenic line
Intestinal epithelium Absorptive cells	Mouse monoclonal antibody, 4E8 *(16)* marks the brush border Na/K-ATPase Alkaline phosphatase	*Tg(gutGFP) (3)*; *Tg(ifabp-GFP) (8, 9)*
Pancreas Exocrine tissue	Elastase A	*Tg(elaA:GFP) (11)*; *Tg(ptf1a:eGFP) (12)*; *Tg(gutGFP) (2, 4)*
Pancreas Islet	Mouse monoclonal antibody against Islet1/2 (Developmental Studies Hybridoma Bank) Guinea pig polyclonal antibody against Insulin (Biomeda) Rabbit polyclonal antibody against Somatostatin	*Tg(nkx2.2a-EGFP) (3)*; *Tg(pdx1-GFP) (13)*; *Tg(gutGFP)(2, 4)*
Liver – hepatocyte		*Tg(gutGFP)(2, 4)* *Tg(lfabp:dsRed (11)*

2. Materials

2.1. General Methods

2.1.1. Embryo/Larval Rearing, Staging, and Fixation

1. Embryo medium (EM): 1.2 g Ocean Nature Sea Salt (Aquasonic) in 20 L deionized water, final concentration 60 μg/L.
2. EM with PTU: 0.003% *N*-phenylthiourea (PTU) (Sigma) in EM. Used to inhibit pigmentation in embryos that will be observed >24 hpf. Critical for bright field and DIC imaging.
3. 9 cm diameter bacterial grade Petri dishes to culture embryos/larvae (Greiner Bio One).
4. Disposable 3-mL graduated polyethylene transfer pipets (Samco).
5. Watchmaker's forceps to dechorionate embryos (Dumont no. 5).
6. 5-mL Technoplas tubes to fix and/or store embryos/larvae.
7. EM with benzocaine (EMB; Sigma): final concentration 0.2 mg/mL to anaesthetize embryos.
8. Bouin's fixative: 75 mL saturated picric acid, 25 mL formalin, 5 mL glacial acetic acid (Merck). Preferred fixative for histology and immunohistochemistry.

9. 1X phosphate buffered saline (PBS): 0.8% *(w/v)* NaCl, 0.02% *(w/v)* KCl, 0.02 M phosphate buffer, pH 7.4.
10. 4% *(w/v)* Paraformaldehyde (PFA) in 1X PBS. Prepared from a 16% *(w/v)* PFA stock solution (Electron Microscopy Services). Provided in sealed ampoules and stored at room temperature . Dilute immediately prior to use. Once diluted, keep at 4°C and use within 1 wk or store frozen at −20°C and thaw once for use. Preferred fixative when only mild cross-linking required (e.g., prior to sectioning on a vibrating microtome).
11. PBST: Phosphate buffered saline (PBS) containing 0.1% *(w/v)* Tween-20. Tween-20 prevents the fixed embryos from sticking together and/or to the transfer pipets.
12. Absolute ethanol.

2.2. Histologic Analysis of the Developing Zebrafish Digestive System

1. Low melting temperature (LMT) agarose (SeaPlaque Agarose, Cambrex Bio Science).
2. 50-mL tubes (Falcon).
3. Cryomolds (Tissue Tek disposable vinyl specimen molds 15 × 15 × 5 mm, (Sakura; cat. no. 4566).
4. Probe 1 (straight 15.5 cm, Fine Science Tools, or similar).
5. Probe 2 (finer, handmade probe – made with either a human eyelash/eyebrow hair, or paintbrush bristle as the probe-tip – attached to a handle or a micropipette tip or similar).
6. Absolute ethanol.
7. Toluene.
8. Paraffin wax (Histosec) (Merck; cat. no. K91354261).
9. Tissue-Tek® TEC™ 5, Tissue Embedding Console System (Sakura) or equivalent.
10. 15 × 15 × 5 mm base molds (Grale Scientific).
11. Tissue Processing Embedding Cassettes (Sakura) or equivalent.
12. Leica TP1020 Automatic Tissue Processor (Leica Instruments; Heidelberg, Germany) or equivalent.
13. Scalpel blades (Swann-Morton: carbon steel surgical blade no. 11).
14. Curved fine forceps (Lawton).
15. Long curved probe (Lawton).
16. Leica RM 2035 microtome (Leica Instruments) or equivalent.
17. Water-bath (Labec) or equivalent.
18. Warming plate (WT1, Ratek) or equivalent.
19. Slide rack and staining dish (polyethylene-terephthalate (PET) plastic or glass, various suppliers).
20. Acid-alcohol: 0.5% hydrochloric acid in 70% ethanol.
21. Mayer's hematoxylin (Fronine).

22. Eosin: 5 g EosinY, 1 g calcium nitrate in 1 L of H_2O (Amber Scientific).
23. Histolene (Fronine).
24. Scott's Tap Water: 20 g $MgSO_4$, 2 g $NaHCO_3$.
25. Safety Mount No.4 (slide mountant) (Fronine), or equivalent.
26. Superfrost Plus slides (Menzel-Glaser).
27. Cover slips: 24 × 50 mm or 24 × 60 mm (HD Scientific).
28. Alcian Blue Solution: (Tube A) Add 3 mL glacial acetic acid stock to 97 mL dH_20 and mix well. (Tube B) Weigh 0.1 g Alcian Blue 8GX (Sigma) powder into a Falcon tube (*see* **Note 1**). Add 10 mL from Tube A to Tube B and stir well (use a small magnetic stirrer). Store at room temperature (can be re-used twice).
29. 1% Periodic Acid Solution: Add 5 g periodic acid (Merck) to 500 mL dH_2O and store at room temperature. Can be re-used twice.
30. Schiff reagent (Sigma) (use neat).

2.3. Immunohistochemistry

2.3.1. Example 1: Detection of Bromodeoxyuridine (BrdU) Incorporation into Cells in S-Phase

1. All the reagents listed in **Subheading 2.1** and **Subheading 2.2, item 1** are required, except EM with PTU (**Subheading 2.1.1, item 2**) and 4% PFA (**Subheading 2.1.1, item 10**).
2. Cell Proliferation Labeling Reagent (aqueous solution of 10 mM 5-bromo-2-deoxyuridine and 5-fluoro-2'-deoxyuridine) (GE Healthcare).
3. Citrate buffer: 2.1 g anhydrous citric acid (Sigma) in 1 L of dH_20, increase pH to 6.0 using 10 M NaOH, for antigen retrieval.
4. Xylene.
5. 30% Hydrogen peroxide.
6. Humidified container (we use: a plastic container with dimensions of approximately 25 × 15 × 8 cm, a slide support grate and a lid (*see* **Note 2** and **Fig. 1a**).
7. Mini PAP pen (Zymed Laboratories).
8. CAS block buffer (Zymed Laboratories).
9. Primary antibody: mouse anti-BrdU antibody (1:100 dilution, BD Bioscience).
10. Liquid DAB Substrate Chromogen System (Dako).

2.3.2. Example 2: Anti-active Caspase 3 for Detection of Apoptosis

1. All reagents listed in **Subheading 2.3.1** are required, except for citrate buffer (**Subheading 2.3.1, item 3**) and the anti-BrdU antibody (**Subheading 2.3.1, item 9**).
2. Target Retrieval Solution (DAKO) for antigen retrieval.
3. Primary antibody: rabbit anti-human/mouse caspase 3 active (1:400 dilution) (R&D Systems, Inc.).
4. Secondary antibody: horseradish peroxidase (HRP)-coupled goat anti-rabbit IgG (1:200 dilution, Zymed Laboratories).

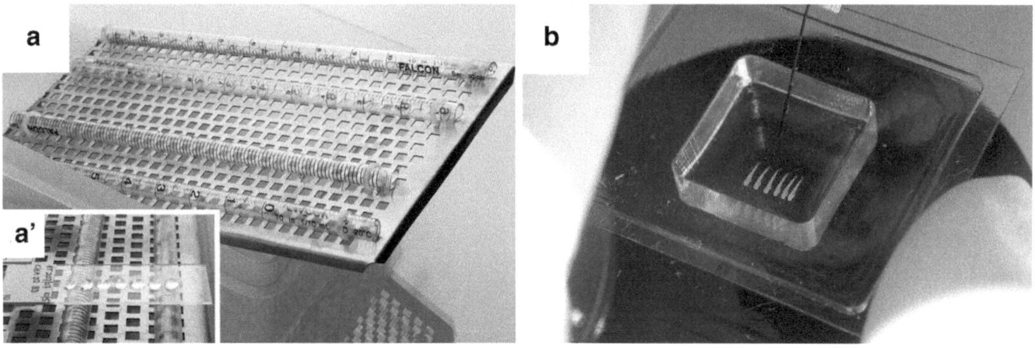

Fig. 1. Equipment required for immunohistochemistry (antibody incubation) and agarose embedding of zebrafish embryo/larvae. (**a**) Humidified container: plastic container with dimensions of approximately 25 cm × 15 cm × 8 cm, a slide support grate (stainless steel mesh with plastic 10 mL Falcon serological pipette attached with cable ties) and a lid. Soaked paper towels are placed under the grate. (**a′**) Slide sitting on a slide support grate incubating in CAS block buffer. (**b**) Positioning of embryos using probe 2 in 1% low melting temperature agarose in a Cryomold (Tissue Tek disposable vinyl specimen mold 15 × 15 × 5 mm). Positioning of embryos/larvae should be horizontal with anterior to the left and as uniform as possible in the horizontal plane to allow comparable sections to be obtained across multiple specimens.

2.4. Imaging of Embryos/Larvae Using Brightfield, Fluorescence/Confocal, and DIC Microscopy

2.4.1. Preparing Embryos for Brightfield, Fluorescence or DIC Imaging

1. All reagents listed in **Subheading 2.1.1, items 1–7**, except **item 6**.
2. 3% methyl cellulose *(w/v)* (Methocel 65 Hg, Fluka).
3. Fine handmade probe (*see* **Subheading 2.2.1, item 4**).

2.4.2. Preparation of Embryos for Confocal Analysis

1. All reagents listed in **Subheading 2.1.1, items 1–5**.
2. FluoroDish™FD35 35-mm cover-glass bottom dish (glass thickness = 0.17 mm) (World Precision Instruments, Inc.) or a Sykes-Moore chamber accommodating a microscope coverslip (glass thickness = 0.17 mm) (Bellco Glass Company).

2.4.3. Confocal Imaging of Live Embryos

1. Nikon C1 laser scanner attached to a Nikon TE-2000-E inverted microscope or equivalent.

2.5. Analysis of Protein Expression Patterns Using Whole-Mount Immunocytochemisty

2.5.1. Wholemount Immunocytochemistry

1. 12-well tissue culture plate with transfer wells (Netwell inserts, membrane diameter 15 mm, mesh size 74 μm; Corning).
2. PBST.
3. PBD: PBST + 0.5% Triton-X.
4. Blocking solution: 1% *(w/v)* bovine serum albumin (BSA), 1% *(v/v)* serum in PBD.
5. Antibody dilution solution: 0.2% *(w/v)* BSA in PBD.
6. Aluminum foil.

2.6. Immunocytochemical Analysis of Protein Localization Patterns in Thick (150–200 µm) Sections

2.6.1. Embryo Embedding and Sectioning

1. Cryovials 2 mL (Nunc).
2. Scalpel blades (Swann-Morton; carbon steel surgical blade no. 11).
3. Superglue.
4. Razor blades.

2.6.2. Immunohistochemistry on Thick (150–200 µm) Sections

1. All reagents listed in **Subheading 2.5.1** except transfer wells.
2. 12-well tissue culture plates.
3. Fine paintbrush.

3. Methods

3.1. General Methods

3.1.1. Embryo/Larvae Rearing, Staging, and Fixation

1. Collect embryos from pairs/groups of breeding fish within 15–30 min of spawning (*see* **Note 3**).
2. Incubate embryos in 9-cm diameter bacterial grade Petri dishes (no more than 40 per dish) containing approximately 30 mL of EM at 28.5°C with a 12 h light:12 h dark photoperiod.
3. If required, dechorionate unhatched embryos using watchmaker's forceps and place embryos in fresh EM (*see* **Note 4**).
4. Anesthetize embryos in EMB to a final concentration of 200 mg/L (*see* **Note 5**). Sort embryos/larvae into groups as required and place up to approximately 100 embryos/larvae in 1 mL EMB into 5 mL Technoplas tubes (or similar).
5. Working in a fume-hood, remove EMB using a transfer pipette and immediately add approximately 1 mL of fixative. For histology and immunohistochemistry, fix overnight at 4°C in Bouin's fixative (**WARNING:** Bouin's fixative is toxic) (*see* **Note 6**). For embryos to be sectioned on a vibrating microtome, lightly fix the embryos either in 4% PFA/PBS (**WARNING:** PFA is toxic) (*see* **Note 7**) at room temperature for 2–3 h or in 2% PFA/PBS overnight at 4°C (*see* **Note 8**).
6. After the required fixation period, use a transfer pipette to remove the fixative and wash embryo/larvae in PBST 3× for 5 min. At this point the embryos/larvae to be used for histology and/or immunohistochemistry are embedded in agarose. If groups of embryos/larvae are to be consolidated for embedding on the same day, they can be transferred to 70% ethanol and stored at 4°C for several months.

3.1.2. Preparing Agarose for Embedding

1. On a hot plate, maintain 600 mL of water at boiling point in a 1 L beaker.

2. Place 25 mL of PBS and 1 g of LMT agarose into a 50 mL Falcon tube to make a solution of 4% LMT agarose (*see* **Note 9**). The percentage of agarose required will vary depending on whether the embryos are to be embedded vertically or horizontally (*see* **Note 10**).
3. Loosen the lid of the Falcon tube and place into the beaker of boiling water.
4. Every 5 min release and retighten the lid of the Falcon tube to release the build up of pressure. Swirl the solution gently to prevent the agarose from forming clumps.
5. Continue to heat the agarose until it has melted completely.

3.2. Using Histology to Visualize the Developing Zebrafish Digestive System

3.2.1. General Histologic Procedure and Hematoxylin and Eosin (H + E) Staining

1. Prepare embryos/larvae up to **Subheading 3.1.1**, **step 6**.
2. Prepare 1% LMT agarose in PBS as described in **Subheading 3.1.2**.
3. Place embryos/larvae into a bacterial grade Petri dish with enough media (ethanol or PBST depending if continuing from **step 5** of **Subheading 3.1.1** or from storage) to prevent them drying out.
4. Put a cryomold (labeled using autoclave tape and a pencil) on the stage of a dissecting microscope.
5. Use a transfer pipet to place embryos/larvae into a cryomold. Remove as much of the transferred medium as possible, immediately prior to adding the molten agarose. The embryos/larvae will almost invariably float to the surface/sides of the agarose. Quickly push them to the bottom/centre of the cryomold with probe 1 and, once approximately positioned, finalize positioning of the embryos/larvae with probe 2. Positioning of embryos/larvae should be parallel to the bottom of the cryomold, right lateral side down and as uniform as possible in the horizontal plane (**Fig. 1b**). This will allow comparable sections to be obtained across multiple specimens and facilitate the production of sagittal sections encompassing the entire digestive system (**Fig. 2a**). If mutants are to be compared to wild-type, embed both genotypes in one block to afford even staining and exact comparisons (*see* **Note 11**).
6. Once the agarose is set, place the entire cryomold in a beaker containing 70% ETOH at room temperature. The blocks can be stored up to 3 d in 70% ETOH. Remove agarose blocks from cryomolds one at a time and trim the blocks with a scalpel blade (with a margin of approximately 1–2 mm around the specimens) and put into a Tissue Tek cassette. Transfer into a tissue basket of a Leica TP1020 Automatic Tissue Processor (or equivalent) inserted in a 2 L plastic beaker containing approximately 1.5 L of 70% ETOH.

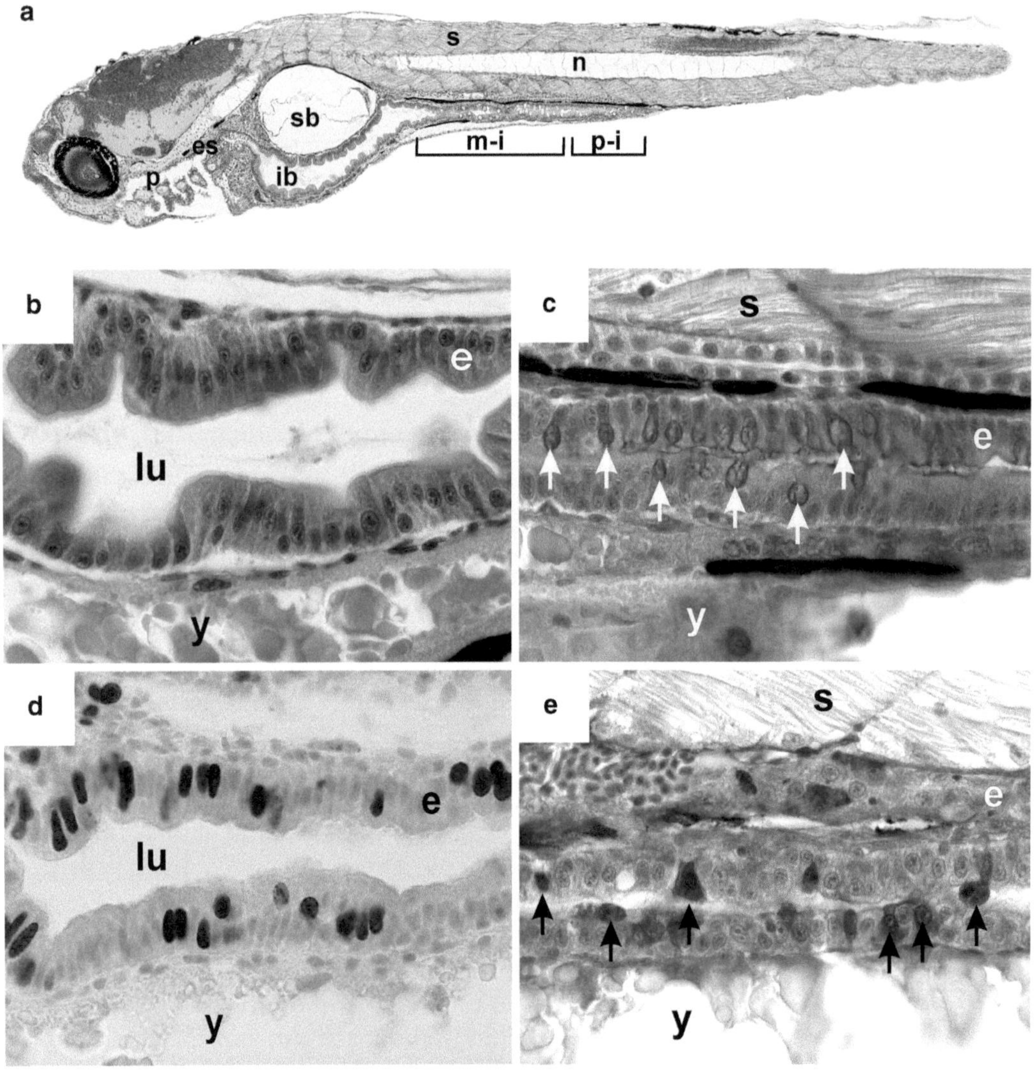

Fig. 2. Morphologic analysis of intestinal epithelium development in zebrafish larvae. (**a**) Full length sagittal section of a wild-type zebrafish larva at 120 hpf stained with hematoxylin and eosin. (**b**) Sagittal section of a wild-type zebrafish embryo stained with hematoxylin and eosin showing the anterior portion of the intestine, the intestinal bulb, at 96 hpf. (**c**) Sagittal section of a wildtype zebrafish embryo stained with alcian blue-periodic acid Schiff reagent showing the mid-intestine region at 96 hpf. *Arrows* indicate turquoise staining of acid mucopolysaccharides within goblet cells in the intestinal epithelium. (**d**) Sagittal section of a wildtype zebrafish embryo at 72 hpf, illustrating BrdU-positive nuclei (dark nuclei) in intestinal epithelial cells in the *S*-phase of the cell cycle. (**e**) Sagittal section of a zebrafish intestinal mutant (*trinculo*^{S451}) embryo depicting active caspase 3-positive cells (*arrows*) in the mid-intestine at 96 hpf. *e* intestinal epithelium; *es* esophagus; *ib* intestinal bulb; *lu* intestinal lumen; *m-i* mid-intestine; *n* notochord; *p* pharynx; *p-i* posterior-intestine *s* somites; *sb* swim bladder *y* yolk.

7. Place the basket on the Leica TP1020 Automatic Tissue Processor for dehydration, clearing, and paraffin infiltration. If an automatic tissue processor is not available, this can be done manually (*see* **Note 12**). The following program is run

overnight: (a) ethanol 70% 30°C for 2 h; (b) ethanol 70% 30°C for 2 h; (c) ethanol 80% 30°C for 1 h; (d) ethanol 90% 30°C for 1 h; (e) ethanol 100% 30°C for 1 h; (f) ethanol 100% 30°C for 1.5 h; (g) 50% ethanol – 50% toluene 30°C for 1 h; (h) 100% toluene 30°C for 1 h; (i) 100% toluene 30°C for 1 h; (j) paraffin wax 60°C for 1 h; (k) paraffin wax 60°C for 1 h (*see* **Note 13**).

8. Embed the samples using a Tissue-Tek® TEC™ 5, Tissue Embedding Console System (or equivalent) (*see* **Note 14**). Maintain the agarose/paraffin blocks at 60°C and one at a time place each block into a base mold also maintained at 60°C and fill with molten paraffin wax (*see* **Note 15**). Immediately mount a Tissue Tek cassette on top of the base mold and fill with molten paraffin wax to form the new block. Store the blocks at RT until required.
9. Prior to sectioning, cool the blocks on ice in order to section properly. Trim blocks using a scalpel blade. A 1–2 mm margin around the periphery of the embryos is desirable; too large a margin around the block will limit the number of sections that can be mounted onto one slide.
10. Place the blocks into the cassette clamp of a Leica RM 2035 Microtome (or equivalent) and section at 5 μm intervals. Trim through the outer layers of wax and look for the first sign of the original agarose block appearing at the cutting face. If unsure, take a ribbon of 5–10 sections and use curved forceps and a long curved probe to transfer the sections from the microtome to a water-bath maintained at 45°C (*see* **Note 16**). Use the forceps to pull the leading section and the probe to pick up the trailing sections and gently place onto the surface of the water. Look for a slightly less transparent and smoother area in the sections. When a section/ribbon containing agarose/tissue is identified, tease it apart from the ribbon with the forceps and probe and immerse a slide into the water behind the section and maneuver it onto the slide next to the label. The section will attach to the slide as they collide at the air-water interface. Complete the attachment by slowly withdrawing the slide from the water (this maneuver will take a little practice).
11. Confirm the presence of tissue by checking under a microscope. Continue to float ribbons of 8 sections onto prelabeled slides (*see* **Note 17**) in 2 rows (16 sections per slide).
12. Drain water off slides by leaning against a vertical surface.
13. When slides are dry, place on a warming plate maintained at approximately 55°C for 15 min to bake the sections onto the slide (aiming to melt the wax a little).

14. When the wax is reset dewaxing can commence by immersing slides in a Histolene bath for 3 min.
15. Complete the de-waxing of sections by immersing slides in a second Histolene bath for 3 min. The following series of steps (**steps 16–27**) are completed by slotting the slides into a slide rack and transferring them between slide staining dishes.
16. Incubate slides in 100% ethanol 3 for 3 min each.
17. Rinse slides in running tap water for 3 min.
18. Stain sections by incubating slides in Mayer's hematoxylin for 5 min.
19. Rinse slides in running tap water for 1.5 min.
20. Dip slides in acid-alcohol (0.5% hydrochloric acid in 70% ethanol) for 2 s.
21. Rinse slides in running tap water for 1.5 min.
22. Incubate slides in Scott's Tap Water for 1.5 min.
23. Rinse slides in running tap water for 1.5 min.
24. Stain sections by incubating slides in eosin for 1.5 min.
25. Rinse slides in running tap water for approximately 30 s.
26. Incubate slides in 100% ethanol × 3 for 1.5 min each .
27. To ensure dehydration of the section, incubate slides in Histolene for 3× for 3 min each.
28. Mount slides in safety mount no. 4.
29. Leave slides to dry in the fume hood overnight at room temperature.

Result: **(Fig. 2a, b)**

3.2.2. Alcian Blue-Periodic Acid Schiff Stain for Visualization of Goblet Cells

The alcian blue-periodic acid Schiff (PAS) stain is a method for detecting neutral and acid polysaccharides. Acidic mucopolysaccharides are stained blue/turquoise with Alcian blue, whereas neutral polysaccharides stain pink with the PAS reaction. The mucins in the goblet cells in the zebrafish intestine contain acid mucopolysaccharides and stain blue/turquoise using this technique *(3)*.

1. Following sectioning, take sections to water (**Subheading 3.2.1, step 19**). Allow Schiff reagent to reach room temperature for 10–20 min prior to use; it works sub-optimally at lower temperatures.
2. Incubate slides in 3% acetic acid for 15 min.
3. Stain slides in 1% alcian blue in 3% acetic acid for 5–15 min (gauge under a microscope).
4. Wash thoroughly with running tap water for 3 min and then rinse by immersing briefly in a bath of H_2O.

5. Incubate in 0.5% periodic acid solution for 15 min.
6. Wash slides thoroughly with running tap water for 5 min and rinse by immersing briefly in a bath of H_2O.
7. Place slides in Schiff reagent for 5–20 min (gauge under a microscope).
8. Wash slides thoroughly with running tap water for 5 min and rinse by immersing briefly in a bath of H_2O.
9. Stain slides for 30 s to 1 min in Mayer's hematoxylin.
10. Rinse for 1.5 min in running tap water and then in Scott's Tap Water for 30 s to 1 min.
11. To complete the protocol continue **steps 26–29** in **Subheading 3.2.1**.

Result: **(Fig. 2c)**

3.3. Immunohistochemistry

Example 1: Anti-BrdU Detection for Assessment of Cell Proliferation

1. Rear embryos as described until **Subheading 3.1.1**, **step 3**.
2. Sort embryos/larvae according to phenotype if required. Incubate embryos in undiluted Cell Proliferation Labeling Reagent for 30 min at 28.5°C in 5 mL Technoplas tubes. Incubate up to 25 embryos in 1 mL of BrdU solution.
3. Wash embryos three times in pre-warmed EM, transfer embryos to a bacterial grade Petri dish and incubate at 28.5°C in EM for 1 h.
4. Anesthetize as in **Subheading 3.1.1**, **step 4** and then follow **steps 5** and **6**.
5. Prepare 1% LMT agarose as described in **Subheading 3.1.2**.
6. Follow steps from **Subheading 3.2.1**, **steps 3–12**.
7. *Preparation of the master slide stained with H + E and unstained slides for immunohistochemistry.* Confirm the presence of tissue by checking under a microscope as in **Subheading 3.2.1**, **step 11**. Place the first section on the master slide; this section is "section 0" of the master slide and the entire series (*see* **Subheading 3.3.1**, **step 9** for use of the master slide). Place the next eight sections on "slide 1" and keep unstained for immunohistochemistry (the first section on slide 1 becomes "section 1 of slide 1"; following section "0" on the master slide). It is important that sections are evenly spaced along the centre of each unstained slide; usually eight sections per slide (*see* **Subheading 3.3.1**, **step 21** below for details) (*see* **Note 18**). Complete the master slide by attaching sections 9, 18, 27, and so on (up to 16 sections in 2 rows can be put on the master slide as in **Subheading 3.2.1**, **step 11**). Put successive groups of eight sections on slide 2, 3, and so on until all the sections are used. Dewax,

stain (H + E) and mount the master slide as in **Subheading 3.2.1, steps 14–29**.

8. Dry and bake unstained sections onto slides as described in **Subheading 3.4.1, steps 12** and **13** (can be stored at room temperature before further processing).
9. Scan the master slide under the microscope to identify the best sections for the region of interest. The next process takes about 7 h.
10. Put the selected unstained slides in a slide rack and incubate in an oven at 60°C for 10 min. Use the slide rack (and staining dish as required) until end of **Subheading 3.3.1, step 19**.
11. Allow slides to cool down at room temperature for 10 min.
12. Dewax slides (xylene 5 min ×2, 100% ethanol 5 min ×2, 70% ethanol 5 min).
13. Wash slides in H_2O for at least 5 min.
14. For antigen retrieval, place slides in 10 mM citrate buffer and microwave on HIGH for 5 min, and then MED-LOW for 5 min.
15. Allow slides to cool for at least 20 min.
16. Wash slides thoroughly in H_2O for 5 min and then in PBS for 1 min.
17. Block endogenous peroxidase activity by incubating slides in 6% H_2O_2 in methanol at room temperature for 30 min with gentle shaking (160 mL methanol + 40 mL H_2O_2) (*see* **Note 19**).
18. Wash slides thoroughly in running tap water for 5 min.
19. Wash slides in PBS for 1 min.
20. Use a Mini PAP pen to draw an impervious barrier around the sections (*see* **Note 20**).
21. Use a micropipet to cover each section with approximately 30 μL of CAS block buffer **(Fig. 1a′)**. Incubate sections for 40 min at room temperature in a humidified container. Ensure that sections do not dry out at any stage during the following process.
22. Remove CAS block by flicking each slide over a sink. Put a slide rack in a staining dish containing PBS and add the slides one-by-one as the CAS block is removed.
23. Use a micropipet to cover each section with approximately 30 μL of anti-BrdU antibody (a 1:100 dilution in PBS). Incubate at RT for 1 h in a humidified container.
24. Remove primary antibody by flicking each slide over a sink. Put a slide rack in a staining dish containing PBS and add the

slides one-by-one as the primary antibody is removed and leave for 5 min. Wash ×2 in PBS for 5 min.

25. Use a micropipet to cover each section with 30 μL secondary antibody (HRP-coupled goat anti-mouse IgG – a 1:200 dilution in PBS). Incubate slides for 1 h at room temperature in a humidified container.
26. Remove secondary antibody by flicking each slide over a sink. Put a slide rack in a staining dish containing PBS and add the slides one-by-one as the secondary antibody is removed and leave for 5 min. Wash ×2 in PBS for 5 min.
27. Place slides under a microscope and flood (~ 0.5 mL) in DAB-chromagen mix (1 drop chromagen plus 1 mL DAB substrate; pre-warm to RT) using a transfer pipet. Carefully watch under the microscope for developing brown color in nuclei of positive cells. Avoid background staining by stopping the reaction at the time of greatest contrast between positive and negative cells. This usually takes no more than 1 min but may take considerably longer.
28. Rinse each slide quickly in PBS in a staining dish, and then hold the slides in a slide rack in a second slide staining dish containing PBS until all slides have been processed through **Subheading 3.3.1, step 28**.
29. Keep slides in slide rack from here until the end of **Subheading 3.3.1, step 34**. Wash slides in running tap water for 5 min.
30. Stain the slides with Mayer's hematoxylin for 30–50 s (start with one slide for 15 s and monitor under the microscope) and then wash in tap water for 2–5 min, or until water just clears.
31. Place slides in Scott's Tap Water for 1 min and then rinse in tap water for 2–5 min and check under microscope before proceeding.
32. Rinse slides in 70% ethanol for 5 min.
33. Incubate slides in 100% ethanol for 5 min ×2.
34. Incubate slides in xylene for 5 min.
35. Mount using safety mount no. 4.
36. Leave slides to dry in a fume hood overnight.

Result: **(Fig. 2d)**

3.3.2. Example 2: Anti-active Caspase 3 for Detection of Apoptosis

1. Rear embryos/larvae as described until **Subheading 3.1.1, step 3**.
2. Sort embryos/larvae according to phenotype if required. Continue from **Subheading 3.3.1, steps 4–14**.

3. Antigen retrieval: Preheat an oven to 97°C. Prepare 1X DAKO Target Retrieval Solution as per manufacturer's instructions (20 mL 10X concentrate DAKO Target Retrieval Solution: 180 mL H_2O). Microwave solution on high for approximately 1.5 min (stop at the first sign of boiling; the solution will go cloudy). Immerse the rack of slides into the hot solution and place into a preheated oven for 20 min.
4. Follow steps **15–22** in **Subheading 3.3.1** and then use a micropipet to cover each section with 30 μL of primary antibody: rabbit anti-human/mouse caspase 3 active (1:400 dilution in PBS), at room temperature for 1 h in humidified container. Apply approximately 30 μL to each section using a micropipette.
5. Continue as in **Subheading 3.3.1, steps 24–25** incubate in secondary antibody (1:100 dilution with 1X PBS) for 1 h and room temperature.
6. Complete protocol by continuing from **Subheading 3.3.1, steps 26–36**.

Result: **(Fig. 2e)**

3.4. Imaging of Embryos/Larvae Using Brightfield, Fluorescence/Confocal, and DIC Microscopy

3.4.1. Preparing Embryos for Brightfield, Fluorescence, or DIC Imaging

1. Rear embryos to the required developmental stage and anesthetize (**Subheading 3.1.1, steps 1–4**).
2. Remove the embryo/larvae from EMB and place on a cavity microscope slide, a Sykes-Moore chamber (Bellco Glass Company) or a clean plastic Petri dish. Remove excess EMB and quickly cover the embryo and surrounding area with 3% methyl cellulose (*see* **Note 21–23**).
3. Push the embryo using a fine handmade probe to the base of the microscope slide or Petri dish for imaging.
4. Remove large bubbles formed in the methyl cellulose with a pipet tip. Fine bubbles can be removed from the field of view with a fine probe.

3.4.2. Preparing Embryos for Confocal Analysis

The methodology to create an agarose mold for an upright microscope is described in Hall et al. (*see* **Note 24**). The methodology described below creates an agarose mold for an inverted microscope. Both microscopes are suitable for imaging the zebrafish digestive system.

1. Rear embryos to the required developmental stage and anesthetize as in **Subheading 3.1.1, step 4**.
2. Mount three embryos in 0.8–1% LMT agarose in EMB in either a 35 mm tissue FluoroDish™ or a Sykes-Moore chamber (*see* **Note 25**). To prevent damage to the embryo, cool the molten agarose for 5–10 min before embedding (*see* **Note 26**).

3. Position the embryos in the centre and on the base of the glass well, lying flat. Separate the embryos from each other to limit photo bleaching that could occur during data acquisition of a neighboring embryo.
4. Once the agarose has set (approximately10–20 min), apply a small volume of EMB to the surface of the agarose to prevent it from drying out.

3.4.3. Confocal Microscopy of Embryos

The settings used for a confocal imaging experiment will differ depending on the equipment used. The following information relates to the use of a Nikon C1 laser scanning confocal microscope attached to a Nikon TE-2000E inverted microscope configured with filters suitable for FITC fluorescence. Record all imaging parameters used during a confocal experiment to allow duplication and appropriate comparisons between test groups. A typical scanning set-up for analysis of the digestive tract in 72–120 hpf embryos is as follows:

1. ×20 multi-immersion lens
2. 20 mW Sapphire solid state laser (Coherent) emitting at 488 nm (for detection of GFP fluorescence)
3. Image format: 512 × 512 for time-lapse or z-series data collection
4. Image format: 1024 × 1024 if higher resolution required
5. Average of 3 images to increase signal to noise (Kalman filter)
6. 20–30 slices for a small 3D reconstruction, no greater than 3 μm apart

The following instructions relate to the EZ-C1 software that controls the Nikon C1 laser scanner unit. However, all confocal software packages allow manipulation of these imaging parameters.

1. Limit the 488 nm laser intensity to the minimum required to obtain a reasonably noise-free image (e.g., 15–30%) (*see* **Note 27** and **28**).
2. Set the pin-hole for the 488 nm channel to medium size (*see* **Note 29**).
3. To avoid over-exposing the image (image saturation) use a look-up-table that will false color the image in the event of pixel saturation.
4. Perform initial scans of the embryo at a low resolution (high scan speed such as 256 × 256 pixels) to permit correct positioning of the embryo. Limit this initial exposure time to minimize photobleaching before image collection.

5. Use minimal laser power and compensate with higher detector gain to avoid photobleaching of fluorophore and possible cell death during image collection (*see* **Note 30**).
6. Raise the resolution of the image for data collection (e.g., 512 × 512 or 1024 × 1024 pixels).

Result: **(Fig. 3a–c)**

3.4.4. Image Analysis and 3D Stack Correction

Image analysis can be performed in the freeware and multiplatform application, ImageJ (http://rsb.info.nih.gov/ij/) (*see* **Note 31**). Slight movements that may occur during time lapse or 3-D projection experiments can usually be corrected with post imaging software using the "StackReg" Plugin (http://bigwww.epfl.ch/thevenaz/stackreg/) (*see* **Note 32** and **Fig. 3d, e**). Record manipulations performed on an image and declare these in publications. The raw data collected from an experiment should be maintained in an unmanipulated state. All alterations should be performed on duplicate image copies.

3.5. Analysis of Protein Expression Patterns Using Wholemount Immunocytochemistry

3.5.1. Wholemount Immunocytochemistry

1. Rear embryos/larvae to the required developmental stage and fix to **Subheading 3.1.1, step 6**.
2. Primary antibody blocking step: Transfer the embryos to transfer wells in a 12-well plate (*see* **Note 33**). Permeablize the embryos in PBD at room temperature with gentle rotation for 2–6 h. Remove this solution and completely cover the embryos with blocking solution (~400 μL of solution/well). Incubate at room temperature for approximately 2 h on a rocking platform.
3. Primary antibody incubation: Dilute primary antibody in antibody dilution solution according to the manufacturer's recommendations (useful dilutions include 1:200, 1:500, 1:1,000, 1:2,000 but must be determined empirically for each antibody). Remove blocking solution and incubate embryos with the primary antibody solution overnight at 4°C on a rocking platform. Apply parafilm around the seal between the lid and plate to prevent evaporation of the antibody solution.
4. Secondary antibody blocking: Wash the embryos three times in PBD for 20 min at room temperature on a rocking platform ×3.
5. Remove PBD and incubate embryos in blocking solution for 1 h at room temperature.
6. Secondary antibody incubation*:* Dilute secondary antibody in dilution solution according to the manufacturer's recommendations (*see* **Note 34**). Secondary antibody dilutions include 1:500, 1:1,000, 1:2,000 but must be determined empirically for each antibody.

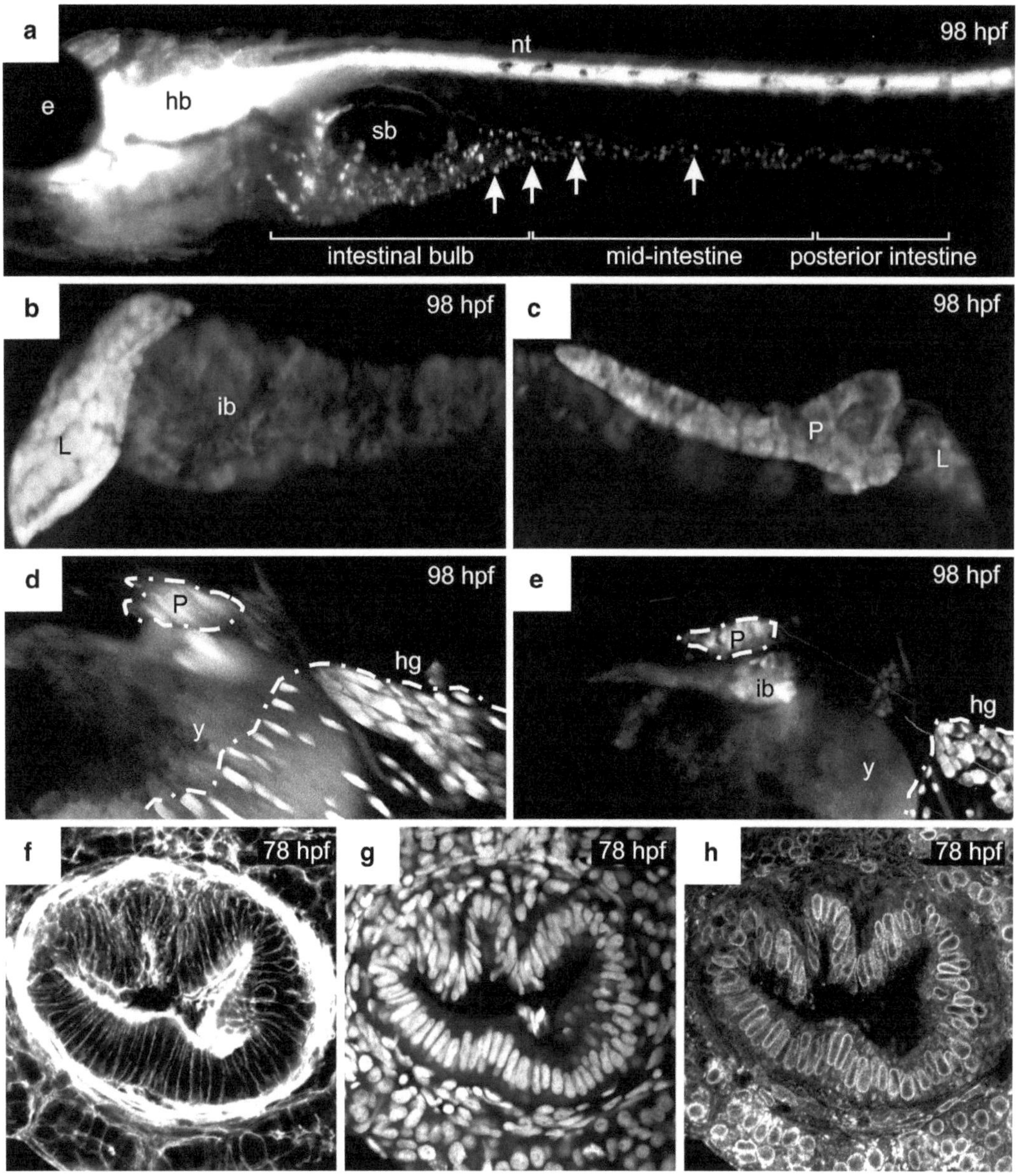

Fig. 3. Analysis of zebrafish intestinal morphogenesis using fluorescent imaging. (**a**) Epi-fluorecence analysis of the *Tg(nkx2.2a:EGFP)* transgenic line *(3)* can be used to explore the differentiation of the enteroendocrine cells *(arrows)* in the zebrafish intestinal epithelium. Cells in the endocrine pancreas, hindbrain, and ventral neural tube also fluoresce brightly. Analysis of the *Tg(gutGFP)* transgenic line reveals GFP expression in the intestinal bulb and liver (**b**) and pancreas (**c**) of the developing zebrafish embryo. (**d**) During analysis of a flotte lotte mutant embryo on the "gutGFP" background the 3% methylcellulose solution in which the embryo was embedded began to flex and cause the embryo to move. Using ImageJ x1.39 (http://rsb.info.nih.gov/ij/) and the "StackReg" Plugin (http://bigwww.epfl.ch/thevenaz/stackreg/) this movement was corrected (**e**) and provided useful analysis of the intestinal and pancreatic defects in this mutant. (**f-h**) Immunocytochemical techniques applied to thick sections of zebrafish embryos permit analysis of the tissue architecture of the intestinal epithelium. (**f**) f-actin staining using rhodamine phalloidin (**g**) demonstrates that the intestinal epithelium is highly polarized at 78 hpf. Application of the Hoechst33342 DNA binding stain demonstrates the basally located nuclei in the intestinal epithelium. (**h**) Immunohistochemistry of nucleoporins, visualized with an Alexa Fluor 488 secondary antibody, reveals the punctate localization of nuclear pore complexes in the nuclear envelope of intestinal epithelial cells. *e* eye; *hb* hindbrain; *hg* hatching gland; *ib* intestinal bulb; *L* liver; *P* pancreas; *nt* neural tube; *sb* swim bladder; *y* yolk.

7. Remove PBD and incubate embryos in secondary antibody solution overnight at 4°C on a rocking platform.
8. Secondary antibody removal: Wash the embryos in PBD for 20 min at room temperature or 4°C on a rocking platform ×3.
9. Imaging of embryos stained with secondary antibodies: If the secondary antibody is conjugated to a fluorophore such as Alex Fluor 488 (Molecular Probes), mount the embryos on a coverslip. Use 1% LMT agarose for longer periods of image acquisition, such as a *z*-series in a confocal microscope. Collected images can be projected using a maximum intensity method. Alternatively, use 3% methyl cellulose to hold the embryos in position for shorter periods of image acquisition (*see* **Note 21**). During image acquisition, keep the remainder of the embryos in complete darkness and/or stored at 4°C. This can be achieved by wrapping the plate in aluminum foil. If the embryos are to be stored for an extended period of time, apply parafilm around the seal between the lid and plate to prevent evaporation.

3.6. Immunocytochemical Analysis of Protein Localization Patterns in Thick (150–200 μm) Sections

3.6.1. Embryo Embedding and Sectioning

Note that the following instructions relate to the preparation of transverse sections, but can be adjusted for other planes of section.

1. Rear and process embryos/larvae to **Subheading 3.1.1, step 6**.
2. Prepare 4% LMT agarose in PBS as described in **Subheading 3.1.2**.
3. Cut 1 cm off the end of a standard 1000 μL micropipet tip (to give a larger aperture) (*see* **Note 35**).
4. Pierce the yolk of the embryos/larvae with sharp fine forceps to allow agarose to infiltrate (*see* **Note 36**).
5. Place embryos/larvae into a 9-cm bacterial grade Petri dish with approximately 20 mL EM to prevent them drying out.
6. Put a cryomold (labeled with autoclave tape and a pencil) on the stage of a dissecting microscope.
7. For transverse sections of embryos >48 hpf, add approximately 1.2 mL of 4% LMT agarose into the cryomold with a modified pipet tip (*see* **Note 37** and **38**). Modified tips can be used once only.
8. Immediately following the addition of the agarose to the cryomold, remove any bubbles that have formed on the surface (*see* **Note 39**).
9. Using a fine handmade probe, drag the embryos away from the cryomold edges into the middle of the mold. Gently push

down the embryos until they are lying horizontally close to the bottom surface of the dish.

10. For transverse sections, manipulate the embryos so that the head of the embryo is resting on the bottom surface with the tail pointing vertically upwards. Evenly space the embryos around the centre of the mold (*see* **Note 40**).
11. Do not move the mold until the agarose solution has started to set (agarose becomes cloudy) as the embryos may move out of position (*see* **Note 41**).
12. To prevent the agarose drying out, place a small amount of PBS onto the surface of the agarose (*see* **Note 42** and **43**).
13. Snap a double-sided razor blade in two and screw one half tightly into the blade holder (*see* **Note 44**).
14. Place the blade holder into position and fill the vibrating microtome buffer tray with PBS.
15. Flip the agarose out of the cryomold with a scalpel blade.
16. Use a scalpel blade to trim all four edges of the block to reduce the amount of agarose to cut away during sectioning (*see* **Note 45**). Remove any meniscus in the agarose to allow for an even cut during sectioning (*see* **Note 46**).
17. Use Superglue to fix the agarose mold to the vibrating microtome stage so that the embryo heads are facing up.
18. Begin sectioning on the vibrating microtome according to the manufacturer's instructions. Suitable parameters to begin with on the Leica VT1000S are: speed and frequency offset at 4, section thickness 150–200 μm. Collect single sections onto a paintbrush and transfer to individual wells of a 24-well plate containing 300 μL PBS (*see* **Note 47**).

3.6.2. Immunohistochemistry Applications on Thick (150–200 μm) Sections

1. Primary antibody blocking step: Remove PBS solution and completely cover the sections with blocking solution (~ 400 μL). Incubate the sections for approximately 2 h at RT on a rocking platform.
2. Primary antibody incubation: Dilute the primary antibody in antibody dilution solution according to the manufacturer's recommendations or determined empirically. Remove the blocking solution and incubate the embryos with 200–250 μL primary antibody solution overnight at 4°C on a rocking platform. Apply parafilm around the seal between the lid and the plate to prevent evaporation of the antibody solution.
3. Wash the embryos in PBST for 20 min at 4°C on a rocker ×3.
4. Remove PBST and incubate embryos in blocking solution for 2 h at 4°C.

5. Secondary antibody incubation: Dilute the secondary antibody in antibody dilution solution according to the manufacturer's recommendations (*see* **Note 34**). Common dilutions of secondary antibodies for this application are 1:500, 1:1,000 and 1:2,000. Remove PBST and incubate embryos in 200–250 μL secondary antibody solution overnight at 4°C on a rocking platform.
6. Secondary antibody removal: Wash the embryos in PBST for 20 min at 4°C on a rocking platform ×3.
7. Mounting: Remove sections required for imaging with a paintbrush and place onto microscope slide for fluorescence analysis (*see* **Note 47**). In order to image both sides of the section, a slide can be placed on top of the section and carefully inverted to release the section from the first slide. The plate containing the remainder of the sections should be kept in complete darkness and stored at 4°C by covering it with aluminum foil.
8. Imaging: Employ imaging strategies as described in **Subheading 3.4.3**, except in the case of multiple fluorophores when an appropriate laser and filter set is required for each fluorophore. Use sequential image acquisition strategies to avoid bleed-through (*see* **Note 48**). For the images shown in **Fig. 3f–h**, the following coherent lasers were used: a 532 nm Compass **(Fig. 3f)**, a 405 nm Cube **(Fig. 3g)**, and a 488 nm Sapphire **(Fig. 3h)**.

4. Notes

1. Or any disposable tube (alcian blue staining is irreversible).
2. Slide support grate is a stainless steel mesh curved at the edges to raise the mesh approximately 1 cm above the bottom of the plastic container. When in use, the space between the mesh and the bottom of the container is filled with water-soaked paper towels. Slides are supported in the humidified atmosphere on two rows of disposable 10-mL plastic pipets attached with cable ties **(Fig. 1b)**.
3. This ensures that collected embryos are synchronized at the same developmental stage.
4. Embryos less than 10 hpf should be fixed prior to dechorionation. For older embryos, dechorionation is easier on unfixed (live) embryos.
5. Benzocaine is a skin, eye and respiratory irritant.

6. Bouin's fixative contains formaldehyde, which is volatile and a carcinogen. Accordingly, fixation must be conducted in a fume hood.
7. PFA is volatile and carcinogenic. Fixation must be conducted in a fume hood.
8. It is important not to over-fix embryos; otherwise they become too brittle to section properly.
9. Do not allow the agarose to clump in the bottom of the 50-mL tube as this will increase the length of time required to dissolve the solution.
10. Embryos/larvae that are to be sectioned vertically should be embedded in 4% LMT agarose as rapid setting is required once the specimens are in position. Successful horizontal embedding can usually be achieved with 1% LMT agarose.
11. When comparing mutant embryos to wildtypes it is best to embed equal numbers of each per block. To array the embryos in a desired pattern, it is advisable to trim the tips of the tails from the wildtypes prior to embedding so that they can be easily identified when the agarose is added and the embryos/larvae move around.
12. Samples can be processed manually in a fume hood, using sealable glass containers, water and the required reagents.
13. The paraffin wax will coat around the agarose block.
14. Following the dehydration, clearing, and paraffin infiltration the agarose/paraffin wax blocks need to be further embedded with paraffin wax.
15. Ensure a thin layer of paraffin wax is set at the bottom of the base mold just before the agarose/paraffin block is positioned, so it does not form the bottom of the new block. Unlike with larger samples, such as mouse tissue for example, this is very important to prevent cutting into the embryos/larvae when the first few sections are cut on the microtome.
16. A black background to the water-bath will allow better visualization of the floating sections. Maintaining the temperature at 45°C is to ensure that the sections will flatten out on the water surface but do not melt.
17. Pre-labeling slides affords more efficient processing of tissue sections. Importantly, slides need to be numbered to keep chronological order.
18. Sagittal sectioning at 5 μm will produce approximately 8–12 slides containing 8 sections per slide.
19. Blocking endogenous peroxidase activity can also be done after the primary antibody incubation (after the running tap water and PBS washes are completed).

20. The wax circle around each section creates a water repellent boundary that allows for the use of small volumes of reagents usually 25–30 μL, held in place by surface tension. Eight sections per slide evenly spaced will allow enough room to comfortably outline sections with the Mini PAP pen.
21. After swim bladder inflation, embryos are difficult to maintain in a set position for even brief periods in EM and a more viscous solution is required. Selecting the correct medium to adequately immobilize embryos depends on the time course of image acquisition. Three percent methyl cellulose is suitable for immobilizing specimens in the desired position for brightfield and fluorescence imaging for periods of several minutes.
22. Methyl cellulose (3%) is kept at −20°C for long-term storage. As frozen samples of methyl cellulose take lengthy periods of time to thaw, defrost overnight. A working aliquot of methyl cellulose can be kept at 4°C for a few weeks.
23. When methyl cellulose is added, bubbles may form. Use a pipet tip to remove the large bubbles and manipulate the fine air bubbles away from the embryo with a fine probe.
24. The growth and expansion of the inverted embryo will be retarded during an extended time-lapse experiment. Using an upright microscope or a paralysed mutant such as *nic1* alleviates this problem (*see* Chapter "Live Cell Imaging of Zebrafish Leukocytes").
25. The production of 3D projection images of the endoderm-derived tissues by confocal analysis requires exposure times of more than 20 min. Time lapse experiments for observing intestinal morphogenesis may run for several hours. Methyl cellulose tends to disperse over extended periods of time and is not suitable for confocal analysis. Instead, embryos are embedded in LMT agarose in EMB.
26. To decrease the time required for the agarose to cool before embedding the embryos, run cold water over the outside of the flask for a short period of time. To maintain the agarose in a molten state while embedding the embryos, keep it in the boiling/hot water.
27. The laser monitor (absolute value) of laser intensity should be turned on during the experiment and the data values recorded. This allows for correction of the laser power used when the intensity of the beam reduces due to the age.
28. The use of high laser powers must be avoided to reduce photo bleaching.
29. The smaller the pinhole, the finer the optical resolution but this also reduces the intensity of the fluorescence. Brighter

fluorescence can be observed with an increase in the size of the pin-hole. This will also increase the out-of-focus blur detected in the image.

30. Saturation of signal in the image should be avoided for all experiments as it reduces the available information for further analysis. This will also cause problems with *z*-series 3-D reconstructions.
31. Although embryo movement during the collection phase of an experiment can often be corrected after image acquisition using image manipulation software such as ImageJ (http://rsb.info.nih.gov/ij/), it is best avoided.
32. You must install the "TurboReg" plugin (http://bigwww.epfl.ch/thevenaz/turboreg/) before the "StackReg" can be used *(17)*.
33. As an alternative to transfer wells (Netwell inserts), use 5-mL Technoplas tubes (Crown Scientific) or 12-well plates without transfer wells. Netwell inserts are short polystyrene tubes fitted with a polyester mesh bottom to permit easy transfer of embryos between solutions.
34. If a fluorescent secondary antibody is used, all solutions containing aliquots of the secondary antibody must be maintained in complete darkness. This can be achieved by covering the plate with aluminum foil.
35. Modified 1000-μL tips afford greater accuracy in filling the cryomold with the required volume of agarose. This is more important when embedding embryos vertically; too little agarose will cause the embryos to tilt, as there will not be enough height in the block to accommodate the entire length of the specimens.
36. Snipping the yolk allows the agarose to infiltrate the embryos. This increases the likelihood that the embryo sections will stay in the agarose following sectioning, rather than popping out.
37. 4% LMT agarose sets relatively quickly (1–3 min) so a maximum of six embryos should be embedded at a time. If more embryos are required, the percentage of LMT agarose can be dropped to no lower than 3%.
38. Embryos older than 48 hpf require 1.2 mL of agarose solution due to their length. If performing transverse sections on embryos less than 48 hpf, a reduced amount of agarose solution can be used (~ 800 μL).
39. This may also be performed using a plastic disposable transfer pipet, or a 200 μL pipet tip.
40. View the mold from the side to determine if all embryos are in the correct position with embryo heads on the bottom

surface, and tails completely vertical. All embryos need to be at the same level in the mold so that comparable tissue structures can be observed in the same section.

41. The embryos can still be manipulated somewhat once the surface of the agarose has hardened a little.
42. If PBS is placed on the embryos before the agarose has set, an impression will form on the surface which will cause problems during sectioning.
43. If sectioning is to be done on another day, place the block in a humidified container at 4°C in the refrigerator.
44. Keep razor blade in wrapping during breaking process, or use a wire cutter to reduce risk of injury.
45. Do not cut the block too close to the embryos as this will increase the risk of losing embryos/larvae during sectioning.
46. Cut one corner off the block so that the orientation of the agarose section can be followed during the imaging of the embryos.
47. This step requires gentle manipulation as the embryo sections can be easily lost if roughly handled.
48. An example of bleed-through is where some of the green fluorescent signal is detected in the red channel. Controls where only one fluorophore is present should be used to determine the extent of bleed-through under the same imaging parameters used for the experiment. Collect imaging data for each fluorophore sequentially with only the appropriate laser switched on.

Acknowledgments

We would like to thank Val Feakes and Stephen Cody for assistance and advice with histology and imaging, respectively. Val Feakes, Annie Ng, Tanya de Jong-Curtain, Heather Verkade and Elsbeth Richardson are acknowledged for their role in protocol development. This work was supported by Project Grants (280916 and 433614) to JKH from the NHMRC, Australia. Adam Parslow is a Ludwig Institute Ph.D student in the Department of Surgery at the Royal Melbourne and Western Hospitals, University of Melbourne and a recipient of an Australian Postgraduate Award.

References

1. Kikuchi, Y., Agathon, A., Alexander, J., Thisse, C., Waldron, S., Yelon, D., Thisse, B., and Stainier, D. Y. (2001). Casanova encodes a novel Sox-related protein necessary and sufficient for early endoderm formation in zebrafish. *Genes Dev* **15**, 1493–1505.
2. Field, H. A., Ober, E. A., Roeser, T., and Stainier, D. Y. (2003). Formation of the digestive system in zebrafish. I. Liver morphogenesis. *Dev Biol* **253**, 279–290.
3. Ng, A. N., de Jong-Curtain, T. A., Mawdsley, D. J., White, S. J., Shin, J., Appel, B., Dong, P. D., Stainier, D. Y., and Heath, J. K. (2005). Formation of the digestive system in zebrafish. III. Intestinal epithelium morphogenesis. *Dev Biol* **286**, 114–135.
4. Field, H. A., Dong, P. D., Beis, D., and Stainier, D. Y. (2003). Formation of the digestive system in zebrafish. II. Pancreas morphogenesis. *Dev Biol* **261**, 197–208.
5. Wallace, K. N., Akhter, S., Smith, E. M., Lorent, K., and Pack, M. (2005). Intestinal growth and differentiation in zebrafish. *Mech Dev* **122**, 157–173.
6. Wallace, K. N., and Pack, M. (2003). Unique and conserved aspects of gut development in zebrafish. *Dev Biol* **255**, 12–29.
7. Pauls, S., Zecchin, E., Tiso, N., Bortolussi, M., and Argenton, F. (2007). Function and regulation of zebrafish nkx2.2a during development of pancreatic islet and ducts. *Dev Biol* **304**, 875–890.
8. Andre, M., Ando, S., Ballagny, C., Durliat, M., Poupard, G., Briancon, C., and Babin, P. J. (2000). Intestinal fatty acid binding protein gene expression reveals the cephalocaudal patterning during zebrafish gut morphogenesis. *Int J Dev Biol* **44**, 249–252.
9. Her, G. M., Chiang, C. C., and Wu, J. L. (2004). Zebrafish intestinal fatty acid binding protein (I-FABP) gene promoter drives gut-specific expression in stable transgenic fish. *Genesis* **38**, 26–31.
10. Her, G. M., Yeh, Y. H., and Wu, J. L. (2004). Functional conserved elements mediate intestinal-type fatty acid binding protein (I-FABP) expression in the gut epithelia of zebrafish larvae. *Dev Dyn* **230**, 734–742.
11. Wan, H., Korzh, S., Li, Z., Mudumana, S. P., Korzh, V., Jiang, Y.-J., Lin, S., and Gong, Z. (2006). Analyses of pancreas development by generation of gfp transgenic zebrafish using an exocrine pancreas-specific elastaseA gene promoter. *Exp Cell Res* **312**, 1526–1539.
12. Davison, J. M., Woo Park, S., Rhee, J. M., and Leach, S. D. (2008). Characterization of Kras-mediated pancreatic tumorigenesis in zebrafish. *Methods Enzymol* **438**, 391–417.
13. Huang, H., Vogel, S. S., Liu, N., Melton, D. A., and Lin, S. (2001). Analysis of pancreatic development in living transgenic zebrafish embryos. *Mol Cell Endocrinol* **177**, 117–124.
14. Dong, P. D., Munson, C. A., Norton, W., Crosnier, C., Pan, X., Gong, Z., Neumann, C. J., and Stainier D. Y. (2007). Fgf10 regulates hepatopancreatic ductal system patterning and differentiation. *Nat Genet* **39**, 397–402.
15. Dong, P. D., Provost, E., Leach, S. D., and Stainier, D. Y. (2008). Graded levels of Ptf1a differentially regulate endocrine and exocrine fates in the developing pancreas. *Genes Dev* **22**, 1445–1450.
16. Crosnier, C., Vargesson, N., Gschmeissner, S., Ariza-McNaughton, L., Morrison, A., and Lewis, J. (2005). Delta-Notch signalling controls commitment to a secretory fate in the zebrafish intestine. *Development* **132**, 1093–1104.
17. Thévanaz, P., Ruttimann, U. E., Unser, M. (1998) IEEE Transactions on Image Processing. A pyramid approach to subpixel registration based on intensity. Volume 7: 27–41.

Chapter 19

In Toto Imaging of Embryogenesis with Confocal Time-Lapse Microscopy

Sean G. Megason

Summary

Microscopy has been one of the most direct and powerful tools since the beginning of biological research. Continued advances such as confocal and two-photon fluorescence microscopy and fluorescent proteins now make imaging useful at a variety of spatial scales (molecules, circuits, cells, tissues, and even whole embryos) and temporal scales (<seconds to days). Zebrafish is uniquely poised to benefit from these continued technological improvements because of its inherent suitability for both imaging and genetics. This chapter presents an approach called "in toto imaging". The goal of in toto imaging is to image and track every single cell movement and division that forms a tissue or organ. This approach is powerful for understanding how cell lineage, shape changes, and movements control the morphogenesis of a tissue. When used with transgenic lines, in toto imaging can be used to "digitize" data at single cell level over time from a living organism. This quantitative, digitized data can then serve as the basis for forming models of how biological circuits orchestrate developmental processes.

Key words: Confocal microscopy, Two-photon, In toto imaging, In vivo, Time lapse, Zebrafish, Image Analysis, Fluorescent, GFP, Live cell imaging.

1. Introduction

In the relatively short period of their use, zebrafish have already proven their power in the field of developmental biology. Their small size and transparency is also allowing zebrafish researchers to ask developmental questions at higher levels of resolution, namely understanding how development works at a cellular level. This single-cell precision has been one of the real hallmarks of research on *Caenorhabditis elegans* allowing it to become a powerful model. We feel that zebrafish can match this ability yet in a vertebrate system. We also feel that the natural progression of

Graham J. Lieschke et al. (eds.), *Zebrafish,* Methods in Molecular Biology, vol. 546
DOI: 10.1007/978-1-60327-977-2_19

many current approaches such as development, genetics, genomics, and imaging is to become more and more quantitative and systematic in our understanding of how the biological circuits encoded by the genome orchestrate the transformation of an egg into an embryo, and that zebrafish is the best model system in which to integrate these various approaches. Towards this goal we have developed an approach called in toto imaging that seeks to extract quantitative cellular and molecular data from intact developing zebrafish embryos in a high-throughput and systematic fashion *(1, 2)*. The goal of in toto imaging is to extract complete cell lineages of how tissues and organs are formed, but also to annotate these lineages with quantitative molecular data suitable for making predictive models of development. In toto imaging presents significant technical challenges, so here we break down the steps of in toto imaging into labeling, mounting, image acquisition, and image analysis to try to make this approach more widely useable.

2. Materials

2.1. Media and Reagents

1. 10X Tricaine: Make a 0.1% solution of Tricaine in egg water. It will be very acidic. Add 1 M Tris, pH 8.0, to adjust the tricaine solution to 10 mM Tris. It will still be a little acidic. Slowly add 0.1 M NaOH to the tricaine solution to bring it to a final pH of 7.0. Store in dark at 4°C for up to 1 wk.
2. 300% Danieau buffer: 174 mM NaCl, 2.1 mM KCl, 1.2 mM $MgSO_4$, 1.8 mM $Ca(NO_3)_2$, 15 mM HEPES, pH 7.6. Use at 30% for raising dechorionated embryos.
3. DNA plasmids: pCS-H2B-EGFP and pCS-membrane-mCherry (available from author).
4. Nucleotide cleanup column (Qiagen).
5. mMessage Machine mRNA synthesis kit (Ambion).
6. RNA cleanup column (Qiagen).

2.2. Instrumentation

1. Laser Scanning Confocal microscope (e.g., Zeiss 510) with a 25 mW 488 nm laser for exciting EGFP, emission filter set to 500–530 nm band pass and a 5 mW 543 nm laser for mCherry, emission filter set to 600 nm long pass.
2. High NA, high working distance objectives: ×40 C-Apochromat 1.2NA (290 μm working distance) and the ×40 LD C-Apochromat 1.1 NA (600 μm working distance) from Zeiss.
3. Incubator box for microscope.
4. Sufficient data storage capacity (e.g., hard drives or DVDs).

5. NanoDrop spectrophotometer (Nano Drop Technologies).
6. Fluorescent dissection microscope with green luorescent protein (GFP) and red fluorescent prtein (RFP) filter sets (e.g., Olympus MVX with ×1 0.25 NA and a ×2 0.50NA objectives.
7. Microinjector: Nanoject with Universal Adaptor and Support Base (Thomas Scientific) or similar.
8. Micromanipulator: Märzhäuser M33 or similar.
9. Standard gel electrophoresis apparatus.

2.3. Software

1. Megacapture (free download from digitalfish.org).
2. GoFigure (free download from digitalfish.org).

2.4. Other Equipment

1. Lucite embryo mount templates (*see* **Subheading 3.2** and **Figs. 1** and **2**).
2. Microcapillary tubes: Drummond (cat. no. 3–000–203-G/X).
3. Coverslips: no. 1 thickness, 43 × 50 mm, 25 × 25 mm (Electron Microscopy Sciences).

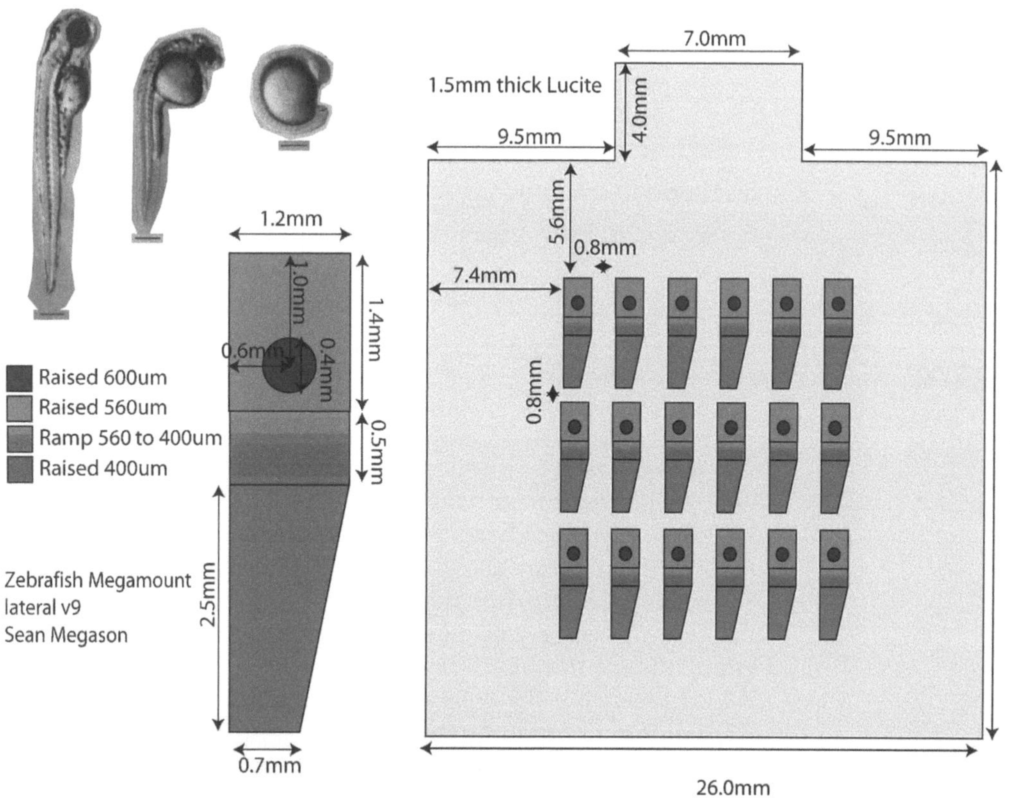

Fig. 1. Design for a lateral embryo array template.

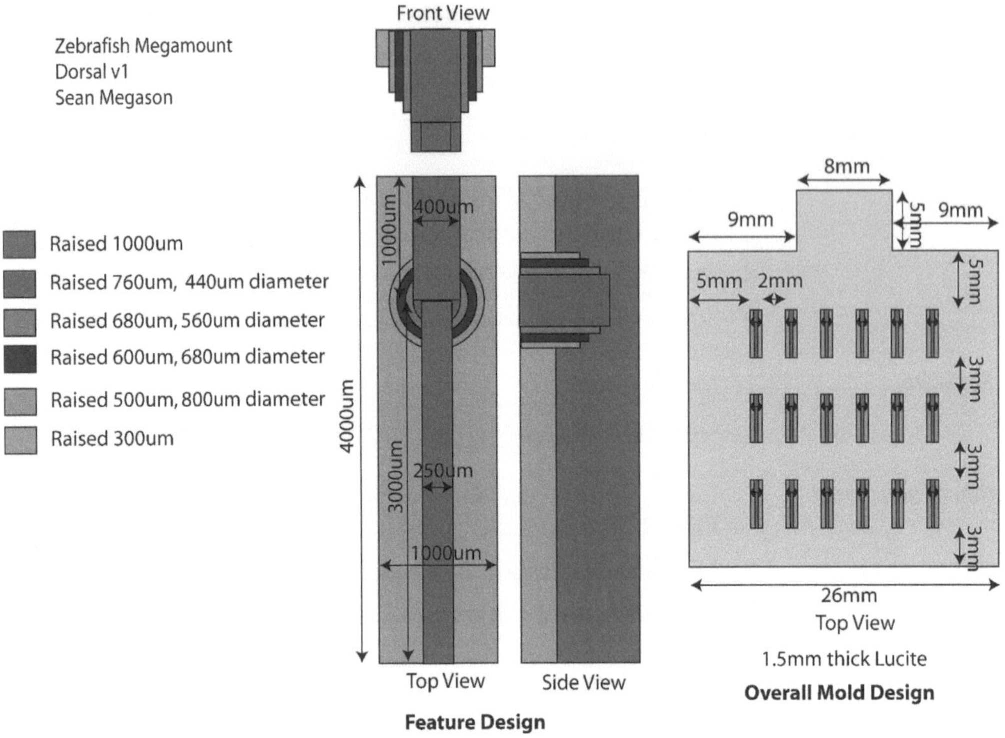

Fig. 2. Design for a dorsal embryo array template.

4. Watchmakers' forceps, no. 5.
5. Hairloop embryo manipulator.
6. Silicone Rubber Aquarium Sealant (Fisher; cat. no. S18180B).

3. Methods

The steps of in toto imaging can be roughly divided into: labeling, mounting, image acquisition, and image analysis. In toto imaging is technically very difficult (barely possible) for most tissues so it is essential that these four steps be considered together and that all the steps are optimized as a whole. For example, the image analysis is only possible if the images were acquired at high enough signal-to-noise, spatial, and temporal resolution, and this is only possible if the embryos were labeled and mounted well. Thus, if you are having trouble in the last step of image acquisition, it could be due to the first step of labeling not being optimized. For in toto imaging to work, every step must be perfected together, not just gotten by.

3.1. Labeling

There are two types of labels used for in toto imaging – segmentation markers and expression markers. Segmentation markers are fluorescent labels that allow all the cells to be individually recognized, tracked, and relevant subcellular compartments to be identified. Expression markers are fluorescent labels that mark some other piece of data pertinent to the specific experiment at hand such as a particular gene expression pattern. For some experiments such as lineage analysis, an expression marker is not required. For segmentation markers, we currently use a combination of a membrane-localized fluorescent protein in one color and a histone-fused fluorescent protein of another color. The histone fusion marks chromatin allowing nuclei to be tracked as cells move. Nuclei tend to be easier to track than whole cells since they have a simpler shape and are more separated. Importantly, the histone fusion also stays localized to chromatin during mitosis allowing daughter nuclei to be linked with their mother. For some tissues such as the neural tube, the nuclei can appear touching even in very thin optical sections making automatic segmentation more difficult. The membrane localized fluorescent protein is useful in such cases because this signal can be used to split touching nuclei. The membrane label is also nice in that it highlights tissue morphology well and along with the histone label allows the nuclear, cytoplasmic, and membrane subcellular compartments to be identified.

Each marker must be a different color fluorescent protein, should be as bright as possible, and should be well separated spectrally *(3)*. If no expression marker is needed, we currently use histone-EGFP and membrane-mCherry. If an expression marker is used, then the expression marker is done with citrine and the segmentation markers are histone-cerulean and membrane-mCherry. It is important to consider the effects of the "tail" of fluorescent emission and the relative levels of expression of each marker. For example, if the segmentation markers are expressed at very high levels, bleed though into the expression marker channel may be a problem unless it is expressed at similar levels or is well separated spectrally.

The most simple method for labeling with segmentation markers is using RNA injection. This approach allows very high levels of label and thus bright signals to be achieved. Because RNA takes a few hours to get expressed and is degraded after a few days, it is only useful if one is interested in the period of development from 5 to 60 h post-fertilization (hpf). For earlier or later stages, a transgenic source of segmentation markers should be used.

Labeling using RNA injection:

1. RNA for histone2B-EGFP and membrane-mCherry can be made using in vitro transcription from the plasmids pCS-H2B-EGFP and pCS-membrane-mCherry available from the author. These plasmids should be linearized with *Not*I,

purified using a Qiagen nucleotide removal spin column, and used as template for in vitro transcription with the Ambion mMessage Machine kit. In vitro transcribed RNA can be purified using a Qiagen RNA cleanup column and quantitated on a NanoDrop spectrophotometer. RNA integrity can also be assessed using gel electrophoresis.

2. The amount of RNA injected should be determined by titration such that the maximal level of labeling that allows normal development is achieved. Make up injection solutions in the range of 20 ng/μL, 40 ng/μL, and 80 ng/μL of RNA individually and together. Inject these solutions into batches of 1-cell stage embryos. Allow these embryos to develop until 24-hpf and check for any signs of developmental retardation or malformation compared with uninjected embryos. Also check embryos for levels of fluorescence using a fluorescent dissecting microscope. Comparing different batches can be done easier if they are photographed under the same conditions on a fluorescent dissecting microscope and these images compared side by side. The combination of RNA concentrations that gives the brightest labeling in both channels yet still gives normal development should be determined and used subsequently. This is typically 40 ng/μL of each RNA.
3. Labeling embryos for in toto imaging: Setup male and female fish in mating cages separated by a divider the night before. The next morning remove the dividers and let the fish mate for 10–15 min before collecting eggs. It is useful to collect the eggs soon after mating so they can be injected early. Image analysis is much easier when the cells are evenly and uniformly labeled. To achieve uniform labeling, inject eggs between 10 and 25 min after fertilization (in the early to middle part of the 1-cell stage). It is also best to inject into the single blastomere or blastodisc rather than into the yolk.
4. Raise embryos to the appropriate stage. Discard any embryos that are not morphologically normal. Screen remaining embryos on a fluorescent dissecting microscope in both channels. Look for embryos with bright, uniformly distributed labeling with the segmentation markers. Discard any embryos that are dim or have patchy/mosaic expression of the segmentation markers. Select the 5–10 best embryos for mounting and transfer to a fresh Petri dish.

3.2. Mounting

Mounting is a critically important step to achieving high-quality images. The goals of mounting are to ensure that the ambryo stays healthy and develops normally; that the tissue of interest is positioned such that it can be viewed as directly as possible (through the least amount of overlying tissue) throughout the time lapse; and that the tissue of interest is positioned as close to the coverslip as possible (important for using high numerical aperture, short working distance objectives). Zebrafish is often touted as being superior for imaging because of the transparency of its embryos. While certainly true, the suitability of zebrafish embryos for mounting is as important. The first goal of normal development is relatively easy to achieve for zebrafish embryos since they are aquatic, normally develop freely outside their mother, and can readily develop outside their chorion. The last two goals of getting the cells of interest as close to the objective as possible and with the least amount of intervening tissue is made easier in zebrafish embryos by their relatively small size. The exact details of mounting may vary depending on the tissue being studied, but the above goals will still apply.

We present below a general approach for mounting we call megamounts or embryo arrays. These mounts are available for both dorsal and lateral views and allow most parts of the embryo to be imaged continuously across most of embryonic development. Embryo arrays are agarose mounts made from a plastic template. The plastic template creates "embryo shaped" wells in defined positions in the agarose that allow the embryo to develop in a defined orientation (lateral or dorsal). Embryos can develop normally in the mounts for at least 3 d and be continuously imaged during this time.

Tricaine solutions should be made fresh weekly from powder that is <6 mo old. The 10X solution must be adjusted to pH 7.0. The final working concentration should be empirically determined by testing different concentrations to find the range that prevents movement but permits normal development. The working concentration is normally approximately 0.013%. If it is much higher, then the tricaine is probably deteriorating and should be replaced.

1. The templates for the mounts are made from Lucite machined with a 500 μm bit on a CNC mill to contain the features as shown on the designs for the dorsal or lateral mounts.
2. Make 100 mL of 0.7% agarose in 30% Danieau by dissolving 0.7 g agarose in 100 mL media by microwave.
3. Let the agarose cool to 65°C. Agarose should be made up and used fresh. Molten agarose will begin to lose its strength if stored for >1 d at 65°C.
4. For use on an upright microscope: Add 1 mL of 10X tricaine to a 60 × 15 mm Petri dish. Add 9 mL of molten agarose to dish and mix with the tricaine. Gently lay the template onto the molten agarose. Avoid trapping air bubbles by laying one edge down first. Allow to set for 1 h.
5. For use on an inverted microscope. Prepare a coverslipped bottom Petri dish ahead of time by affixing a large coverslip with silicone rubber to the bottom of a 60-mm Petri dish that has

had a window cut out of its bottom using a flame heated scalpel. Apply 0.5 mL of 0.7% agarose in embryo media with 1X tricaine to the coverslip. Place the template on top and center it. Put a weight (I use a glass vial filled with sand) on top of the template to hold it against the coverslip. Allow to set 1 h.

6. Add 10 mL of 30% Danieau with 1X tricaine. Allow to set for 30 min (this helps loosen the template and makes sure the agarose is well set).
7. Gently remove the template from the agarose by slowly prying up the tabbed edge using forceps. For inverted mounts, it is useful to gently press on the bottom of the coverslip from below with a Kimwipe such that the coverslip and the agarose will slightly bow. Because the template will not bow, it will become loose from the agarose.
8. Mounting the embryos: For lateral mounts, the embryos can be mounted anytime after shield stage. For dorsal mounts, anytime after epiboly. Manually dechorionate embryos in embryo media using sharp forceps. Petri dishes coated with agarose should be used for embryos <24 hrs old to avoid them sticking.
9. Transfer the embryos to the mounting chamber. It is easiest to see the wells of the mounting chamber using oblique transillumination.
10. Under a dissecting scope, position one embryo in each well using a hairloop. Orient the embryos such that the animal pole is pointing where the head should go and the shield or somites is pointing where the back should go. With lateral mount, this is most easily accomplished by first positioning the embryos animal side up in the circular depression within each well. The embryos can then be rotated about their animal-vegetal axis such that the shield is toward the side it should be **(Fig. 3)**. The embryos can then be rotated along the *D-V* axis so that the animal pole is pointing where it should be. With the dorsal mounts, the embryos do not spin easily within the hemispherical depression in each well so the embryos should be dropped into the hemisphere already in

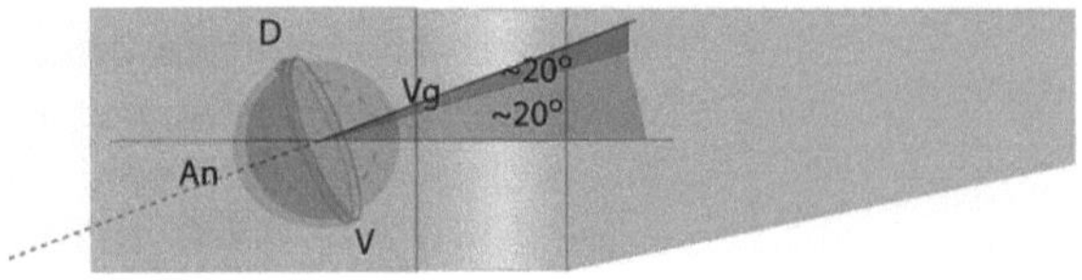

Fig. 3. Correct initial position for shield stage embryos in lateral mounts. The vegetal pole should be pointing approximately 20° above the horizontal and approximately 20° out of the page.

the proper orientation. This is most easily accomplished by positioning the embryos in the groove just posterior to the hemisphere with their animal side up and their dorsal facing the posterior end of the well. The embryos can then be rolled forward into the hemisphere such that they undergo a quarter rotation as they fall into the hemisphere resulting in their dorsal side being up and their animal pole facing the anterior of the well.

11. Coverslipping: A coverslip is placed within the square created by the template in order to hold the embryos in place. A coverslip should be used that is 1 mm smaller than the width of the template such that it barely fits within the square depression created by the template (25 mm^2). The coverslip should be prewetted by soaking it in embryo media for 10 min to reduce is hydrophobicity. Properly applying the coverslip without moving the embryos requires practice, especially with the lateral mounts where the embryos can move much more easily. To apply the coverslip, rotate the mount such that the tab that sticks off the square in the agarose depression is to your right. Pick up the wetted coverslip with a forcep in your right hand. Under a dissecting scope, position the left edge of the coverslip in the bottom left edge of the agarose square depression. Lower the right edge (the side being gripped by the forceps) of the coverslip to the surface of the media such that the edge is right at the surface so that the coverslip is half sunk and half floating. At this point, the coverslip should be partly submerged with its left edge in position under the media along the left edge of the agarose depression and its right edge stuck to the surface of the media/floating. You can the release the coverslip and it should stay in this half sunk/half floating position. You should be able to position the coverslip in this position without disturbing the embryos and you can even reposition some of the embryos if you're careful.
12. Lowering the coverslip: Now take a pair of forceps in both hands. Put the closed point of the forceps in the right hand under the floating edge of the coverslip with fingers resting on the edge of the Petri dish for support. With the forceps in the left hand, press the floating edge of the coverslip under the media such that it sinks onto the forceps in the right hand. Very carefully and slowly lower the forceps in the right hand down and into the agarose such the coverslip will be lowered into the square. Slowly withdraw the forceps through the tabbed area (it is OK to pit the agarose in this area). Check all four corners of the coverslip to make sure that they are within the square agarose depression and that the coverslip is lying flat against the agarose.

13. Lowering the coverslip can also be done using a micromanipulator. After the coverslip is in the half floating/half sunk arrangement, position a pulled glass pipet attached to an *xyz* micromanipulator under the floating edge of the coverslip. Using forceps press the floating edge under the surface so that it sinks onto the pulled glass pipet. Using the micromanipulator, slowly lower the coverslip.
14. It is now ready to image! For an upright compound scope, the objectives can be dipped into the media in the Petri dish. For an inverted compound scope, a drop of water can be placed on a water immersion objective and the objective raised to meet the coverslip as normal. For long-term imaging on a dissecting scope, the Petri dish can be covered with a watch glass to prevent evaporation and the embryos observed through the watch glass or it can be covered with a Petri dish lid containing a coverslip "skylight" for imaging.

3.3. Image Acquisition

After the zebrafish have been mounted the next step is to acquire images. For tracking all the cells in a tissue it is necessary to capture a complete volume of the tissue of interest (an xyz image) and to repeat this over time to create an *xyzt* image set. Capturing volumetric images of tissues requires the use a confocal or multiphoton microscope. We typically use a laser-scanning confocal for imaging (*see* **Note 1**). To segment cells in 3-D (across the *z*-axis) you need high z-resolution (1 μm), which requires thin optical sections and a close spacing between optical sections (*see* **Note 2**), which in turn requires the use of high numerical aperture objectives (*see* **Note 3**). Tracking cells as they move and divide also requires high temporal resolution (~2 min). Achieving high signal to noise, high spatial resolution, and high-temporal resolution requires a difficult balancing act. It is easy to achieve one at the expense of another but can be challenging to achieve all these things and a healthy embryo at the same time.

Actually acquiring all the images can also be difficult because of the sheer number of images and the amount of disk space they require. An in toto image set can be >100 Gb so it is much easier if not essential that these images be stored as multiple files rather than one huge file. It is also quite useful to be able to adjust the imaging parameters (e.g., laser power, gain, location of *z*-stack) during image acquisition to account for photobleaching and drift of the embryo. Some image acquisition routines that come with microscopes block out user interaction during acquisition.

We have developed an image acquisition macro for Zeiss microscopes called Megacapture that automates the collection of in toto image sets. This macro allows you to set the number of time points, time interval, number of *z*-sections, and *z*-interval. It saves each *xy* image as a separate file which makes file management of huge image sets easier, and can also optionally save images compressed saving valuable disk space. Megacapture can also be used with a motorized stage to automate tiling across embryos and for capturing multiple embryos in a row/column format such as with embryo arrays. And finally, Megacapture allows the image acquisition parameters to be altered during an experiment and separately records the imaging parameters for each image. It can be downloaded from the software section of digitalfish.org.

1. Install the proper objective onto the microscope (*see* **Note 3**).
2. Pre-warm the microscope to 28°C for 2 h prior to starting imaging. The use of an incubator box that encloses the entire microscope is preferable because it is easier to maintain a uniformly controlled temperature which is essential for proper embryonic development and to prevent focal drift caused by thermal expansion/contraction.
3. Mount embryos as discussed above and transfer the dish to the microscope. The microscope should have a stage insert capable of securely holding a Petri dish. For an inverted microscope, add water to the objective and then place the Petri dish on top with your embryo directly over the front lens of the objective. For an upright microscope, the 30% Danieau in the Petri dish can serve as the immersion media.
4. Find the embryo. This can be done using bright field and manually adjusting the focus while looking through the oculars until the embryo is in focus. If multiple embryos were mounted, scan through all the embryos to find the one that is mounted best and has the brightest and most uniform labeling with the segmentation markers. Center the tissue of interest in the field of view.
5. Setup image acquisition parameters. We have been successful on a Zeiss 510 with the following: 25 mW 488 nm laser for exciting EGFP set to a power of <5%, 5 mW 543 nm for mCherry set to a laser power of <15%, the emission filter for EGFP set to 500–530 nm band pass, the emission filter for mCherry set to a 600-nm long pass, and both pinholes set to 1 Airy unit. We do "single tracking" which means both channels are captured simultaneously. It is more difficult to get cleanly separated and balanced channels using single tracking but it allows the imaging to be done twice as quickly. We also use bidirectional scanning in which pixels are captured on both the forward and backward pass of the laser. This can cause aliasing artifacts but also doubles the speed of acquisition. We use 1024 × 1024 pixel images and scan at the maximal scan rate which results in a speed of just under 1 s per *xy* image. The gain and background should be adjusted such that the full dynamic range of the signal is captured – the gain should be increased until a small percentage of pixels in the brightest parts of the

image have maxed out and the background should be set such that only pixels in area where there is no signal have a minimal value. A pixel depth of 8-bits is sufficient to capture the informational content of fluorescent images taken under these conditions because of their inherent "shot noise" (Poisson noise).

6. Through the microscope control software, set the top and bottom limits of the *z*-stack. To achieve a 2 min time resolution with a 1 s frame rate, one is limited to about 100 images per *z*-stack after some time for microscope movement and saving images is accounted for. At 1 μm *z*-spacing, this gives a maximal *z*-stack size of 100 μm. In setting the upper and lower limits of the *z*-stack, one should bracket your tissue of interest by several microns to allow for focal drift of the microscope and normal morphogenetic movements. One should also be able to predict the direction and amount of any tissue movements caused by morphogenesis and set the block of space that you image such that your tissue grows within this space.
7. Click start on Megacapture and hope for the best. Watch the first few *z*-stacks being captured and make sure they are being saved properly.
8. "Babying" the microscope. To play it safe, it is a good idea to check on the imaging every hour or two throughout the time lapse. It might be necessary to add immersion water to make up for any evaporation. On an upright microscope, water can just be added to the Petri dish with a pipet. On an inverted microscope, water can be added between the objective and the coverslipped bottom Petri dish using a flame-pulled P1000 (blue) plastic pipet tip. Laser power and gain can be adjusted to make up for any photobleaching/degradation of the segmentation markers or changes in the expression levels of expression markers. The xy stage position and *z*-stack limits can also be adjusted to account for morphogenetic movement. Ideally, all these adjustments should be made in the short gap in between time points.
9. At the end of the time lapse, make sure all the images are safely stored to hard drive or DVD. Clean the objective front lens with fresh water, blot dry, and store safely. The option exists to unmount the embryo that was imaged by carefully removing the upper coverslip and then raise the embryo long enough to convince yourself that it is healthy and did not suffer any ill effects from imaging. Catch up on sleep before proceeding.

3.4. Image Analysis

The goal of image analysis for in toto imaging is to make biologic sense of the huge amount of image data. There are two important components to image analysis. The first is visualization. In toto imaging generates huge image sets that are four-dimensional. There is a basic need just to be able to picture how the tissue is moving and changing shape within the large image sets over the course of the experiment. The second component of image analysis is the actual quantitative analysis. This comes down to converting the pixel-based data that is in the images into cell-based data that is useful for a biologist. This process of recognizing and demarcating cells, cell tracks, and cell lineages within images is called "segmentation". The enormity of the number of images and cells makes manual segmentation unfeasible. Automatic segmentation is the process in which a computer program automatically parses an image set to recognize and demarcate objects of interest. It has been a very active area of research in computer science and artificial intelligence for several decades because it turns out that a human's ability to recognize objects in their surroundings is quite complex and very difficult to mimic with a computer. Cell segmentation in particular can now be done fairly routinely for well isolated cells such as those grown in flat culture and imaged in 2-D. However, automatic segmentation still poses a significant challenge for densely packed cells in real tissue and for performing segmentation in 3-D and in 4-D, as is needed for in toto imaging.

We have invested a lot of effort into developing a software package called GoFigure to address the image analysis needs of in toto imaging. GoFigure allows for visualization of large 4-D image sets. It allows the user to work with image sets larger than the physical memory of the computer (which is typically 30× smaller) by providing different methods of dynamically subsetting the data. The default is to load all the *z*-sections for the current time point to form an *xyz* subset and all the time points for the current focal plane to form an *xyt* subset. GoFigure can also load, for example, every other time point or *z*-section, downsample the images in *xy* (**Fig. 4**), or crop the loaded images along the *z*-axis or time axis to facilitate working with large image sets.

GoFigure allows segmentation to be done manually as well as automatically using several different segmentation algorithms. GoFigure breaks down segmentation into several steps corresponding to the different dimensions of the image set: namely, figures correspond to the 2-D objects such as sections through cells, meshes correspond to 3-D objects such as a whole cell, tracks correspond to 4-D objects such as the movement of a cell over time, and lineages represent the branched trees that represent cell divisions. GoFigure provides different algorithms for automatically generating each of these levels of segmentation.

1. Download and install GoFigure. GoFigure currently runs on Windows but we will soon release a cross-platform version of GoFigure that will run on Windows, Mac, and Unix. GoFigure can be downloaded from the software section of digitalfish.org.

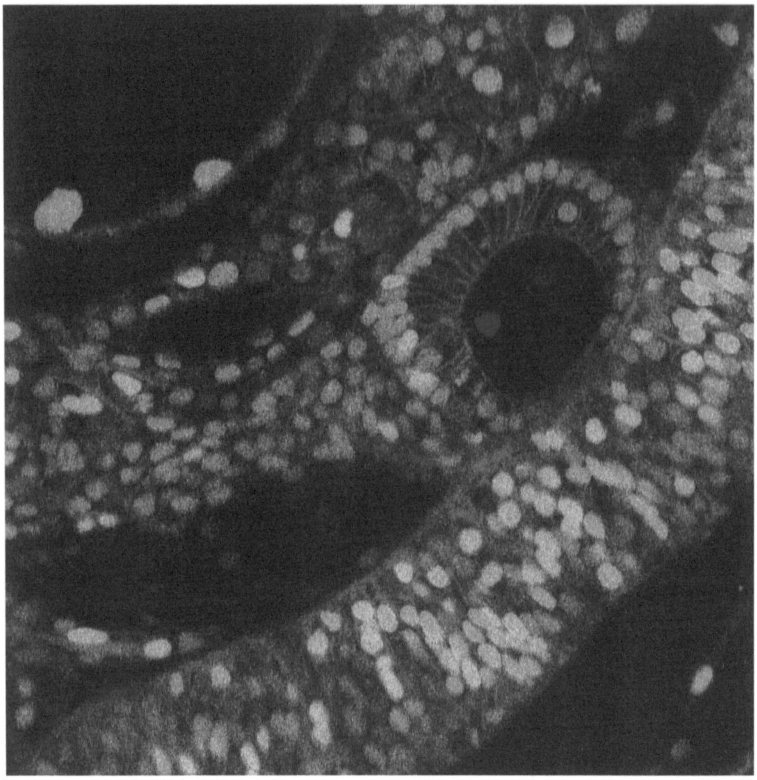

Fig. 4. A single *xy* image extracted from a very large in toto image to show a typical image. This image is a dorsal-lateral view of the otic vesicle and hindbrain. Nuclei are marked in green with H2B-EGFP and cell membranes are marked in red with a membrane localized mCherry (*see Color Plates*).

2. Import images into GoFigure. GoFigure provides several methods for importing images. If images were captured using MegaCapture, then they can be automatically imported with all of their proper dimensions and image capture settings into GoFigure. GoFigure can also import collections of RAW images that have their coordinates in *z* and time indicated in their filenames using a wild card in the name. We are currently developing import filters for Zeiss and other format image collections.
3. Visualize your image set to ensure you captured the complete volume and complete time window. Areas where the embryo moved can be identified as discontinuities in the *xz, yz, xt,* or *yt* computed cross-sections.
4. Manually segment approximately 10 representative cells. This can be done using the pencil tool and drawing outlines around each cell using the mouse.
5. Select these cells by clicking on them and use them to train the automatic segmentation tool. GoFigure uses the shape and

size of what cells should look like to help guide the segmentation process.

6. Click on the automatic segmentation button to segment other cells in the image set. Depending on the segmentation algorithm this procedure may need to be repeated to build progressively higher dimension segmented objects.
7. Manually edit the automatically segmented objects. Automatic segmentation saves a great deal of time but it can make both false positives and false negatives so the results should be verified by a well trained human eye.
8. Once the data is segmented, it is possible to perform cell based quantitative analysis. The cell size, fluorescent content, location, speed, shape, and many other parameters is displayed in the "list views" and can be used to generate interactive histograms and scattergrams within GoFigure.

4. Notes

1. A spinning disc confocal allows too much cross talk between pinholes with a widely labeled specimen such as is used in in toto imaging causing image degradation. Two-photon microscopy can be used but it is more expensive, more difficult to maximally excite multiple fluorescent proteins simultaneously (because there is typically just a single tunable laser line), and the resolution is twice as bad (becausee the wavelength is twice as long). Laser scanning confocal can typically achieve good enough quality images for cell tracking to a depth of 100 μm. For deeper tissues, two photon may be preferable. Two photon also can provide less photobleaching and photodamage, which are typically much less of a problem in zebrafish than other systems with properly done confocal imaging.
2. Achieving thin optical sections requires the use of a tight pinhole (we aim for 1 Airy unit) and a high numerical aperture objective (>1.0). The use of a high numerical aperture objective is absolutely imperative. While lateral resolution is inversely proportional to the NA, axial resolution is inversely proportional to the square of the NA. In practice, resolution is often limited by signal-to-noise ratio in addition to diffraction, but NA also helps here since signal is directly proportional to the square of the NA. Optical section spacing should ideally be set to half of the optical section thickness to achieve proper sampling across the *z*-axis according to Nyquist criteria. However, in practice there is often just not enough time in between

time points to capture that many *z*-sections so we often settle for a z-section spacing of 1 μm.

3. There are only a few objectives that are really suitable for in toto imaging. A suitable objective should have an NA of >1.0, enough working distance to reach the cells being studied (200–600 μm), and be water immersion. Zeiss makes two ×40 objectives that meet these criteria – the ×40 C-Apochromat 1.2NA (290 μm) and the ×40 LD C-Apochromat 1.1 NA (600 μm working distance). Zeiss, Leica, and Olympus have also all come out with 20× objectives with NAs of 0.95–1.05 and long working distances.

References

1. Megason SG, Fraser SE. (2003) Digitizing life at the level of the cell: high-performance laser-scanning microscopy and image analysis for in toto imaging of development. *Mech Dev.* 120(11):1407–1420.
2. Megason SG, Fraser SE. (2007) Imaging in Systems Biology. *Cell*, 130:784–795.
3. Shaner NC, Steinbach PA, Tsien RY. (2005) A guide to choosing fluorescent proteins. *Nature Methods* 2:905–909.

INDEX

S

T

U

W

X

Z

MIX
Papier aus verantwortungsvollen Quellen
Paper from responsible sources
FSC® C105338

If you have any concerns about our products,
you can contact us on
ProductSafety@springernature.com

In case Publisher is established outside the EU,
the EU authorized representative is:
Springer Nature Customer Service Center GmbH
Europaplatz 3, 69115 Heidelberg, Germany

Printed by Libri Plureos GmbH
in Hamburg, Germany